Assessing Complexity in Physiological Systems through Biomedical Signals Analysis II

Assessing Complexity in Physiological Systems through Biomedical Signals Analysis II

Guest Editors

Paolo Castiglioni
Luca Faes
Gaetano Valenza
Andrea Faini

MDPI

Basel • Beijing • Wuhan • Barcelona • Belgrade • Novi Sad • Cluj • Manchester

Guest Editors

Paolo Castiglioni
Department of Biotechnology and Life Science (DBSV)
Università degli Studi dell'Insubria
Varese
Italy

Luca Faes
Department of Energy, Information Engineering and Mathematical models (DEIM)
University of Palermo
Palermo
Italy

Gaetano Valenza
Department of Information Engineering and Research Center "E. Piaggio"
University of Pisa
Pisa
Italy

Andrea Faini
Department of Cardiovascular, Neural and Metabolic Sciences
Istituto Auxologico Italiano
Milan
Italy

Editorial Office
MDPI AG
Grosspeteranlage 5
4052 Basel, Switzerland

This is a reprint of the Special Issue, published open access by the journal *Entropy* (ISSN 1099-4300), freely accessible at: https://www.mdpi.com/journal/entropy/special_issues/NQ3RZ17964.

For citation purposes, cite each article independently as indicated on the article page online and as indicated below:

Lastname, A.A.; Lastname, B.B. Article Title. *Journal Name* **Year**, *Volume Number*, Page Range.

ISBN 978-3-7258-6366-2 (Hbk)
ISBN 978-3-7258-6367-9 (PDF)
https://doi.org/10.3390/books978-3-7258-6367-9

Cover image courtesy of Paolo Castiglioni

Contents

About the Editors vii

Paolo Castiglioni, Luca Faes, Gaetano Valenza and Andrea Faini
Assessing Complexity in Physiological Systems Through Biomedical Signals Analysis II
Reprinted from: *Entropy* **2025**, *27*, 1185, https://doi.org/10.3390/e27121185 1

Francesco Riganello, Martina Vatrano, Paolo Tonin, Antonio Cerasa and Maria Daniela Cortese
Heart Rate Complexity and Autonomic Modulation Are Associated with Psychological Response Inhibition in Healthy Subjects
Reprinted from: *Entropy* **2023**, *25*, 152, https://doi.org/10.3390/e25010152 5

Paolo Castiglioni, Giampiero Merati, Gianfranco Parati and Andrea Faini
Sample, Fuzzy and Distribution Entropies of Heart Rate Variability: What Do They Tell Us on Cardiovascular Complexity?
Reprinted from: *Entropy* **2023**, *25*, 281, https://doi.org/10.3390/e25020281 17

Alejandro Muñoz-Diosdado, Éric E. Solís-Montufar and José A. Zamora-Justo
Visibility Graph Analysis of Heartbeat Time Series: Comparison of Young vs. Old, Healthy vs. Diseased, Rest vs. Exercise, and Sedentary vs. Active
Reprinted from: *Entropy* **2023**, *25*, 677, https://doi.org/10.3390/e25040677 34

Jacques-Olivier Fortrat and Guillaume Ravé
Autonomic Nervous System Influences on Cardiovascular Self-Organized Criticality
Reprinted from: *Entropy* **2023**, *25*, 880, https://doi.org/10.3390/e25060880 56

Laurent M. Arsac
Entropy-Based Multifractal Testing of Heart Rate Variability during Cognitive-Autonomic Interplay
Reprinted from: *Entropy* **2023**, *25*, 1364, https://doi.org/10.3390/e25091364 64

Juan Ruiz de Miras, Chiara-Camilla Derchi, Tiziana Atzori, Alice Mazza, Pietro Arcuri, Anna Salvatore, et al.
Spatio-Temporal Fractal Dimension Analysis from Resting State EEG Signals in Parkinson's Disease
Reprinted from: *Entropy* **2023**, *25*, 1017, https://doi.org/10.3390/e25071017 75

Jiawen Li, Guanyuan Feng, Chen Ling, Ximing Ren, Xin Liu, Shuang Zhang, et al.
A Resource-Efficient Multi-Entropy Fusion Method and Its Application for EEG-Based Emotion Recognition
Reprinted from: *Entropy* **2025**, *27*, 96, https://doi.org/10.3390/e27010096 87

Yuan Lu and Jingying Chen
Improved EEG-Based Emotion Classification via Stockwell Entropy and CSP Integration
Reprinted from: *Entropy* **2025**, *27*, 457, https://doi.org/10.3390/e27050457 108

Rosa Molina, Yasmina Crespo-Cobo, Francisco J. Esteban, Ana Victoria Arias, Javier Rodríguez-Árbol, Maria Felipa Soriano, et al.
Dynamic Evolution of EEG Complexity in Schizophrenia Across Cognitive Tasks
Reprinted from: *Entropy* **2025**, *27*, 226, https://doi.org/10.3390/e27030226 123

Brenda Y. Angulo-Ruiz, Elena I. Rodríguez-Martínez, Francisco J. Ruiz-Martínez, Ana Gómez-Treviño, Vanesa Muñoz, Sheyla Andalia Crespo and Carlos M. Gómez
Brain Complexity and Parametrization of Power Spectral Density in Children with Specific Language Impairment
Reprinted from: *Entropy* **2025**, *27*, 572, https://doi.org/10.3390/e27060572 . . . . **137**

Irene Franzone, Yuri Antonacci, Fabrizio Giuliano, Riccardo Pernice, Alessandro Busacca, Luca Faes and Giuseppe Costantino Giaconia
Time-Resolved Information-Theoretic and Spectral Analysis of fNIRS Signals from Multi-Channel Prototypal Device
Reprinted from: *Entropy* **2025**, *27*, 694, https://doi.org/10.3390/e27070694 . . . . **161**

Andrea Faini, Laurent M. Arsac, Veronique Deschodt-Arsac and Paolo Castiglioni
Multifractal Multiscale Analysis of Human Movements during Cognitive Tasks
Reprinted from: *Entropy* **2024**, *26*, 148, https://doi.org/10.3390/e26020148 . . . . **179**

Stelios M. Potirakis, Fotios K. Diakonos and Yiannis F. Contoyiannis
A Spike Train Production Mechanism Based on Intermittency Dynamics
Reprinted from: *Entropy* **2025**, *27*, 267, https://doi.org/10.3390/e27030267 . . . . **193**

About the Editors

Paolo Castiglioni

Paolo Castiglioni is an Associate Professor at the "Università degli Studi dell'Insubria" in Varese, Italy, and a researcher at the Laboratory of Movement Analysis and Rehabilitation Bioengineering (Lamobir), IRCCS Fondazione Don Gnocchi, in Milan, Italy. He received his M.Sc. in Electronic Engineering (1987) and his Ph.D. in Bioengineering (1993) from Politecnico di Milano, Italy. He is the author of 170 scientific papers published in peer-reviewed journals and 11 book chapters, a Guest Editor of the journal *Entropy*, and a member of the scientific committee of the "European Study Group on Cardiovascular Oscillations" (ESGCO). His research activities include the analysis of biological signals, mainly cardiovascular signals, EMG, and EEG; advanced algorithms for biosignal analysis, including advanced spectral techniques; time-frequency distributions; and complexity-based methods for time-series analysis. His research interests also focus on modeling humoral and neural mechanisms for the control of the cardiovascular system; the development of portable systems for the long-term monitoring of physiological signals; and gravitational and sleep physiology.

Luca Faes

Luca Faes is Full Professor of Biomedical Engineering at the Department of Engineering of the University of Palermo, Italy, where he leads the Biosignals and Information Theory Laboratory (BIT lab) and teaches courses on Statistical Analysis of Biomedical Signals, Biosensors, and Biomedical Devices. He received his Master's degree from the University of Padova, Italy, in 1998, and his PhD in Electronic Devices from the University of Trento, Italy, in 2003. He has been a Visiting Researcher at the State University of New York (Stony Brook, USA, 2007), at the Worcester Polytechnic Institute (MA, USA, 2010), at the University of Gent (Belgium, 2013), at the Federal University of Minas Gerais (Brazil, 2015), and at Boston University (USA, 2016). Dr. Faes is a Senior Member of the IEEE, a member of the IEEE Engineering in Medicine and Biology Society (IEEE-EMBS) and of the Technical Committee of Biomedical Signal Processing of the IEEE-EMBS, and a member of the European Study Group on Cardiovascular Oscillations (ESGCO). He is Specialty Chief Editor for the journal *Frontiers in Network Physiology* and Associate Editor for the journals *IEEE Transactions on Biomedical Engineering* and *Entropy*. He has been involved in several nationally and internationally funded research projects. His research focuses on multivariate time series analysis and information theory applied to cardiovascular neuroscience, brain connectivity, brain–heart interactions, and network physiology. Within these fields, he has co-authored 8 book chapters and more than 250 peer-reviewed publications, receiving >9000 citations (h-index: 56; font: Scholar, December 2025). Thanks to his scientific output, in 2021–2024, he was included in the list of World's Top 2% Scientists provided by Stanford University for the discipline Biomedical Engineering.

Gaetano Valenza

Gaetano Valenza is an Associate Professor of Bioengineering and is head of the NeuroCardiovascular Intelligence Lab at the Research Center "E. Piaggio" and Department of Information Engineering, University of Pisa, Italy. He received his Ph.D. in Automatics, Robotics, and Bioengineering from the University of Pisa, Italy. His research interests include statistical and nonlinear biomedical signal and image processing, cardiovascular and neural modeling, physiologically interpretable AI systems, and wearable systems for physiological monitoring.

Andrea Faini

Andrea Faini is a researcher at the Department of Cardiovascular, Neural and Metabolic Sciences, I.R.C.C.S. Istituto Auxologico Italiano, Milan, Italy. His scientific interests span signal processing and statistical analysis of clinical data, with a special focus on cardiovascular physiology, heart-rate and blood-pressure variability, sleep-related breathing disorders, and the cardiovascular effects of hypoxia, particularly during high-altitude exposure. He has been actively involved in the HIGHCARE project, contributing to investigations on cardiovascular adaptation to altitude and hypoxic environments. His research activities include studies on autonomic cardiovascular regulation, the impact of altitude on vascular function and hemodynamics, and the use of wearable and non-invasive technologies for physiological monitoring. Over the years, he has authored and co-authored more than a hundred peer-reviewed articles, advancing understanding of how the cardiovascular system responds to hypoxia and sleep-related disorders.

entropy

MDPI

Editorial

Assessing Complexity in Physiological Systems Through Biomedical Signals Analysis II

Paolo Castiglioni [1,2,*], Luca Faes [3], Gaetano Valenza [4] and Andrea Faini [5,6]

1 Department of Biotechnology and Life Sciences (DBSV), University of Insubria, 21100 Varese, Italy
2 IRCCS Fondazione don Carlo Gnocchi, 20148 Milan, Italy
3 Department of Engineering, University of Palermo, 90128 Palermo, Italy; luca.faes@unipa.it
4 Department of Information Engineering and Research Center "E. Piaggio", University of Pisa, 56122 Pisa, Italy; g.valenza@ing.unipi.it
5 Department of Cardiovascular, Neural and Metabolic Sciences, Istituto Auxologico Italiano, IRCCS, 20149 Milan, Italy; a.faini@auxologico.it
6 Department of Electronics Information and Bioengineering, Politecnico di Milano, 20156 Milan, Italy
* Correspondence: paolo.castiglioni@uninsubria.it

Physical systems are widely characterized in terms of their complex dynamics in physiology and medicine to understand the ability of a living system to adapt to external perturbations. Complex dynamics are typical of systems characterized by physical structures with fractal geometry and nonlinear components, often composed of numerous interdependent elements that self-organize, interacting at different hierarchical levels and time scales. This allows homeostasis to be maintained even in the presence of external perturbations and interactions with other physiological systems, through neural and humoral networks.

The variables that fully describe the states of physiological systems often remain inaccessible, but it is possible to observe the temporal evolution of related biosignals. Their analysis may theoretically allow for the extraction of information about the complex dynamics of the system under study. Typical measures are related to the unpredictability and self-similarity of the recorded time series, as well as the amount of information exchanged between biosignals. These complexity measures are proving increasingly valuable for better understanding the functioning of a physiological system, monitoring its health over time, or predicting pathological events.

Therefore, the lively interest in the complexity analysis of biomedical signals has prompted us to propose this Special Issue, "Assessing Complexity in Physiological Systems through Biomedical Signals Analysis II", as a continuation of the Special Issue on complexity that we published in this same journal a few years ago [1]. The present Special Issue brings together thirteen contributions that address, from different methodological and applicative perspectives, the topic of complexity in physiological systems. The systems studied in this issue are the cardiovascular and central nervous systems. However, a clear separation between systems is impossible because psychological factors, cognitive loads, and the heart–brain connection via the autonomic nervous system, for example, produce important interactions between these systems, as well as with other regulatory systems in the body, such as those responsible for motor and respiratory control.

A total of five contributions concern heart rate variability (HRV). The paper "Heart Rate Complexity and Autonomic Modulation Are Associated with Psychological Response

Inhibition in Healthy Subjects" [2] studied HRV indices linked to entropy and sympathovagal balance in healthy volunteers during the execution of cognitive tests to assess flexibility, inhibition abilities, and rule learning. The authors explored the relationship between complexity, sympathovagal balance, and age in regulating impulsive reactions during cognitive tests, suggesting future applications of this methodology in assessing age-related cognitive decline and the brain–heart interaction.

In "Sample, Fuzzy, and Distribution Entropies of Heart Rate Variability: What Do They Tell Us about Cardiovascular Complexity?" [3], the authors compared three entropy metrics, often used in the HRV literature without properly considering their respective characteristics. The three metrics—Sample, Fuzzy, and Distribution Entropy—were applied to quantify the degree of unpredictability of heart rate series recorded in able-bodied and spinal cord-injured participants, separately in two postures. The authors compared entropy changes associated with the expected variations in HRV unpredictability that follow the postural shift, and with the expected alterations in cardiovascular complexity induced by the spinal lesion. The article discusses the robustness, sensitivity, and clinical interpretability of these entropy measures, emphasizing that, taken together, they complement each other.

An application of complex network analysis is presented in "Visibility Graph Analysis of Heartbeat Time Series" [4]. The authors applied a graph-based approach to HRV analysis. This method associates a complex network to the heart rate series, where a node of the network corresponds to a value of the series. The degree of connectivity among nodes may reveal subtle properties of the time-series dynamics, and the authors showed that the visibility graphs discriminate between young and old participants, between active and sedentary individuals, or between healthy subjects and heart failure patients.

The concept of self-organized criticality has recently been introduced to describe cardiovascular complexity. The paper "Autonomic Nervous System Influences on Cardiovascular Self-Organized Criticality" discusses how the autonomic system influences the properties of Zip's law, a feature of self-organized criticality [5]. The authors studied beat-by-beat heart rate recordings in professional soccer players during a training session, revealing an association between Zip's law characteristics and autonomic adjustments.

Finally, the study "Entropy-Based Multifractal Testing of Heart Rate Variability during Cognitive-Autonomic Interplay" applies a multifractality analysis to HRV recordings of volunteers undertaking three cognitive tasks (Stroop color and word task, stop-signal, and go/no-go) [6]. This contribution discusses the implications of cardiac complexity changes in relation to the interaction between the cognitive tasks and autonomic modulation.

On the central nervous system side, five studies use the electroencephalogram (EEG) as a window into brain complexity. The study "Spatio-Temporal Fractal Dimension Analysis from Resting State EEG Signals in Parkinson's Disease" performed a four-dimensional fractal analysis of the EEG tracings in Parkinson's disease patients and healthy controls [7]. This work provides evidence that such a complexity approach may characterize the specific changes in brain dynamics associated with Parkinson's disease.

In "A Resource-Efficient Multi-Entropy Fusion Method and Its Application for EEG-Based Emotion Recognition", the authors propose an entropy-based method for classifying emotional states with low computational costs [8]. Such an approach is promising for portable applications in mental health monitoring, human–computer interaction, and affective computing. The study "Improved EEG-Based Emotion Classification via Stockwell Entropy and CSP Integration" also uses complex EEG analysis for emotion recognition [9]. This work analyzed a publicly available dataset of EEG traces in volunteers while they watched videos encompassing positive, neutral, and negative emotions to induce coherent emotional responses. The authors demonstrate that the integration between Stockwell entropy and spatial filtering allows for obtaining a satisfactory affective classification.

In "Dynamic Evolution of EEG Complexity in Schizophrenia Across Cognitive Tasks", the authors employed Higuchi Fractal Dimension, a measure of signal complexity, to examine the temporal dynamics of EEG activity across five cortical regions during an attentional and a memory-based task in individuals diagnosed with schizophrenia and healthy controls [10]. They found that, while a consistent pattern of higher neural complexity characterizes the attentional task across the different brain regions in controls, by contrast, the EEG complexity varies in a task-dependent manner in schizophrenic patients, highlighting the need for dynamic rather than static approaches.

The study "Brain Complexity and Parametrization of Power Spectral Density in Children with Specific Language Impairment" compared measures of EEG complexity in children diagnosed with specific language impairment and in normo-developed children of the same age [11]. The results suggested alterations associated with the cognitive and linguistic characteristics of this disease, in the excitatory–inhibitory balance and intra- and inter-hemispheric connectivity.

Alongside EEG, the interactions between cerebral hemodynamics and respiration were explored in "Time-Resolved Information-Theoretic and Spectral Analysis of fNIRS Signals from Multi-Channel Prototypal Device" [12]. The authors applied informational and spectral measures to multichannel fNIRS signals during a breath-holding task to characterize the impact of respiratory activity on scalp hemodynamics within the framework of Network Physiology. They highlighted distinct informational dynamics across the breathing and apnea phases, opening new perspectives on the use of complexity for optical neuroimaging.

Another complex physiological system considered in this Special Issue concerns motor control and movement. The study "Multifractal Multiscale Analysis of Human Movements during Cognitive Tasks" explored the regularity of cycling movements by multiscale multifractality during progressive cognitive loads [13]. This study revealed how cognitive load alters the complexity of motor control, suggesting new perspectives for assessing coordination structures between higher neural centers and movement.

Finally, at the theoretical level, "A Spike Train Production Mechanism Based on Intermittency Dynamics" introduces a generative spike-train model of the neuron that reproduces high-frequency spontaneous membrane potential fluctuations by coupling intermittent maps, which are nonlinear first-order difference equations [14]. Such a model may prove to be useful as a test bed for complexity metrics in neurology.

Overall, this Special Issue presents a variety of applications and methodologies: from the heart to the brain, from electrical to hemodynamic and motion signals, and from entropic to multifractal measures. Emerging challenges include the need for robust metrics for short and noisy series, the standardization of the procedures, and the physiological interpretability of results, a prerequisite for clinical translation. The contributions collected here, however, demonstrate that complexity is not just an abstract concept but an attribute of living systems, which, if correctly described and quantified, might allow a better understanding of the physiological processes, possibly becoming a practical tool for aiding clinical interventions.

Acknowledgments: We express our thanks to the authors of the above contributions and to the editorial staff of the journal *Entropy* for their support during this work.

Conflicts of Interest: The authors declare no conflicts of interest.

References

1. Castiglioni, P.; Faes, L.; Valenza, G. Assessing Complexity in Physiological Systems through Biomedical Signals Analysis. *Entropy* **2020**, *22*, 1005. [CrossRef] [PubMed]
2. Riganello, F.; Vatrano, M.; Tonin, P.; Cerasa, A.; Cortese, M.D. Heart Rate Complexity and Autonomic Modulation Are Associated with Psychological Response Inhibition in Healthy Subjects. *Entropy* **2023**, *25*, 152. [CrossRef] [PubMed]
3. Castiglioni, P.; Merati, G.; Parati, G.; Faini, A. Sample, Fuzzy and Distribution Entropies of Heart Rate Variability: What Do They Tell Us on Cardiovascular Complexity? *Entropy* **2023**, *25*, 281. [CrossRef] [PubMed]
4. Muñoz-Diosdado, A.; Solís-Montufar, É.E.; Zamora-Justo, J.A. Visibility Graph Analysis of Heartbeat Time Series: Comparison of Young vs. Old, Healthy vs. Diseased, Rest vs. Exercise, and Sedentary vs. Active. *Entropy* **2023**, *25*, 677. [CrossRef] [PubMed]
5. Fortrat, J.-O.; Ravé, G. Autonomic Nervous System Influences on Cardiovascular Self-Organized Criticality. *Entropy* **2023**, *25*, 880. [CrossRef] [PubMed]
6. Arsac, L.M. Entropy-Based Multifractal Testing of Heart Rate Variability during Cognitive-Autonomic Interplay. *Entropy* **2023**, *25*, 1364. [CrossRef] [PubMed]
7. Ruiz De Miras, J.; Derchi, C.-C.; Atzori, T.; Mazza, A.; Arcuri, P.; Salvatore, A.; Navarro, J.; Saibene, F.L.; Meloni, M.; Comanducci, A. Spatio-Temporal Fractal Dimension Analysis from Resting State EEG Signals in Parkinson's Disease. *Entropy* **2023**, *25*, 1017. [CrossRef] [PubMed]
8. Li, J.; Feng, G.; Ling, C.; Ren, X.; Liu, X.; Zhang, S.; Wang, L.; Chen, Y.; Zeng, X.; Chen, R. A Resource-Efficient Multi-Entropy Fusion Method and Its Application for EEG-Based Emotion Recognition. *Entropy* **2025**, *27*, 96. [CrossRef] [PubMed]
9. Lu, Y.; Chen, J. Improved EEG-Based Emotion Classification via Stockwell Entropy and CSP Integration. *Entropy* **2025**, *27*, 457. [CrossRef] [PubMed]
10. Molina, R.; Crespo-Cobo, Y.; Esteban, F.J.; Arias, A.V.; Rodríguez-Árbol, J.; Soriano, M.F.; Ibáñez-Molina, A.J.; Iglesias-Parro, S. Dynamic Evolution of EEG Complexity in Schizophrenia Across Cognitive Tasks. *Entropy* **2025**, *27*, 226. [CrossRef] [PubMed]
11. Angulo-Ruiz, B.Y.; Rodríguez-Martínez, E.I.; Ruiz-Martínez, F.J.; Gómez-Treviño, A.; Muñoz, V.; Andalia Crespo, S.; Gómez, C.M. Brain Complexity and Parametrization of Power Spectral Density in Children with Specific Language Impairment. *Entropy* **2025**, *27*, 572. [CrossRef] [PubMed]
12. Franzone, I.; Antonacci, Y.; Giuliano, F.; Pernice, R.; Busacca, A.; Faes, L.; Giaconia, G.C. Time-Resolved Information-Theoretic and Spectral Analysis of fNIRS Signals from Multi-Channel Prototypal Device. *Entropy* **2025**, *27*, 694. [CrossRef] [PubMed]
13. Faini, A.; Arsac, L.M.; Deschodt-Arsac, V.; Castiglioni, P. Multifractal Multiscale Analysis of Human Movements during Cognitive Tasks. *Entropy* **2024**, *26*, 148. [CrossRef] [PubMed]
14. Potirakis, S.M.; Diakonos, F.K.; Contoyiannis, Y.F. A Spike Train Production Mechanism Based on Intermittency Dynamics. *Entropy* **2025**, *27*, 267. [CrossRef] [PubMed]

entropy

MDPI

Article

Heart Rate Complexity and Autonomic Modulation Are Associated with Psychological Response Inhibition in Healthy Subjects

Francesco Riganello [1,*], Martina Vatrano [1], Paolo Tonin [1], Antonio Cerasa [1,2,3] and Maria Daniela Cortese [1]

1 S. Anna Institute, Via Siris 11, 88900 Crotone, Italy
2 Institute for Biomedical Research and Innovation (IRIB), National Research Council of Italy (CNR), 98100 Messina, Italy
3 Pharmacotechnology Documentation and Transfer Unit, Preclinical and Translational Pharmacology, Department of Pharmacy, Health Science and Nutrition, University of Calabria, 87036 Rende, Italy
* Correspondence: f.riganello@isakr.it

Abstract: Background: the ability to suppress/regulate impulsive reactions has been identified as common factor underlying the performance in all executive function tasks. We analyzed the HRV signals (power of high (HF) and low (LF) frequency, Sample Entropy (SampEn), and Complexity Index (CI)) during the execution of cognitive tests to assess flexibility, inhibition abilities, and rule learning. Methods: we enrolled thirty-six healthy subjects, recording five minutes of resting state and two tasks of increasing complexity based on 220 visual stimuli with 12 × 12 cm red and white squares on a black background. Results: at baseline, CI was negatively correlated with age, and LF was negatively correlated with SampEn. In Task 1, the CI and LF/HF were negatively correlated with errors. In Task 2, the reaction time positively correlated with the CI and the LF/HF ratio errors. Using a binary logistic regression model, age, CI, and LF/HF ratio classified performance groups with a sensitivity and specificity of 73 and 71%, respectively. Conclusions: this study performed an important initial exploration in defining the complex relationship between CI, sympathovagal balance, and age in regulating impulsive reactions during cognitive tests. Our approach could be applied in assessing cognitive decline, providing additional information on the brain-heart interaction.

Keywords: heart rate variability; inhibitory control; complexity index; entropy

1. Introduction

Heart rate variability (HRV) is a well-known index of autonomic control of the heart, but it is also linked to cognitive and emotional control [1,2] as well as autonomic dysfunctions that may precede the cognitive impairment [3]. The heart rate is directly controlled by the brain through the sympathetic and parasympathetic branches of the autonomic nervous system (ANS), defining the well-known central nervous system-ANS axis, also known as central autonomic network (CAN) [4–7]. The HRV then represents the output of the complex brain-heart interaction [5,8,9]. Inhibition, also known as inhibitory control, is the ability to suppress or regulate impulsive (or automatic) reactions, producing responses via attention and reasoning. This mental ability is part of executive functions and aids in goal-setting and anticipatory planning. Inhibitory control stops inappropriate behaviors and spontaneous reactions and replaces them with a more appropriate, well-thought-out response.

Response inhibition has been identified as common factor underlying the performance in all executive function tasks [10]. They are generally divided into inhibition and interference control, working memory, and cognitive flexibility [11,12]. The last is also called set-shifting, mental flexibility, or mental set-shifting.

Alerting, orienting, and executive monitoring of actions involve salient autonomic responses observable in pupillary reactivity [13], skin conductance [14], and heart rate variability [15,16]. Response inhibition is one of the main cognitive tasks that require the ability to suppress a pre-potent motor response by adjusting it rapidly as a function of environmental changes.

The high-frequency band of the HRV spectrum correlates to parasympathetic heart activity, which is crucial for the individual to have an efficient adaptability to changing environmental demands. The vagal response is also correlated to respiration. Changes in breathing patterns can impact heart rate (HR) and HRV, with a general decrease in respiratory frequency associated with an increase in heart period (i.e., a decrease in heart rate). The respiratory characteristic linked to HR is known as respiratory sinus arrhythmia (RSA) [17,18]. A decrease in vagal control (i.e., reduced HF-HRV) may suggest a lack of flexibility in responding to changing demands, limiting the spectrum of possible responses and hence restricting the person's capacity to create appropriate responses and inhibit incorrect ones [5].

Forte and colleagues, in their review [3], refer that the most-reported HRV analysis during cognitive tests regards the high frequency (HF), low frequency (LF), and its LF/HF ratio. Moreover, studies showed that higher values in HRV entropy were related to cognitive performances [19,20].

Since the heart rhythm is not regular, the entropy analysis better represents its complexity and unpredictable variability [21,22]. Higher and lower entropy rates correlate to a more complex heartbeat sequence and a more regular and predictable heartbeat, respectively. Given the complexity of the brain-heart two-way interaction described by the CAN [7], the HRV entropy may be used to assess the system's health. In fact, a more complex heartbeat sequence was found to indicate better reactivity to the external/internal stimulus [19,23–25].

In the frame of the HRV entropy analysis, the Multiscale Entropy (MSE) [26] was developed to investigate the information content in non-linear signals at different temporal scales (coarse-graining), generally using the Sample Entropy [27]. The Complexity Index (CI), a scalar score that permits gaining insights into the integrated complexity of the measuring system, is the sum of the entropies calculated for several coarse-graining scales. [26]. It was found that the heart rate complexity correlates with brain activity [19,28,29], and that the complexity in heartbeat dynamics grows with it and vanishes with stress.

This work aims to define, for the first time, the overall characteristics of HRV signals in the frequency and non-linear domains during the execution of a Go/NoGo task. This task is used to assess someone's ability to regulate an inappropriate response among interfering visual stop signal (NoGo) by pressing a button whenever a Go signal stimuli is made [30]. The presented test is based on an adaptation of the rule shift card test. This well-known cognitive task assesses flexibility, inhibition abilities, and rule learning [31,32]. We want to assess how the response inhibition in the rule shift card test [31,32] is linked to a higher brain-heart interaction. In particular, we sought to determine the following: (1) if there are significant differences within and between baseline and task phases, in terms of entropy and spectral parameters, (2) the correlations between the errors in the two Go/noGo tasks and physiological parameters, and (3) the accuracy to predict the poor or good performance of the subjects using recorded HRV parameters in the baseline.

2. Materials and Methods

2.1. Subjects

Participants were recruited from Institute S.Anna, Crotone, Italy. The chosen participants had never used any drugs before. Before the experiment, all participants were instructed to refrain from smoking and consuming caffeine for four to six hours. The following criteria were specifically required for inclusion: (1) no evidence of dementia or depression symptoms according to DSM-V criteria; (2) no use of antidepressant, anxiolytic, or antipsychotic drugs that could affect cerebral blood flow; (3) right-handedness;

and (4) absence of chronic medical conditions (heart disease, hypertension, or diabetes). According to these criteria, 36 healthy graduate subjects were enrolled in this study with median age 41 and interquartile range 16 (21 females age 36 ± 11 and 15 males age 43 ± 9). No significant differences were between males and females for age (Mann-Whitney test: $Z = -1.912$, $p = 0.056$). All participants had normal or corrected to normal vision and normal color vision. All the participants gave written informed consent. In accordance with the Helsinki Declaration, the study was approved by the Ethical Committee of Regione Calabria (n.ro 172 17 July 2020).

2.2. Procedure

The experiment was conducted in a dimly lighted, soundproof room. Participants completed a modified version of the traditional Go/NoGo activity based on the Rule Shift Cards [31–33], designed using Biotrace+, while seated in a comfortable chair (https://www.mindmedia.com/Herten/Nederland/ accessed on 5 November 2022).

In the last task, the subject must modify the strategy learned in the first task, memorizing the element that previously appeared and inhibiting the response when the square color is different from that of the previous square.

The protocol study consisted of a baseline (resting state condition) lasting 5 min and two different tasks. The tasks consisted of 220 visual stimuli with 12 × 12 cm red, white, and chess pattern squares on a black background. The first task included three distinct visual stimuli: red squares (frequent stimulus n.154—70%), white squares (rare stimulus n.44—20%), and a chess pattern with squares (distractor n.22—10%). It was based on fundamental response inhibition. A working memory response inhibition task made up the second. There were just red (common stimulus n.176—80%) and white (rare stimulus n.44—20%) squares in this instance. Each stimulus had a 1500 ms gap in between them, and each stimulus lasted 500 ms. The subject had to press the spacebar on the keyboard during the first task when the white square appeared, and during the second task when two successive visual stimuli of the same color were shown (Figure 1).

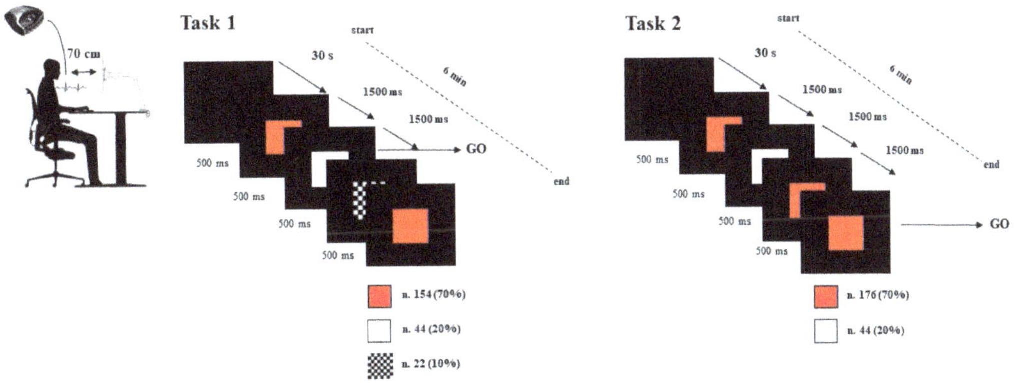

Figure 1. *Experimental detail*: Subject comfortably sits in front of the screen at 70 cm of distance from the monitor with the hand positioned near the spacebar of the computer keyboard. In the first task, the GO is represented by the white square, and the subject must hit the spacebar when it appears, and inhibit the response when the red square appears (the chess pattern with squares was the distractor). In the second task, the GO is represented by the appearance of a square with color equal to the previous square, while the NoGo is represented by a change in the appearance of the square color. The first square of the tasks appeared after 30 s of a black image. The stimulus duration was 500 ms and the interval of time between the stimuli was 1500 ms.

The subject was comfortably seated at a distance of 70 cm from the 24-inch monitor where the sequence of stimuli was displayed. There was no transient noise, and the environment was always the same temperature and brightness. The participant received assignment instructions prior to each activity.

2.3. Data Acquisition

The ECG recording was performed by NEXUS-32, and the stimuli were presented by Biotrace+ software (https://www.mindmedia.com/Herten/Nederland accessed on 5 November 2022). Since the ECG sample frequency rate of the signal acquisition can affect the HRV analysis, and a minimum sample of 250 Hz or higher are suggested [34,35], the signal was acquired at 256 Hz, and the 4 Hz cubic spline interpolation was applied for a correct extraction of the R-peaks.

2.4. Data Analysis

ECG was analyzed by Kubios advanced software for HRV analysis (v 3.1/Kuopio, Finland). The interpolation approach was used to eliminate artifacts and ectopic beats from the data. The CI (i.e., summation of the MSE from 1 to 3) and the natural logarithm of the HRV power of HF (0.15–0.5 Hz) (LnHF), LF (0.04–0.15 Hz) (LnLF), and LF/HF ratio were calculated. Heartbeat, like most physiological signals, is non-stationary due to the complex nature of the biological systems. It frequently contains either slow trends or very slow frequency oscillations. Since the HRV parameters may be affected by the non-stationarity of the signal, a quadratic polynomial model detrended the RR series to reduce the influence of lower frequencies in the power spectral density (PSD) results. The HF and LF PSD were calculated using the Fast Fourier Transform method (Welch's PSD; windows width: 150 s). The transformation of the spectral power in their natural logarithm was applied because the measures showed a skewed distribution (i.e., skewness 2.48 ± 0.39; kurtosis 7.07 ± 0.77). The CI was based on the multiscale entropy (MSE) approach quantifying the degree of irregularity over a range of coarse-grained scales (τ) from 1 to 3. The interval between two consecutive R peaks of the QRS ECG complex represents the data points in the entropy analysis (Figure 2). The original series represents the scale for $\tau = 1$. The coarse-grained scales 2 and 3 were constructed by averaging the IBI/tachogram's data points within non-overlapping windows of increasing length τ (Figure 2). The sample entropy (SampEn) was calculated for each coarse-grained scale, and the CI was extracted by summing the sample entropy for each coarse-grained scale. Given a sample of length N, the SampEn is defined as the negative natural logarithm of the probability that if two sets of simultaneous data points of length m have distance $< r$, then two sets of simultaneous data points of length $m + 1$ also have distance $< r$ [27]. Considering the length of our samples (i.e., original series: baseline $N = 423 \pm 101$, task 1 $N = 509 \pm 83$, task 2 $N = 525 \pm 102$), the parameters m (i.e., embedding dimension; length of the window of the different vector comparisons) and r (i.e., level of tolerance: generally ranging from 0.1 to 0.25, corresponding to the 10–25% of the standard deviation of the series of data onto analysis) of SampEn were set to 2 and 0.2, respectively [27,36], for the original and the coarse-grained scale 2 and 3 (i.e., r was the 20% of the standard deviation of the original and rescaled time series).

2.5. Statistical Analysis

The non-parametric exact test was used for the statistical analysis. This approach provides more accurate results when the sample size is small or in the case of tables sparse or imbalanced. By the Wilcoxon exact test, the HRV parameters (LnLF, LnHF, LF/HF, SampEn for the original and the CI were compared in the different tasks. The effect size r was calculated as the absolute value of $Z\sqrt{(2N)}$ (Wilcoxon's test), where Z is the Z statistic of the statistical test and N is the total number of subjects. The effect size results were considered as follows: $r < 0.1$ not significant; $0.1 \leq r < 0.3$ low; $0.3 \leq r < 0.5$ medium; $r \geq 0.5$ high. The level of significance was set at $p \leq 0.05$.

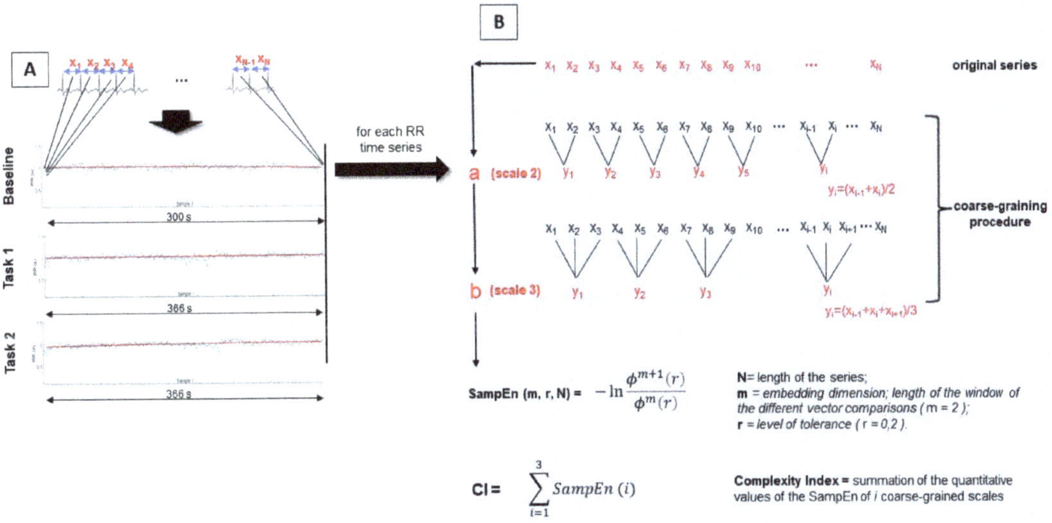

Figure 2. *Multiscale Entropy and Complexity Index Scheme.* (**A**) For each subject the tachogram was extracted in baseline, task 1, and task 2. N represents the length of each original detrended (trend line in red) time series (baseline N = 422 ± 100; task 1 N = 509 ± 83; task 2 N = 524 ± 102). (**B**) The sample entropy (SampEn) was calculated for the original and coarse-grained series A and B, setting the parameters m and r at 2 and 0.2, respectively. The complexity index (CI) was calculated as the sum of the SampEn of the scales 1 (original series), 2 (a), and 3 (b).

The Pearson correlation test analyzed the correlation between HRV parameters and performance levels (errors and Reaction Time).

Because of outliers, the median of total errors was used to divide the subjects into Good Performance (GP) and Poor Performance (PP) groups. Good or poor performance was predicted by binary logistic regression. The backward approach in the logistic regression was used to select the regressors in the model, inserting the performance (i.e., GP and PP classification) as dichotomic variable, and age and HRV parameters (i.e., C.I., LnLF, LnHF, and LF/HF) as independent variables. The backward approach uses the Wald tests' results for the regressor's elimination, removing the variable with the least significant effect that does not meet the level for staying in the model [37]. The removed effect is excluded from the model, and the process is repeated until no other effect in the model meets the specified level for removal. The model's accuracy was also checked, controlling the variables for collinearity by tolerance and its reciprocal variance inflation factor (VIF) [38].

3. Results

Behavioral Data

The inhibitory task was successfully completed by all subjects. Table 1 reports the response time (RT) for GO trials and the percentage of errors for GO and NOGO trials. The RT and numbers of errors in Go and NoGo conditions raised in Task 2, as aspected.

The number of errors were lower in task 1, with a decrease over time. Conversely, they increased over time in task 2 (Figure 3).

In the group, significant differences between baseline and Task 2, with a decreasing trend, were found for LnLF ($Z = -3.189$, $p = 0.001$, $r = 0.38$), LnHF ($Z = -2.726$, $p = 0.005$, $r = 0.32$) and CI ($Z = -2.317$, $p = 0.02$, $r = 0.27$). The same parameters showed significant differences comparing task 1 and task 2 ($-3.473 \leq Z \leq -2.789$, $0.0001 \leq p \leq 0.004$, $0.33 \leq r \leq 0.41$) (Figure 4).

Table 1. Behavioral Variables.

	Task 1	Task 2
	Mean ± SD	Mean ± SD
RT during GO trials (s)	0.37 ± 0.08	0.49 ± 0.11
% errors during GO trials	0.04%	1.38%
% errors during NoGo trials	1.81%	2.93%

RT: Reaction Time; SD: Standard Deviation.

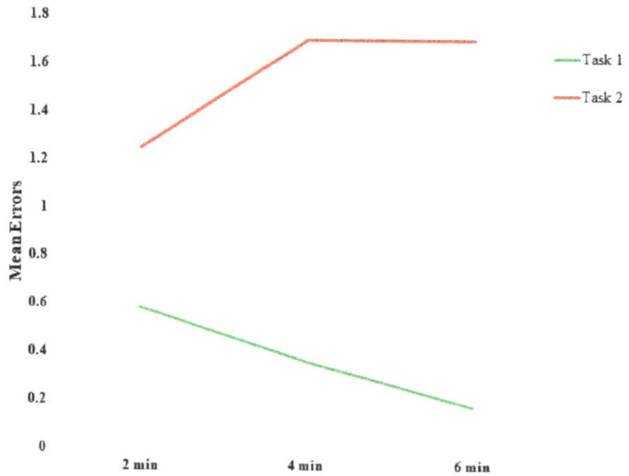

Figure 3. Mean errors during the tasks: number of errors during task 1 (green line) and task 2 (red line) along the time.

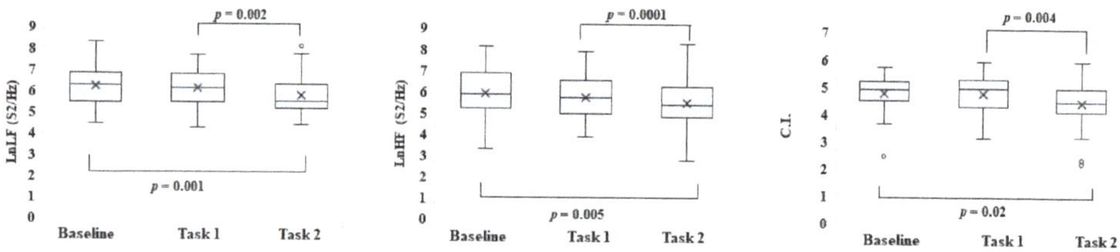

Figure 4. *Boxplot of the HRV parameters*: Boxplot of the natural logarithm of low-frequency power (LnLF) and high-frequency power (LnHF), and Complexity Index (CI) in resting-state (baseline) and during task 1 and task 2.

In baseline, CI had a negative correlation with age (Rho = -0.373, $p = 0.026$) and the SampEn with the LnLF (Rho = -0.351, $p = 0.036$). In Task 1, the SampEn had a positive correlation with LnHF (Rho = 0.495, $p = 0.002$) and negative correlation with LF/HF (Rho = -0.589, $p = 0.0001$), while CI and SampEn had a negative correlation with the errors (Rho = -0.455, $p = 0.005$ and Rho = -0.419, $p = 0.01$, respectively) and positive correlation with LF/HF (Rho = 0.448, $p = 0.006$). In Task 2, LF/HF had a negative correlation with SampEn (Rho = -0.387, $p = 0.02$) and positive correlation with the errors (Rho = 0.350, $p = 0.036$), while the Reaction Time had a negative correlation with the LnLF (Rho = -0.522, $p = 0.001$) and positive correlation with the CI (Rho = 0.378, $p = 0.02$). Excluding the outlier from the analysis overall patterns of significant findings did not change (Figure 5).

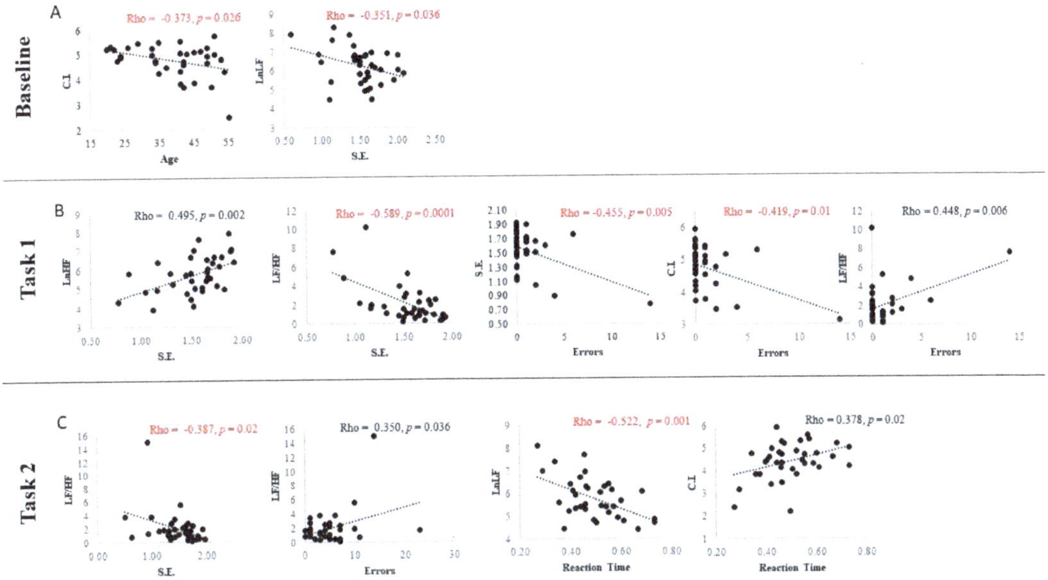

Figure 5. *Correlations among HRV parameters and performance:* Correlation among HRV parameters (SampEn (SE), Complexity Index (CI), HF power (LnHF), LF power (LnLF), LF/HF), performance levels (errors and Reaction Time) and age of the group. In red and black are the negative and positive correlations, respectively. (**A**) baseline; (**B**) task 1; (**C**) task 2.

Because of the presence of outliers by the median of the total errors (median of errors = 5) the subjects were divided in two sample groups: Good Performance (GP) (20 subjects; seven males aged 44 $\pm$ 7; 13 females age 35 $\pm$ 11; median of errors $\leq$ 5) and Poor Performance (PP) (16 subjects; eights males aged 43 $\pm$ 10; eight female age 39 $\pm$ 11; median of errors > 5).

The binary logistic regression was used to observe the probability of predicting the performances (i.e., GP/PP) from the HRV recorded values in the baseline. By backward stepwise binary logistic regression, age, CI, and LF/HF ratio were selected as regressors to classify GP and PP. No collinearity was among the selected variables (i.e., tolerance $\geq$ 0.96, VIF $\leq$ 1.8). The extracted model (GP/PP = $-9{,}35 + 0.036 \times$ age $+ 1.375 \times$ CI $+ 0.597 \times$ LF/HF; significance of the model (omnibus test) $p = 0.032$; Cox & Snell's R2 = 0.22; Naghelkerte's R2 = 0.29; Hosmer & Lemeshow's test $p = 0.32$) classified correctly the GP (16/20; 80%) and PP (10/16; 63%) with a sensitivity of 73%, specificity of 71%, and balanced accuracy of 71%.

4. Discussion

We demonstrated that during response inhibition tasks, HRV metrics (LF, HF, SampEn, CI, and LF/HF ratio) are characterized by significant changes as a function of performance. In particular, we found that the following: (1) the CI and the logarithmic values of LF and HF decreased from baseline to task 2; (2) in baseline, the CI correlated negatively with age; (3) in task 2 the CI is positively correlated with the reaction time, and the LF/HF ratio is positively correlated with the errors; finally, (4) in baseline age, CI and LF/HF ratio are predictors of better performance (i.e., fewer numbers of errors) in the tests.

Several works report higher levels of HRV associated with better emotional responses [39], attentional [2,40], and executive performances [1,32,41,42], while lower HRV was associated with dysfunctional attitudes [43]. Studies found that in healthy subjects, a greater HRV in the resting state predicted better performance in tasks related to executive brain function [1,44,45], with similar results in psychopathological populations [1,45]. In resting-state conditions, less efficient task performance and lower response accuracy in pleasant and unpleasant Go trials were associated with lower HRV values [46]. Ottaviani and colleagues [32] demonstrated that higher parasympathetic levels predicted performance during the Rule Shift Cards and the Hayling Sentence Completion inhibitory tasks.

We found that LnLF and LnHF power and CI decreased from baseline to task 2. The results are consistent with the reactivity or the possible state of stress during mental fatigue [22,29,47–49].

Several studies reported the correlation between autonomic function and cognition, observing a higher parasympathetic modulation associated with better performance on tasks involving executive function [1,50–52]. However, knight and colleagues suggest that both sympathetic and parasympathetic should be considered together to observe changes in cognitive functions [53]. Indeed, a better cognitive performance was associated with increased sympathetic activity but not decreased vagal activity in older patients [54]. In our study, in baseline LnLF was negatively correlated with the entropy, and in task 1 and 2 the LF/HF ratio positively correlated with the errors (i.e., increases in the parasympathetic tone).

A decrease in the HRV complexity was correlated to major depressive disorders [20] and during cognitive tasks in anxious subjects [55]. Cardiac complexity was found to be particularly effective in differentiating between active, effortful emotion regulation and less effortful control and dysregulation [56]. In the field of disorders of consciousness, the CI was found to be a useful marker to discriminate unresponsive wakefulness syndrome/vegetative state patients from minimally conscious state patients [25], with higher CI correlated with a higher level of consciousness. Again, HRV complexity diminishes with age and disease [57,58], while MSE research in healthy persons over 40 revealed an age-related fall in heart rate complexity [59,60].

We found the CI negatively correlated with the errors in task 1 and positively correlated with the reaction time in task 2. Bakhchina and colleagues found that higher HRV entropy, measured by the Sample Entropy, is associated with a more complex response in goal-directed behavioral [61]. In particular, our results highlight that a higher CI is linked to less impulsivity and fewer errors, indicating a more rich complexity in the brain-heart interaction [62]. Moreover, we found that higher reaction time is associated with higher CI and lower LF. These results suggest that a higher brain-heart interaction is related to higher flexibility and adaptation at the rule shift. Moreover, we found that the CI is negatively correlated with age.

These findings supported the relationship between sympathovagal control, CI, and age, due to a reduction in the complexity of the physiologic dynamics associated with aging [63,64] and their involvement in the test performance. Moreover, the negative correlation between the SE with LF/HF ratio highlights the effect of the sympathetic response on the heart rate complexity. Porta and colleagues correlated the increase in sympathetic activity with decreased HRV entropy [65]. The sympathetic activity was found to increase during mental stress induced by various methods in healthy subjects [66,67]. The effect induced by the increased sympathetic activity is an increase in the heart rate and regularity of the heartbeats producing a decrease in the heart rate complexity [68,69].

Furthermore, CI, age, and LF/HF ratio were the regressors involved in the logistic model, which allowed in baseline to predict the performances with sensitivity and specificity of 73% and 72%, respectively.

To the best of our knowledge, in this study, a significant correlation between CI and inhibitory performance has been described for the first time. In particular, we found that the increased HRV entropy was linked to increased reaction time and decreased LF/HF ratio. The HRV complexity better identifies considerable changes in autonomic regulation providing new insight into changes under various physiological and pathological conditions, complementing the analysis in the time and frequency domain [70–73]. It has been demonstrated that the CI provided novel information on the brain-heart interaction in patients with disorders of consciousness [25,74,75]. Our data could provide new markers to assess the performances in cognitive tests in healthy and pathological subjects.

However, the relatively small sample size represents the limitations of this study. Moreover, the HRV decreases with the age [76,77], and as well as it was documented, the menstrual cycle can influence the HRV in women and also its non-linear property [78]. Again, a recording of 300 s for the baseline and 366 s for the tasks could represent a limit in calculating the multiscale entropy. However, the sample entropy has also been calculated in short data points (i.e., N = 100; [36]). A bigger sample size could better explain the outlier observed in Figure 2, where a male subject over 40 increased the LF/HF values from 5.5 in baseline to 14.9 in task 2. The outlier does not change the significance of the results. However, it is interesting to observe that the high error numbers are associated with increased HR from baseline to task 2 and that while LF did not change, the HF decreased. This highlights the importance of HF autonomic modulation during cognitive tests [79].

On the other hand, the strength of this study was the evaluation of the non-linear proprieties of the HRV during all phases (baseline, tasks 1 and 2). In this way, it has been possible to observe differences in the performance non-related to a significant difference in the vagal modulation of the ANS but to a different modulation in the brain-heart two-way interaction. Further studies should consider the ECG and EEG simultaneous recording to understand the brain-heart two-way interaction better. Moreover, the use of novel methods in the entropy analysis, such as Refined MSE, the Linear MSE, or the Modified MSE [80,81], could better explain this complex interaction. Moreover, the analysis of the respiratory sinus arrhythmia could provide further information about how vagal system mediates the cognitive performance.

Our study performed an important initial exploration in defining the complex relationship between CI, sympathovagal balance, and age in inhibitory function.

Our approach could be applied in assessing cognitive decline and as a complement to the EEG and fMRI analysis, providing additional information on the brain-heart interaction.

Author Contributions: M.D.C. developed the experimental design; F.R. and M.V. performed the analysis; F.R. and M.V. acquired ECG data; F.R., M.D.C., M.V. and A.C. write the paper. F.R., M.V., P.T., A.C. and M.D.C., contribute to the final version of the paper. All authors have read and agreed to the published version of the manuscript.

Funding: This research received no external funding.

Institutional Review Board Statement: The study was approved by the Ethical Committee of Regione Calabria (n.ro 172 17 July 2020).

Informed Consent Statement: Informed consent was obtained from all subjects involved in the study.

Data Availability Statement: The data presented in this study are available on request from the corresponding author.

Conflicts of Interest: The authors declare no conflict of interest.

References

1. Hansen, A.L.; Johnsen, B.H.; Thayer, J.F. Vagal Influence on Working Memory and Attention. *Int. J. Psychophysiol.* **2003**, *48*, 263–274. [CrossRef] [PubMed]
2. Duschek, S.; Muckenthaler, M.; Werner, N.; del Paso, G.A.R. Relationships between Features of Autonomic Cardiovascular Control and Cognitive Performance. *Biol. Psychol.* **2009**, *81*, 110–117. [CrossRef] [PubMed]
3. Forte, G.; Favieri, F.; Casagrande, M. Heart Rate Variability and Cognitive Function: A Systematic Review. *Front. Neurosci.* **2019**, *13*, 710. [CrossRef] [PubMed]
4. Napadow, V.; Dhond, R.; Conti, G.; Makris, N.; Brown, E.N.; Barbieri, R. Brain Correlates of Autonomic Modulation: Combining Heart Rate Variability with FMRI. *Neuroimage* **2008**, *42*, 169–177. [CrossRef] [PubMed]
5. Thayer, J.F.; Lane, R.D. Claude Bernard and the Heart–Brain Connection: Further Elaboration of a Model of Neurovisceral Integration. *Neurosci. Biobehav. Rev.* **2009**, *33*, 81–88. [CrossRef]
6. Tonhajzerova, I.; Ondrejka, I.; Turianikova, Z.; Javorka, K.; Calkovska, A.; Javorka, M. Heart Rate Variability: An Index of the Brain–Heart Interaction. In *Tachycardia*; InTechOpen: London, UK, 2012; pp. 185–202.

7. Benarroch, E.; Singer, W.; Mauermann, M. *Autonomic Neurology*; Oxford University Press: New York, NY, USA, 2014; ISBN 978-0-19-992019-8.
8. Palma, J.-A.; Benarroch, E.E. Neural Control of the Heart: Recent Concepts and Clinical Correlations. *Neurology* **2014**, *83*, 261–271. [CrossRef]
9. Porges, S.W. The Polyvagal Theory: New Insights into Adaptive Reactions of the Autonomic Nervous System. *Clevel. Clin. J. Med.* **2009**, *76*, S86–S90. [CrossRef]
10. Yeh, C.-H.; Juan, C.-H.; Yeh, H.-M.; Wang, C.-Y.; Young, H.-W.V.; Lin, J.-L.; Lin, C.; Lin, L.-Y.; Lo, M.-T. The Critical Role of Respiratory Sinus Arrhythmia on Temporal Cardiac Dynamics. *J. Appl. Physiol.* **2019**, *127*, 1733–1741. [CrossRef]
11. Lewis, G.F.; Furman, S.A.; McCool, M.F.; Porges, S.W. Statistical Strategies to Quantify Respiratory Sinus Arrhythmia: Are Commonly Used Metrics Equivalent? *Biol. Psychol.* **2012**, *89*, 349–364. [CrossRef]
12. Miyake, A.; Friedman, N.P. The Nature and Organization of Individual Differences in Executive Functions: Four General Conclusions. *Curr. Dir. Psychol. Sci.* **2012**, *21*, 8–14. [CrossRef]
13. Miyake, A.; Friedman, N.P.; Emerson, M.J.; Witzki, A.H.; Howerter, A.; Wager, T.D. The Unity and Diversity of Executive Functions and Their Contributions to Complex "Frontal Lobe" Tasks: A Latent Variable Analysis. *Cogn. Psychol.* **2000**, *41*, 49–100. [CrossRef] [PubMed]
14. Lehto, J.E.; Juujärvi, P.; Kooistra, L.; Pulkkinen, L. Dimensions of Executive Functioning: Evidence from Children. *Br. J. Dev. Psychol.* **2003**, *21*, 59–80. [CrossRef]
15. Geva, R.; Zivan, M.; Warsha, A.; Olchik, D. Alerting, Orienting or Executive Attention Networks: Differential Patters of Pupil Dilations. *Front. Behav. Neurosci.* **2013**, *7*, 145. [CrossRef] [PubMed]
16. Howells, F.M.; Stein, D.J.; Russell, V.A. Perceived Mental Effort Correlates with Changes in Tonic Arousal during Attentional Tasks. *Behav. Brain Funct.* **2010**, *6*, 39. [CrossRef]
17. Ramírez, E.; Ortega, A.R.; Reyes Del Paso, G.A. Anxiety, Attention, and Decision Making: The Moderating Role of Heart Rate Variability. *Int. J. Psychophysiol.* **2015**, *98*, 490–496. [CrossRef]
18. Sørensen, L.; Wass, S.; Osnes, B.; Schanche, E.; Adolfsdottir, S.; Svendsen, J.L.; Visted, E.; Eilertsen, T.; Jensen, D.A.; Nordby, H.; et al. A Psychophysiological Investigation of the Interplay between Orienting and Executive Control during Stimulus Conflict: A Heart Rate Variability Study. *Physiol. Behav.* **2019**, *211*, 112657. [CrossRef]
19. Aboy, M.; Cuesta-Frau, D.; Austin, D.; Mico-Tormos, P. Characterization of Sample Entropy in the Context of Biomedical Signal Analysis. In Proceedings of the 29th Annual International Conference of the IEEE Engineering in Medicine and Biology Society, Lyon, France, 22–26 August 2007; pp. 5942–5945.
20. Bouny, P.; Deschodt-Arsac, V.; Touré, E.; Arsac, L. Entropy and Multifractality in Heart Rate Dynamics as Markers of Specific Brain-Heart Coordinations When Adapting to Cognitive Tasks. In Proceedings of the ACAPS 2021 Neurosciences—Contrôle moteur—Methodology and Technology, Montpellier, France, 27–29 October 2021.
21. Bornas, X.; Llabrés, J.; Noguera, M.; López, A.M.; Gelabert, J.M.; Vila, I. Fear Induced Complexity Loss in the Electrocardiogram of Flight Phobics: A Multiscale Entropy Analysis. *Biol. Psychol.* **2006**, *73*, 272–279. [CrossRef]
22. Voss, A.; Heitmann, A.; Schroeder, R.; Peters, A.; Perz, S. Short-Term Heart Rate Variability–Age Dependence in Healthy Subjects. *Physiol. Meas.* **2012**, *33*, 1289–1311. [CrossRef]
23. Riganello, F.; Larroque, S.K.; Bahri, M.A.; Heine, L.; Martial, C.; Carrière, M.; Charland-Verville, V.; Aubinet, C.; Vanhaudenhuyse, A.; Chatelle, C.; et al. A Heartbeat Away from Consciousness: Heart Rate Variability Entropy Can Discriminate Disorders of Consciousness and Is Correlated with Resting-State FMRI Brain Connectivity of the Central Autonomic Network. *Front. Neurol.* **2018**, *9*, 769. [CrossRef]
24. Deschodt-Arsac, V.; Blons, E.; Gilfriche, P.; Spiluttini, B.; Arsac, L.M. Entropy in Heart Rate Dynamics Reflects How HRV-Biofeedback Training Improves Neurovisceral Complexity during Stress-Cognition Interactions. *Entropy* **2020**, *22*, 317. [CrossRef]
25. Costa, M.; Goldberger, A.L.; Peng, C.-K. Multiscale Entropy Analysis of Biological Signals. *Phys. Rev. E* **2005**, *71*, 021906. [CrossRef] [PubMed]
26. Delgado-Bonal, A.; Marshak, A. Approximate Entropy and Sample Entropy: A Comprehensive Tutorial. *Entropy* **2019**, *21*, 541. [CrossRef] [PubMed]
27. Young, H.; Benton, D. We Should Be Using Non-linear Indices When Relating Heart-Rate Dynamics to Cognition and Mood. *Sci. Rep.* **2015**, *5*, 16619. [CrossRef]
28. Dimitriev, D.A.; Saperova, E.V.; Dimitriev, A.D. State Anxiety and Nonlinear Dynamics of Heart Rate Variability in Students. *PLoS ONE* **2016**, *11*, e0146131. [CrossRef] [PubMed]
29. Byun, S.; Kim, A.Y.; Jang, E.H.; Kim, S.; Choi, K.W.; Yu, H.Y.; Jeon, H.J. Entropy Analysis of Heart Rate Variability and Its Application to Recognize Major Depressive Disorder: A Pilot Study. *Technol. Health Care* **2019**, *27*, 407–424. [CrossRef]
30. Georgiou, G.; Essau, C.A. Go/No-Go Task. In *Encyclopedia of Child Behavior and Development*; Goldstein, S., Naglieri, J.A., Eds.; Springer: Boston, MA, USA, 2011; pp. 705–706, ISBN 978-0-387-79061-9.
31. Espinosa, A.; Alegret, M.; Boada, M.; Vinyes, G.; Valero, S.; Martínez-Lage, P.; Peña-Casanova, J.; Becker, J.T.; Wilson, B.A.; Tárraga, L. Ecological Assessment of Executive Functions in Mild Cognitive Impairment and Mild Alzheimer's Disease. *J. Int. Neuropsychol. Soc.* **2009**, *15*, 751–757. [CrossRef]
32. Ottaviani, C.; Zingaretti, P.; Petta, A.M.; Antonucci, G.; Thayer, J.F.; Spitoni, G.F. Resting Heart Rate Variability Predicts Inhibitory Control above and beyond Impulsivity. *J. Psychophysiol.* **2019**, *33*, 198–206. [CrossRef]

33. Antonucci, G.; Spitoni, G.F.; Orsini, A.; D'Olimpio, F.; Cantagallo, A. Taratura Italiana Della Batteria per La Valutazione Della Sindrome Disesecutiva: BADS. Available online: https://iris.unicampania.it/handle/11591/159383?mode=full.274 (accessed on 22 April 2022).
34. Task Force of the European Society of Cardiology and the North American Society of Pacing and Electrophysiology. Heart Rate Variability: Standards of Measurement, Physiological Interpretation and Clinical Use. *Circulation* **1996**, *93*, 1043–1065. [CrossRef]
35. Singh, B.; Singh, M.; Banga, V.K. Sample Entropy Based HRV: Effect of ECG Sampling Frequency. *Biomed. Sci. Eng.* **2014**, *2*, 68–72. [CrossRef]
36. Richman, J.S.; Moorman, J.R. Physiological Time-Series Analysis Using Approximate Entropy and Sample Entropy. *Am. J. Physiol. -Heart Circ. Physiol.* **2000**, *278*, H2039–H2049. [CrossRef]
37. Bursac, Z.; Gauss, C.H.; Williams, D.K.; Hosmer, D.W. Purposeful Selection of Variables in Logistic Regression. *Source Code Biol. Med.* **2008**, *3*, 17. [CrossRef] [PubMed]
38. Midi, H.; Sarkar, S.K.; Rana, S. Collinearity Diagnostics of Binary Logistic Regression Model. *J. Interdiscip. Math.* **2010**, *13*, 253–267. [CrossRef]
39. Appelhans, B.M.; Luecken, L.J. Heart Rate Variability as an Index of Regulated Emotional Responding. *Rev. Gen. Psychol.* **2006**, *10*, 229–240. [CrossRef]
40. Williams, D.P.; Thayer, J.F.; Koenig, J. Resting Cardiac Vagal Tone Predicts Intraindividual Reaction Time Variability during an Attention Task in a Sample of Young and Healthy Adults. *Psychophysiology* **2016**, *53*, 1843–1851. [CrossRef] [PubMed]
41. Hansen, A.L.; Johnsen, B.H.; Sollers, J.J.; Stenvik, K.; Thayer, J.F. Heart Rate Variability and Its Relation to Prefrontal Cognitive Function: The Effects of Training and Detraining. *Eur. J. Appl. Physiol.* **2004**, *93*, 263–272. [CrossRef]
42. Colzato, L.S.; Steenbergen, L. High Vagally Mediated Resting-State Heart Rate Variability Is Associated with Superior Action Cascading. *Neuropsychologia* **2017**, *106*, 1–6. [CrossRef]
43. Beevers, C.G.; Ellis, A.J.; Reid, R.M. Heart Rate Variability Predicts Cognitive Reactivity to a Sad Mood Provocation. *Cogn. Ther. Res.* **2011**, *35*, 395–403. [CrossRef]
44. Hansen, A.L.; Johnsen, B.H.; Thayer, J.F. Relationship between Heart Rate Variability and Cognitive Function during Threat of Shock. *Anxiety Stress Coping* **2009**, *22*, 77–89. [CrossRef]
45. Hovland, A.; Pallesen, S.; Hammar, Å.; Hansen, A.L.; Thayer, J.F.; Tarvainen, M.P.; Nordhus, I.H. The Relationships among Heart Rate Variability, Executive Functions, and Clinical Variables in Patients with Panic Disorder. *Int. J. Psychophysiol.* **2012**, *86*, 269–275. [CrossRef]
46. Moretta, T.; Sarlo, M.; Buodo, G. Problematic Internet Use: The Relationship Between Resting Heart Rate Variability and Emotional Modulation of Inhibitory Control. *Cyberpsychol. Behav. Soc. Netw.* **2019**, *22*, 500–507. [CrossRef]
47. Thayer, J.F.; Friedman, B.H.; Borkovec, T.D. Autonomic Characteristics of Generalized Anxiety Disorder and Worry. *Biol. Psychiatry* **1996**, *39*, 255–266. [CrossRef] [PubMed]
48. Verkuil, B.; Brosschot, J.F.; Borkovec, T.D.; Thayer, J.F. Acute Autonomic Effects of Experimental Worry and Cognitive Problem Solving: Why Worry about Worry? *Int. J. Clin. Health Psychol.* **2009**, *9*, 439–453.
49. Melo, H.M.; Nascimento, L.M.; Takase, E. Mental Fatigue and Heart Rate Variability (HRV): The Time-on-Task Effect. *Psychol. Neurosci.* **2017**, *10*, 428–436. [CrossRef]
50. Barber, A.D.; John, M.; DeRosse, P.; Birnbaum, M.L.; Lencz, T.; Malhotra, A.K. Parasympathetic Arousal-Related Cortical Activity Is Associated with Attention during Cognitive Task Performance. *Neuroimage* **2020**, *208*, 116469. [CrossRef]
51. Barber, A.D.; Gallego, J.A.; DeRosse, P.; Birnbaum, M.L.; Lencz, T.; Ali, S.A.; Moyett, A.; Malhotra, A.K. Contributions of Parasympathetic Arousal-Related Activity to Cognitive Performance in Patients with First-Episode Psychosis and Control Subjects. *Biol. Psychiatry Cogn. Neurosci. Neuroimaging* **2021**, ahead of print. [CrossRef]
52. Schaich, C.L.; Malaver, D.; Chen, H.; Shaltout, H.A.; Zeki Al Hazzouri, A.; Herrington, D.M.; Hughes, T.M. Association of Heart Rate Variability with Cognitive Performance: The Multi-Ethnic Study of Atherosclerosis. *J. Am. Heart Assoc.* **2020**, *9*, e013827. [CrossRef]
53. Knight, E.L.; Giuliano, R.J.; Shank, S.W.; Clarke, M.M.; Almeida, D.M. Parasympathetic and Sympathetic Nervous Systems Interactively Predict Change in Cognitive Functioning in Midlife Adults. *Psychophysiology* **2020**, *57*, e13622. [CrossRef]
54. Dalise, A.M.; Prestano, R.; Fasano, R.; Gambardella, A.; Barbieri, M.; Rizzo, M.R. Autonomic Nervous System and Cognitive Impairment in Older Patients: Evidence from Long-Term Heart Rate Variability in Real-Life Setting. *Front. Aging Neurosci.* **2020**, *12*, 40. [CrossRef]
55. Blons, E.; Arsac, L.M.; Gilfriche, P.; McLeod, H.; Lespinet-Najib, V.; Grivel, E.; Deschodt-Arsac, V. Alterations in Heart-Brain Interactions under Mild Stress during a Cognitive Task Are Reflected in Entropy of Heart Rate Dynamics. *Sci. Rep.* **2019**, *9*, 18190. [CrossRef]
56. Berry, K.J.; Mielke, P.W. Exact and Monte Carlo Resampling Procedures for the Wilcoxon-Mann-Whitney and Kruskal-Wallis Tests. *Percept. Mot. Ski.* **2000**, *91*, 749–754. [CrossRef]
57. Lipsitz, L.A. Aging as a Process of Complexity Loss. In *Complex Systems Science in Biomedicine*; Deisboeck, T.S., Kresh, J.Y., Eds.; Topics in Biomedical Engineering International Book Series; Springer: Boston, MA, USA, 2006; pp. 641–654, ISBN 978-0-387-33532-2.
58. Goldberger, A.L.; Peng, C.-K.; Lipsitz, L.A. What Is Physiologic Complexity and How Does It Change with Aging and Disease? *Neurobiol. Aging* **2002**, *23*, 23–26. [CrossRef]

59. Beckers, F. Aging and Nonlinear Heart Rate Control in a Healthy Population. *AJP Heart Circ. Physiol.* **2006**, *290*, H2560–H2570. [CrossRef] [PubMed]
60. Pikkujamsa, S.M.; Makikallio, T.H.; Sourander, L.B.; Raiha, I.J.; Puukka, P.; Skytta, J.; Peng, C.-K.; Goldberger, A.L.; Huikuri, H.V. Cardiac Interbeat Interval Dynamics from Childhood to Senescence: Comparison of Conventional and New Measures Based on Fractals and Chaos Theory. *Circulation* **1999**, *100*, 393–399. [CrossRef] [PubMed]
61. Bakhchina, A.V.; Arutyunova, K.R.; Sozinov, A.A.; Demidovsky, A.V.; Alexandrov, Y.I. Sample Entropy of the Heart Rate Reflects Properties of the System Organization of Behaviour. *Entropy* **2018**, *20*, 449. [CrossRef]
62. Manor, B.; Lipsitz, L.A. Physiologic Complexity and Aging: Implications for Physical Function and Rehabilitation. *Prog. Neuro-Psychopharmacol. Biol. Psychiatry* **2013**, *45*, 287–293. [CrossRef] [PubMed]
63. Takahashi, A.C.M.; Porta, A.; Melo, R.C.; Quitério, R.J.; da Silva, E.; Borghi-Silva, A.; Tobaldini, E.; Montano, N.; Catai, A.M. Aging Reduces Complexity of Heart Rate Variability Assessed by Conditional Entropy and Symbolic Analysis. *Intern. Emerg. Med.* **2012**, *7*, 229–235. [CrossRef] [PubMed]
64. Geovanini, G.R.; Vasques, E.R.; de Oliveira Alvim, R.; Mill, J.G.; Andreão, R.V.; Vasques, B.K.; Pereira, A.C.; Krieger, J.E. Age and Sex Differences in Heart Rate Variability and Vagal Specific Patterns—Baependi Heart Study. *Glob. Heart* **2020**, *15*, 71. [CrossRef] [PubMed]
65. Porta, A.; Gnecchi-Ruscone, T.; Tobaldini, E.; Guzzetti, S.; Furlan, R.; Montano, N. Progressive Decrease of Heart Period Variability Entropy-Based Complexity during Graded Head-up Tilt. *J. Appl. Physiol.* **2007**, *103*, 1143–1149. [CrossRef]
66. Castaldo, R.; Melillo, P.; Pecchia, L. Acute Mental Stress Detection via Ultra-Short Term HRV Analysis. In Proceedings of the World Congress on Medical Physics and Biomedical Engineering, IFMBE, Toronto, Canada, 7–12 June 2015; Jaffray, D.A., Ed.; Springer: Berlin/Heidelberg, Germany, 2015; Volume 51, pp. 1068–1071, ISBN 978-3-319-19386-1.
67. Kim, H.-G.; Cheon, E.-J.; Bai, D.-S.; Lee, Y.H.; Koo, B.-H. Stress and Heart Rate Variability: A Meta-Analysis and Review of the Literature. *Psychiatry Investig.* **2018**, *15*, 235–245. [CrossRef]
68. Weippert, M.; Behrens, M.; Rieger, A.; Behrens, K. Sample Entropy and Traditional Measures of Heart Rate Dynamics Reveal Different Modes of Cardiovascular Control During Low Intensity Exercise. *Entropy* **2014**, *16*, 5698–5711. [CrossRef]
69. Porta, A.; Bari, V.; Ranuzzi, G.; De Maria, B.; Baselli, G. Assessing Multiscale Complexity of Short Heart Rate Variability Series through a Model-Based Linear Approach. *Chaos Interdiscip. J. Nonlinear Sci.* **2017**, *27*, 093901. [CrossRef]
70. Costa, M.; Goldberger, A.L.; Peng, C.-K. Multiscale Entropy Analysis of Complex Physiologic Time Series. *Phys. Rev. Lett.* **2002**, *89*, 068102. [CrossRef] [PubMed]
71. Voss, A.; Schulz, S.; Koschke, M.; Bär, K.J. Linear and Nonlinear Analysis of Autonomic Regulation in Depressed Patients. In Proceedings of the 30th Annual International Conference of the IEEE Engineering in Medicine and Biology Society, Vancouver, BC, Canada, 20–25 August 2008; pp. 2653–2656. [CrossRef]
72. Voss, A.; Schulz, S.; Schroeder, R.; Baumert, M.; Caminal, P. Methods Derived from Nonlinear Dynamics for Analysing Heart Rate Variability. *Philos. Trans. R. Soc. A Math. Phys. Eng. Sci.* **2009**, *367*, 277–296. [CrossRef] [PubMed]
73. Cysarz, D.; Porta, A.; Montano, N.; Van Leeuwen, P.; Kurths, J.; Wessel, N. Quantifying Heart Rate Dynamics Using Different Approaches of Symbolic Dynamics. *Eur. Phys. J. Spec. Top.* **2013**, *222*, 487–500. [CrossRef]
74. Riganello, F.; Chatelle, C.; Schnakers, C.; Laureys, S. Heart Rate Variability as an Indicator of Nociceptive Pain in Disorders of Consciousness? *J. Pain Symptom Manag.* **2018**, *57*, 47–56. [CrossRef] [PubMed]
75. Cortese, D.; Riganello, F.; Arcuri, F.; Lucca, L.; Tonin, P.; Schnakers, C.; Laureys, S. The Trace Conditional Learning of the Noxious Stimulus in UWS Patients and Its Prognostic Value in a GSR and HRV Entropy Study. *Front. Hum. Neurosci.* **2020**, *14*, 97. [CrossRef] [PubMed]
76. Spangler, D.P.; McGinley, J.J. Vagal Flexibility Mediates the Association Between Resting Vagal Activity and Cognitive Performance Stability Across Varying Socioemotional Demands. *Front. Psychol.* **2020**, *11*, 2093. [CrossRef]
77. Garavaglia, L.; Gulich, D.; Defeo, M.M.; Mailland, J.T.; Irurzun, I.M. The Effect of Age on the Heart Rate Variability of Healthy Subjects. *PLoS ONE* **2021**, *16*, e0255894. [CrossRef]
78. Kumral, D.; Schaare, H.L.; Beyer, F.; Reinelt, J.; Uhlig, M.; Liem, F.; Lampe, L.; Babayan, A.; Reiter, A.; Erbey, M.; et al. The Age-Dependent Relationship between Resting Heart Rate Variability and Functional Brain Connectivity. *NeuroImage* **2019**, *185*, 521–533. [CrossRef]
79. Bai, X.; Li, J.; Zhou, L.; Li, X. Influence of the Menstrual Cycle on Nonlinear Properties of Heart Rate Variability in Young Women. *Am. J. Physiol.-Heart Circ. Physiol.* **2009**, *297*, H765–H774. [CrossRef]
80. Valencia, J.F.; Porta, A.; Vallverdu, M.; Claria, F.; Baranowski, R.; Orlowska-Baranowska, E.; Caminal, P. Refined Multiscale Entropy: Application to 24-h Holter Recordings of Heart Period Variability in Healthy and Aortic Stenosis Subjects. *IEEE Trans. Biomed. Eng.* **2009**, *56*, 2202–2213. [CrossRef]
81. Humeau-Heurtier, A. The Multiscale Entropy Algorithm and Its Variants: A Review. *Entropy* **2015**, *17*, 3110–3123. [CrossRef]

entropy

MDPI

Article

Sample, Fuzzy and Distribution Entropies of Heart Rate Variability: What Do They Tell Us on Cardiovascular Complexity?

Paolo Castiglioni [1,2,*], Giampiero Merati [1,2], Gianfranco Parati [3,4] and Andrea Faini [4,5]

1 Department of Biotechnology and Life Sciences (DBSV), University of Insubria, 21100 Varese, Italy
2 Laboratory of Movement Analysis and Bioengineering of Rehabilitation (Lamobir), IRCCS Fondazione Don Carlo Gnocchi ONLUS, 20148 Milan, Italy
3 Department of Medicine and Surgery, University of Milano-Bicocca, 20126 Milan, Italy
4 Department of Cardiovascular, Neural and Metabolic Sciences, Istituto Auxologico Italiano, IRCCS, 20145 Milan, Italy
5 Department of Electronics, Information and Bioengineering (DEIB), Politecnico di Milano, 20131 Milan, Italy
* Correspondence: pcastiglioni@dongnocchi.it

Abstract: Distribution Entropy (DistEn) has been introduced as an alternative to Sample Entropy (SampEn) to assess the heart rate variability (HRV) on much shorter series without the arbitrary definition of distance thresholds. However, DistEn, considered a measure of cardiovascular complexity, differs substantially from SampEn or Fuzzy Entropy (FuzzyEn), both measures of HRV randomness. This work aims to compare DistEn, SampEn, and FuzzyEn analyzing postural changes (expected to modify the HRV randomness through a sympatho/vagal shift without affecting the cardiovascular complexity) and low-level spinal cord injuries (SCI, whose impaired integrative regulation may alter the system complexity without affecting the HRV spectrum). We recorded RR intervals in able-bodied (AB) and SCI participants in supine and sitting postures, evaluating DistEn, SampEn, and FuzzyEn over 512 beats. The significance of "case" (AB vs. SCI) and "posture" (supine vs. sitting) was assessed by longitudinal analysis. Multiscale DistEn (mDE), SampEn (mSE), and FuzzyEn (mFE) compared postures and cases at each scale between 2 and 20 beats. Unlike SampEn and FuzzyEn, DistEn is affected by the spinal lesion but not by the postural sympatho/vagal shift. The multiscale approach shows differences between AB and SCI sitting participants at the largest mFE scales and between postures in AB participants at the shortest mSE scales. Thus, our results support the hypothesis that DistEn measures cardiovascular complexity while SampEn/FuzzyEn measure HRV randomness, highlighting that together these methods integrate the information each of them provides.

Keywords: multiscale entropy; spinal cord injury; posture; autonomic nervous system; SampEn; FuzzyEn; DistEn

1. Introduction

In the last decades, the interest in the entropy of heart rate time series has risen steadily due to the possibility of obtaining information on complexity aspects of cardiovascular dynamics and their alterations with disease [1]. This interest was ignited by the work of S.M. Pincus [2,3] who in 1991 proposed a computationally practical way to estimate the Kolmogorov–Sinai (K-S) entropy, which is the rate of information produced by dynamical systems. This method, called approximate entropy (ApEn), was aimed at overcoming the limits related to the computational demands and strong dependence on noise of the Grassberger–Procaccia [4] and Takens [5] formula by approximating their calculus (hence the ApEn name). The method considered segments of m samples as the coordinates of points in an m-dimensional space and evaluated how many segments were similar to each other, which means that they appeared as points closer than a given distance r. Then, the method evaluated how many similar segments remained similar when the dimension increased to $m + 1$. The number of similar segments can only decrease (or remain the same) from m to $m + 1$ and ApEn was calculated from the rate of this decrease. In this way, ApEn measured how unpredictable the value of a new sample is, given the m values of its preceding samples. This approach made possible the practical estimation of entropy from a relatively short series (hundreds of beats). However, ApEn was a biased estimator. In 2000, J.S. Richman and J.R. Moorman corrected this problem by introducing the Sample Entropy (SampEn) method which excludes self-matches from the count of similar segments [6]. ApEn and SampEn provide similar estimates for relatively long series while for short series ApEn gives lower entropy values than SampEn.

Later, W. Chen et al. abandoned the dichotomous classification of "similar" or "dissimilar" segments introducing less rigid criteria based on fuzzy functions, mitigating the arbitrariness of the threshold choice, and making the estimates statistically more stable. They called their method Fuzzy entropy (FuzzyEn) [7,8]. SampEn and FuzzyEn, like ApEn from which they are derived, estimate the K-S entropy of physiological systems by evaluating a conditional probability and by approximating the original formulas to analyze short series.

More recently, P. Li et al. proposed a different approach called Distribution Entropy (DistEn) [9]. Their motivation was to remove the arbitrary choice of the distance threshold and to improve further the statistical consistency for analyzing shorter series. Like SampEn and FuzzyEn, DistEn considers segments of m samples as points in the m-dimensional space and evaluates the distances between points. However, differently from SampEn and FuzzyEn, neither a threshold nor an additional space dimension is considered because DistEn is obtained from the relative frequencies of the probability distribution function of the distances.

This procedure makes DistEn intrinsically different from SampEn or FuzzyEn. While SampEn and FuzzyEn estimate the entropy of the time series, DistEn estimates the entropy of a structure of the phase space of the system, i.e., the distances between points of the phase space. This suggested that DistEn could be a better estimator of the "complexity structure" of the cardiovascular system than SampEn or FuzzyEn, by contrast, with more focus on the unpredictability of the series [9,10]. In this regard, the difference between "time-series unpredictability" and "system complexity", although not rigorously defined in mathematical terms, was clearly enunciated by Costa et al. [11]. This was done to describe the apparent inconsistency of certain diseases, such as atrial fibrillation, that have similar or even higher heart-rate entropy than healthy systems, while it is expected that these disease conditions should be associated with a lower complexity of the cardiovascular system. The inconsistency could be explained by the higher randomness of heart rate during atrial fibrillation, similar to the increase of entropy in surrogate data when time series generated by complex dynamical systems are randomized. To solve the inconsistency, these authors proposed a multiscale evaluation of entropy which takes into account the entropic structure of the temporal fluctuations [11]. Successively, DistEn was proposed as an alternative approach to distinguish between time-series randomness and system complexity based on the spatial structure of the vector distances rather than on the temporal structure investigated by multiscale entropy [9].

In a previous study, DistEn performed better than SampEn in distinguishing the expected alterations of cardiovascular complexity associated with aging or heart rhythm disturbances from short data segments [12], a result that may depend on more efficient estimates because DistEn exploits all the data (not only those with distance closer than the selected threshold) and is less sensitive than SampEn to the choice of the analysis parameters. However, this result might also indicate that DistEn is better focused than SampEn on the features of the complex cardiac dynamics altered by aging or cardiac diseases.

Our study aims to explore the nature of the information on heart rate variability (HRV) provided by DistEn and the differences with SampEn or FuzzyEn, over multiple scales. In perspective, this may allow the early diagnosis of alterations of circulatory regulation due to disease and the monitoring of the dynamics of the cardiovascular system during treatments or rehabilitation programs. For this aim, we consider two experimental conditions, one in which we expect changes in the randomness of HRV not associated with structural alterations in the generating cardiovascular system; the other, in which we expect differences in the HRV complexity associated with a structurally impaired integrative autonomic control of the circulation. By autonomic integrative control, we mean the role played by the autonomic nervous system to coordinate the heart action and the vasomotor changes in response to the ever-changing organs and body needs, and the way it interfaces with the centers of the brain, medulla, and spinal cord to organize and control body reactions and behavior [13]. Therefore, we consider the dataset of heart rate recordings in supine and sitting postures that we collected in able-bodied (AB) and spinal cord injured (SCI) individuals with a low lesion level previously [14]. To understand the rationale for the choice of this dataset, we should consider that the spinal cord injury represents a clear alteration of the overall cardiovascular control. By interrupting the flow of information between vascular districts below the lesion level and the higher brain centers, the mechanisms of integrative control are weakened allowing local regulations of individual vascular districts to prevail. Actually, we previously demonstrated that the complexity structure of HRV, as quantified by multifractality, is altered in SCI individuals [14]. By contrast, the postural shift from supine to sitting does not induce any structural alteration in the cardiovascular control and its influence on the HRV dynamics is the consequence of a relatively modest modulation of the cardiac autonomic control, consisting of increased sympathetic tone and decreased vagal tone. Selective pharmacological autonomic blockade in humans demonstrated that the short-term self-similar structure of vagal driven fluctuations of heart rate resembles the unpredictable white noise dynamic while sympathetic driven fluctuations of heart rate show much more predictable, Brown-noise like, long-term correlations [15]. Thus, the level of unpredictability of HRV in our participants should reflect the mixture of vagal and sympathetic cardiac outflows. Consequently, the HRV randomness is expected to decrease from supine to sitting, in parallel with the decrease of vagal tone and the increase in sympathetic tone, with negligible effects on the system complexity.

Therefore, this dataset allows testing two working hypotheses separately, possibly understanding the nature of the information provided by these entropy estimators better. First, assuming that a postural shift from supine to sitting does not alter the complexity structure of the cardiovascular control but only the irregularity/unpredictability of the heart-rate series by modulating the autonomic tone, we will test the hypothesis that postural differences will be clearer for SampEn or FuzzyEn than for DistEn (preliminary results on this issue have been presented at the 12th ESGCO Conference in 2022 [16]). Second, assuming that in SCI individuals, previously associated with an altered multifractality, the impaired integrative autonomic regulation influences the cardiovascular complexity, we will test the hypothesis that DistEn reveals more pronounced and significant differences between AB and SCI groups than SampEn or FuzzyEn.

2. Materials and Methods

2.1. Synthetized Series

To compare the general characteristics of SampEn, FuzzyEn, and DistEn, we generated 30 series for each of 3 stochastic processes: Gaussian white, pink, and brown noises. Furthermore, by using the logistic recursion $x(n+1) = \omega \times x(n) \times (1 - x(n))$ we generated 30 chaotic ($\omega = 4$) and 30 periodic ($\omega = 3.5$) series. Each of the 5×30 synthesized series was composed of 512 samples.

2.2. Subjects and Data Collection

We considered the 14 SCI individuals with a complete lesion (ASIA scale A) at a level between the 12th thoracic and 4th lumbar vertebra and 34 AB controls who participated in our previous study for evaluating spectral and fractal HRV components [17]. Paraplegic individuals had complete traumatic lesions with no current or previous history of overt dysautonomia (details on exclusion criteria are reported in [18]). Due to the low level of the lesion, the paraplegic participants had impaired autonomic control of organs innervated by the pelvic nerve and by efferent pathways from the mesenteric ganglia but intact autonomic cardiac control. The experimental protocol, which is described in detail in [18], consisted in recording the ECG at 200 Hz sampling rate for 10 min in sitting and 10 min in supine postures after an adaptation period of 5 min, in a quiet environment (for technical reasons, the supine recording was missing in two SCI participants). The study was approved by the ethics committee of IRCCS Don C. Gnocchi Foundation, Milan (Italy) and each subject gave informed consent before the start of the experiment.

AB and SCI groups were matched by age (AB, 39.3 ± 12.1 years; SCI, 40.9 ± 10.0 years; mean ± SD), body mass index (AB, 24.7 ± 2.6; SCI, 25.5 ± 4.8 kg·m^{-2}) and female/male ratio (AB, 5/29; SCI: 2/12). As demonstrated previously [17] (p. 6), the mean R-R interval (RRI) did not differ significantly between groups (supine AB = 958 ± 26 ms, SCI = 885 ± 40 ms; sitting AB = 864 ± 23 ms, SCI = 853 ± 34 ms, median ± SE median); furthermore, AB and SCI participants were also matched in terms of the time-domain vagal indexes RMSSD (supine AB = 30 ± 4 ms^2, SCI = 28 ± 6 ms^2; sitting AB = 27 ± 3 ms^2; SCI = 28 ± 5 ms^2) and pNN50 (supine AB = 7.8 ± 3.6%, SCI = 8.8 ± 3.3%; sitting AB = 6.2 ± 2.1%, SCI = 9.9 ± 3.7%), as well as frequency-domain vagal and sympatho/vagal indexes, HF power (supine AB = 258 ± 90 ms^2, SCI = 132 ± 84 ms^2; sitting AB = 194 ± 47 ms^2, SCI = 179 ± 80) and LF/HF powers ratio (supine AB = 2.3 ± 0.3, SCI = 2.9 ± 1.2; sitting AB = 3.9 ± 0.4, SCI = 3.5 ± 0.7).

2.3. Entropy Estimators

Premature beats visually identified on the RRI time series from the ECG were removed. The HRV entropy was estimated over a segment of 512 consecutive beats, with embedding dimensions $m = 1$ and $m = 2$.

As to SampEn and FuzzyEn, we set the threshold r equal to 20% of the standard deviation of the series. SampEn was calculated as in [6]: given the N samples $\mathbf{X} = \{x_1\ x_2\ \ldots\ x_N\}$ we constructed the template vectors for the dimension m,

$$\mathbf{X}_i^m = [x_i, x_{i+1}, \ldots, x_{i+m-1}]^T \quad 1 \le i \le N - m, \tag{1}$$

calculated the infinity norm distance between vectors

$$d_{ij}^m = \|\mathbf{X}_i^m - \mathbf{X}_j^m\|_\infty \quad 1 \le i, j \le N - m,\ j \ge i + 1, \tag{2}$$

and counted the pairs of vectors with a distance lower than r, $n_p(m,r)$. We repeated the same steps for $m + 1$ obtaining:

$$SampEn(X, N, m, r) = -ln\frac{n_p(m+1, r)}{n_p(m, r)} \tag{3}$$

FuzzyEn was calculated as in [8]: first, we obtained the average of the similarity degree among vectors $\mathbf{X}_i^m$ as

$$\phi^m = \sum_{i=1}^{N-m}\left(\frac{1}{N-m-1}\sum_{j=1, j\neq i}^{N-m} exp\left(-\frac{d_{ij}^{mn}}{r}\right)\right) \tag{4}$$

with $n = 2$. In Equation (4) we employed an exponential membership function following [19] but different choices might imply different behaviors of the estimator [8]. Then we calculated the same quantity for $m + 1$ and Fuzzy Entropy as:

$$FuzzyEn(X, N, m, r, n) = ln[\phi^m(n, r)] - ln\left[\phi^{m+1}(n, r)\right] \tag{5}$$

It should be noted that the original definition of FuzzyEn [7] removes local trends making FuzzyEn not directly comparable to SampEn at the same m [20]. For this reason, we estimated the global FuzzyEn without trend removals as introduced in [8].

DistEn was obtained as in [9] by calculating the empirical probability distribution function (ePDF) of the d_{ij}^{m} distances. We calculated ePDF on M = 512 equispaced bins over the range of the distance values, its Shannon entropy ShEn as

$$ShEn = -\sum_{t=1}^{M} p_t log_2(p_t) \tag{6}$$

and

$$DistEn(m) = \frac{ShEn(m, M)}{log_2(M)} \tag{7}$$

Figure 1 schematizes the steps for the calculus of these three entropy methods.

Multiscale estimates of SampEn (mSE), FuzzyEn (mFE), and DistEn (mDE) at scales τ between 2 and 20 beats were obtained after low-pass filtering the RRI series at each τ by a zero-phase Butterworth filter with a cut-off frequency equal to $0.5/\tau$ as in [21]. We used this filter because of its better transition band compared to the moving average originally proposed for coarse-graining [22]. Then the coarse-graining decimation (consisting in taking one sample for every τ samples) was not applied to use all the samples of the low-pass filtered series to improve the quality of the estimate [21], and we calculated mSE, mFE, and mDE setting a delay equal to τ between consecutive samples of the template vectors as defined in [23].

2.4. Statistics

As stated in the introduction, the aim is to evaluate whether Sample, Fuzzy and Distribution entropy identify or not the expected HRV changes due to (1) the postural autonomic activation; and (2) the impaired integrative autonomic control. This logically leads to testing two null hypotheses for each entropy estimator, one regarding the significance of the factor "posture" comparing supine and sitting positions, the other regarding the significance of the factor "case" comparing AB and SCI participants.

To test these hypotheses for SampEn, FuzzyEn, and DistEn, we applied a Linear Mixed-Effects Model that simultaneously provides the statistical significance of the factors Case (AB vs. SCI) and Posture (Supine vs. Sitting) and their interaction. We tested if the residuals were normally and equally distributed and rank-transformed the data if the assumptions were not satisfied. When one of the factors or their interaction was significant at $p < 0.05$, we tested the differences between Supine and Sitting for each group and the differences between AB and SCI participants in each posture with a-posteriori contrasts, accounting for multiple comparisons with the false discovery rate correction.

As to the multiscale entropies, the statistical tests should regard each of the 19 scales from $\tau = 2$ to $\tau = 20$ separately. For conciseness, we employed a non-parametric approach (Wilcoxon–Mann–Whitney test) to check the assumption of Linear Mixed-Effects Models for each scale. A null hypothesis was tested separately at each scale and thus corrections for multiple comparisons were not applied. The ability of mSE, mFE, and mDE to detect HRV changes associated with posture was assessed by the Wilcoxon Signed Rank test comparing Supine and Sitting positions by groups at each τ; the ability to detect HRV changes associated with the impaired integrative autonomic regulation was assessed by the Wilcoxon Rank Sum test comparing AB and SCI groups by posture at each τ.

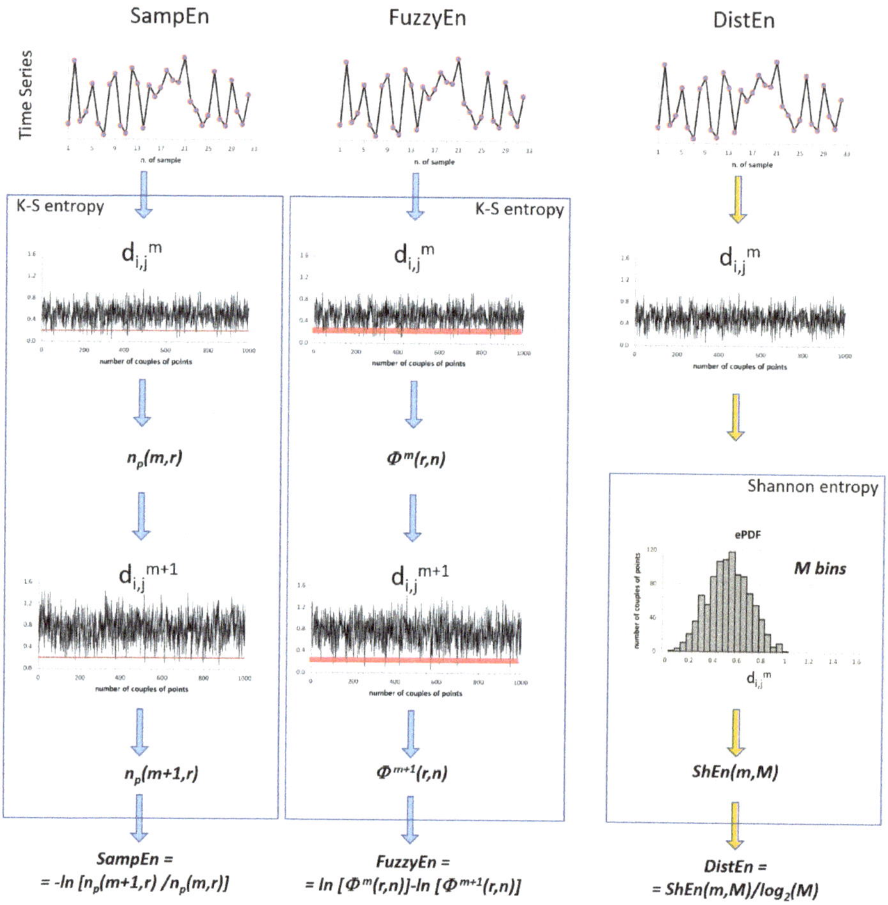

Figure 1. Scheme of Sample (SampEn, **left**), Fuzzy (FuzzyEn, **center**), and Distribution Entropy (DistEn, **right**) calculation. SampEn and FuzzyEn estimate the K-S entropy of the series by representing segments of m and $m + 1$ samples as points in spaces of m and $m + 1$ dimensions and calculating the distances between points in both spaces, $d_{i,j}^{m}$ and $d_{i,j}^{m+1}$, based on the conditional probability that similar segments of m samples remain similar when the segment length increases to $m + 1$. SampEn is the negative logarithm of the ratio between the number of points n_p closer than the threshold r (red line) in spaces of $m + 1$ and m dimensions. FuzzyEn is the difference between the logarithms of the degree of similarity among points in the m and $m + 1$ spaces, $\Phi^{m}(r,n)$ and $\Phi^{m+1}(r,n)$, with r the fuzzy distance (red bar) and n the exponent that defines the similarity function. Unlike SampEn and FuzzyEn, DistEn does not calculate the entropy of the series but the Shannon entropy (ShEn) of the distances, $d_{i,j}^{m}$, by estimating the empirical probability distribution function (ePDF) of $d_{i,j}^{m}$ over M bins and ShEn as $-\sum_{t=1}^{M} p_t log_2(p_t)$, with p_t the relative frequency of bin t.

3. Results

3.1. Synthetized Series

Figure 2 compares the entropy estimates for the synthesized series. Synthetized series allow a better understanding of the role of SampEn and FuzzyEn as estimators of time series irregularity and DistEn as an estimator of system complexity. In fact, the figure highlights a very different behavior of the two estimators of the K-S entropy, SampEn, and FuzzyEn, compared to DistEn. SampEn and FuzzyEn provide the largest entropy for white noise (the more unpredictable series) and lower values for pink and brown noises, reflecting the increase in the long-range correlation of these series. The entropy of the chaotic series is in between pink- and brown-noise entropies, and the periodic series has the lowest entropy, close to zero.

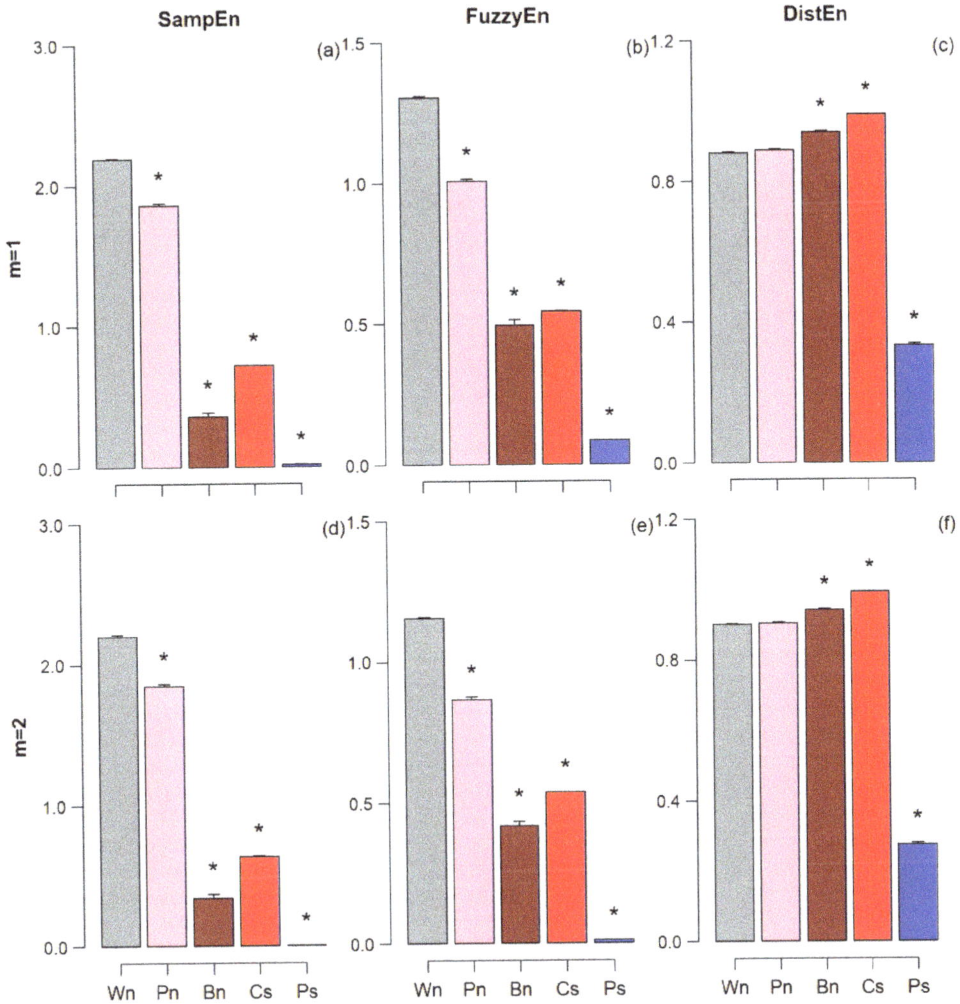

Figure 2. Mean and SEM of SampEn (panels (**a**,**d**)), FuzzyEn (panels (**b**,**e**)), and DistEn (panels (**c**,**f**)) over groups of 30 synthesized series for dimensions $m = 1$ and $m = 2$. Wn = white noise; Pn = pink noise; Bn = Brown noise; Cs = chaotic series; Ps = periodic series. The * indicates differences vs. Wn at $p < 0.05$.

By contrast, DistEn provides a different picture and distribution entropy is the largest for the chaotic series. Furthermore, DistEn is greater (and not lower) for Brown than white noise. Moreover, pink and white noises virtually have the same DistEn. The periodic series has the lowest DistEn, as for SampEn/FuzzyEn. However, while SampEn/FuzzyEn of the periodic series is just 1% of the value of white noise, DistEn of the periodic series is around one-third of the white-noise DistEn.

As to the multiscale entropy (Figure 3), mSE and mFE show almost the same trends: white and pink noises and the chaotic series monotonically decrease with the scale for all $\tau > 2$, with a steeper decrease for mFE; the periodic series is close to zero; and the Brown noise entropy increases with τ, although it happens at all the scales for mSE and up to $\tau = 10$ only for mFE. The estimates appear slightly more stable for mFE than mSE, and for $m = 1$ than $m = 2$. However, there are differences between mSE and mFE in the relative entropy values among signals. For instance, mSE is greater for pink than Brown noise at all the scales while for mFE the difference disappears at $\tau > 10$ for $m = 2$ and is even reversed (with mFE greater for Brown than pink noise) for $m = 1$. Furthermore, white noise and the chaotic series have the same mSE at $\tau > 5$ for mSE, while mFE is consistently greater for white noise than the chaotic series at all τ.

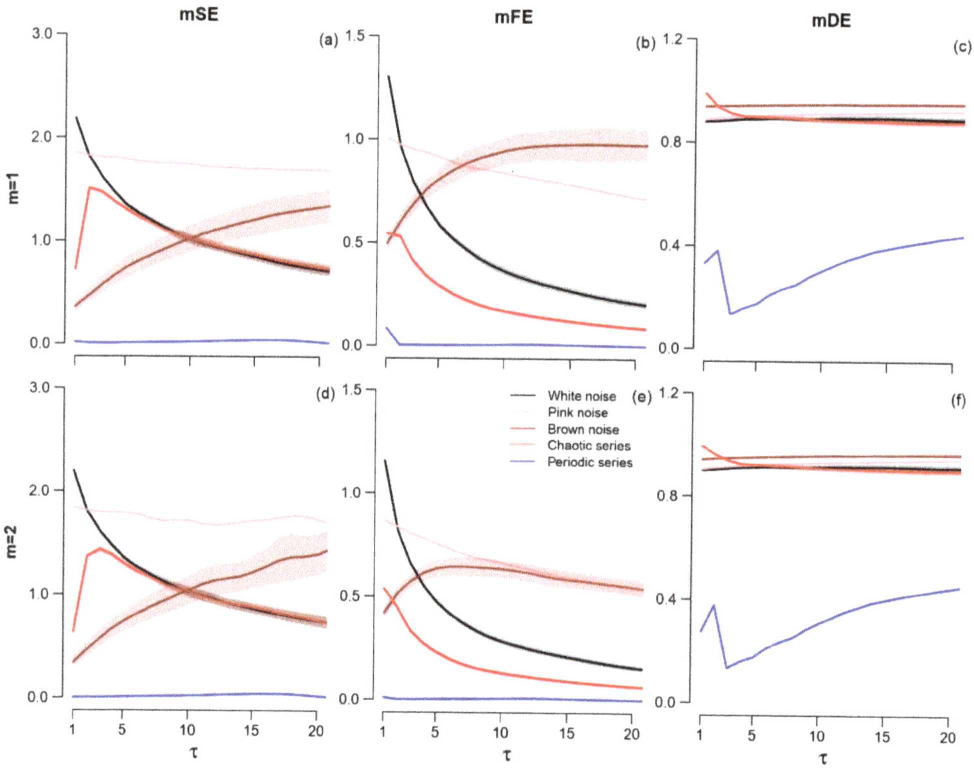

Figure 3. Multiscale Sample (mSE, panels (**a**,**d**)), Fuzzy (mFE, panels (**b**,**e**)), and Distribution (mDE, panels (**c**,**f**)) Entropies of synthesized signals. Values as mean and 95% confidence intervals.

The profile of mDE as a function of the scale appears rather different. The mDE of the chaotic series decreases very quickly with τ so that Brown and pink noises have a greater mDE than the chaotic series at the larger scales and Brown noise has the largest mDE at all $\tau > 2$. The periodic series has the lowest mDE at all scales. It should be noted that the mDE of the periodic series generated by the logistic recursion with $\omega = 3.5$ was calculated in two previous works [24,25] that reported a systematic periodicity at scale multiples of $\tau = 4$ which, however, is absent in our estimates of Figure 3c,f. The reason is likely due to the poor transition band of the moving average filter employed in the previous works before coarse-graining as demonstrated in Appendix A.

3.2. Real Beat-to-Beat Series

3.2.1. SampEn, FuzzyEn, and DistEn

Table 1 reports the significance of "Posture" and "Case" factors and their interaction. Only "Posture" is significant for SampEn and FuzzyEn. In particular, Figure 4 shows lower SampEn and FuzzyEn in Sitting. The percent decrease of entropy from Supine to Sitting is more pronounced for the AB group both when quantified by SampEn (m = 1: AB = −9.9%, SCI = −6.2%; m = 2: AB = −11.8%, SCI = −5.1%) and FuzzyEn (m = 1: AB = −19%, SCI = −0.9%; m = 2: AB = −17.6%, SCI = −2.2%). The differences reach statistical significance in the AB group only.

Table 1. Significance p of Posture and Case factors and their interaction after linear mixed-effects model analysis for two embedding dimensions m.

Factor	SampEn	FuzzyEn	DistEn
m = 1			
Posture	**0.029** *	**0.024** *	0.15
Case	0.17	0.30	**0.020** *
Interaction	0.22	0.13	0.94
m = 2			
Posture	**0.007** *	**0.038** *	0.25
Case	0.20	0.28	**0.026** *
Interaction	0.23	0.23	0.90

Bold and * highlight statistical significances at $p < 0.05$.

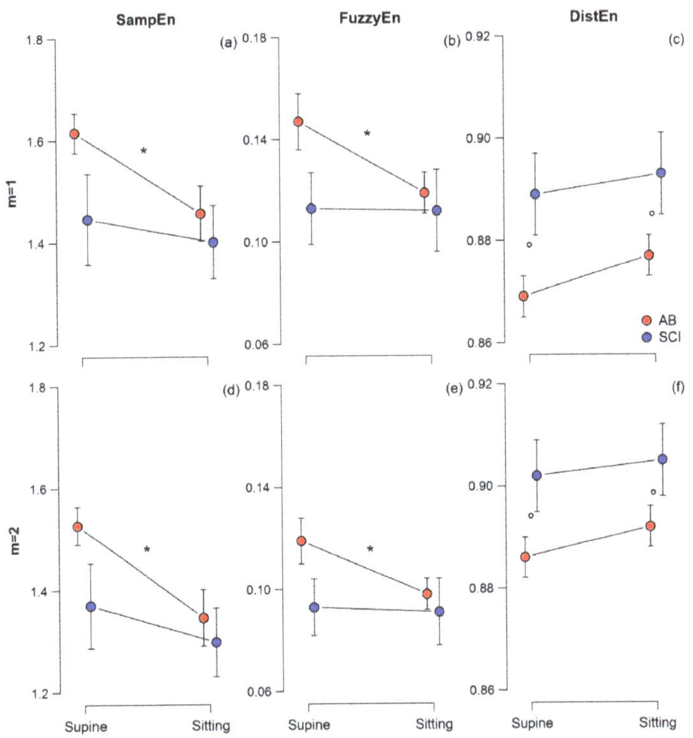

Figure 4. Mean and SEM of SampEn (panels (**a**,**d**)), FuzzyEn (panels (**b**,**e**)), and DistEn (panels (**c**,**f**)) in AB and SCI participants for m = 1 and 2. The * indicates significant differences between postures or groups at $p \leq 0.05$; the ° indicates statistical trends at $p < 0.10$.

By contrast, Table 1 reports that neither "Posture" nor the interaction of "Posture" with "Case" is a significant factor for DistEn. However, differently from SampEn or FuzzyEn, the "Case" factor is significant for DistEn demonstrating that the low-level spinal cord lesion influences cardiovascular complexity. In particular, Figure 4 provides evidence of higher DistEn in SCI than AB participants. The difference between SCI and AB groups expressed as a percentage of the AB value was similar in the two postures (m = 1: Supine = +2.3%, Sitting = +1.8%; m = 2: Supine = +1.8%, Sitting = +1.5%).

3.2.2. mSE, mFE and mDE

Figures 5 and 6 show the profiles of multiscale entropies separately in the two groups and postures, respectively, for embedding dimensions 1 and 2. The capability of DistEn to discriminate between AB and SCI groups is almost lost at $\tau > 1$ and mDE shows just a weak tendency to separate the AB and SCI groups at $\tau < 8$. At greater mDE scales, the two groups overlap each other.

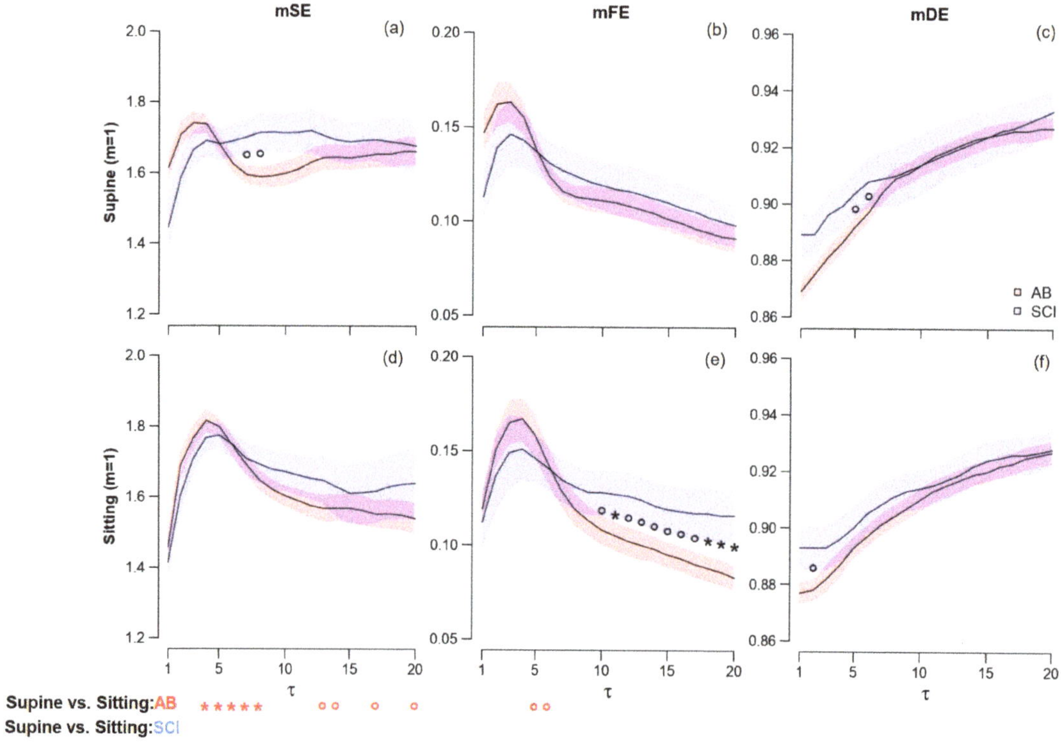

Figure 5. Multiscale Sample (panels **a**,**d**), Fuzzy (**b**,**e**), and Distribution Entropies (**c**,**f**) for m = 1 in supine and sitting postures. Values as mean ± SEM over SCI and AB controls. The symbols refer to the scale-by-scale statistics for $2 \leq \tau \leq 20$: the * and ° in the panels a–f indicate significant differences ($p \leq 0.05$) or trends ($p < 0.10$) between AB and SCI groups; and in the panels (**d**–**f**) indicate significant differences ($p \leq 0.05$) or trends ($p < 0.10$) between postures.

By contrast, the multiscale Fuzzy Entropy allows distinguishing between the AB and SCI groups at scales $\tau > 10$ but only in the sitting position. Differences between groups that are significant (m = 2) or close to the significance level (m = 1) appear for mSE too around $\tau = 7$: in this case, however, the trends characterize the supine position only.

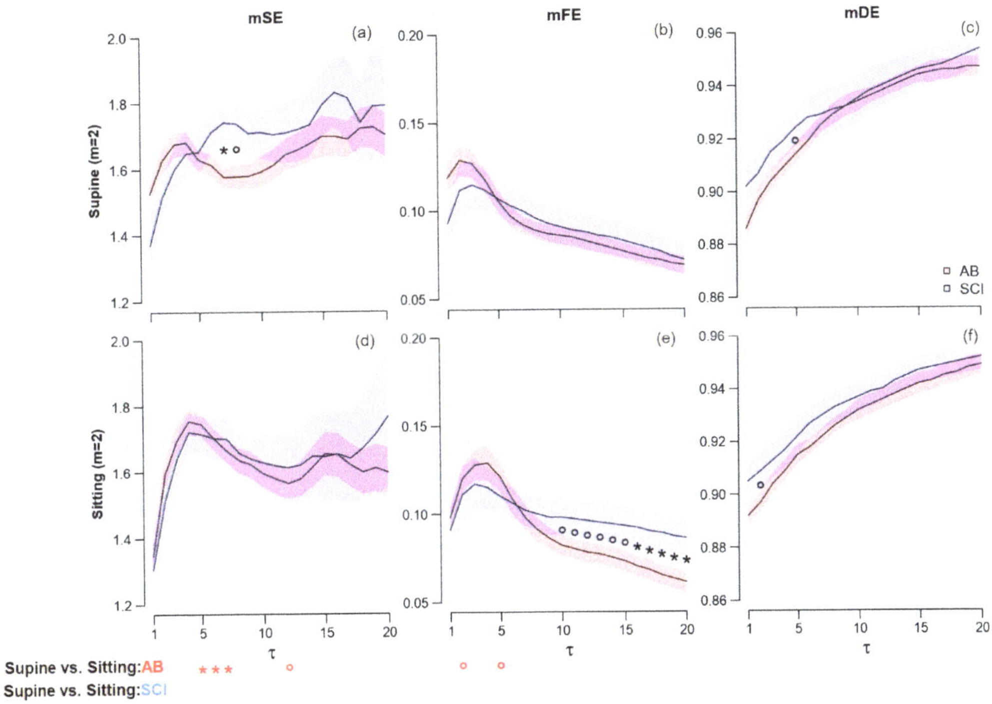

Figure 6. Multiscale Sample (panels **a**,**d**), Fuzzy (**b**,**e**), and Distribution Entropies (**c**,**f**) for $m = 2$ in supine and sitting postures. Values as mean ± SEM over SCI and AB controls. The symbols refer to the scale-by-scale statistics for $2 \leq \tau \leq 20$: the * and ° in the panels a–f indicate significant differences ($p \leq 0.05$) or trends ($p < 0.10$) between AB and SCI groups; and in the panels (**d**–**f**) indicate significant differences ($p \leq 0.05$) or trends ($p < 0.10$) between postures.

Significant differences between postures are detected only by mSE: they regard the AB group only and scale $\tau \leq 8$.

4. Discussion

Distribution entropy has been originally proposed to describe complexity aspects of the physiological systems generating HRV because Sample entropy was considered a measure more focused on the heart-rate irregularity [9,10]. Our study contributes to clarifying this aspect. It is designed to compare how Distribution, Sample and Fuzzy entropies quantify HRV changes having different origins: one is a posture change that modifies HRV without affecting the physiological structure underlying the cardiovascular complexity; the other is a condition anatomically characterized by an impaired integrative regulation.

This comparison was possible considering a human model of impaired integrative autonomic control: the paraplegic individual with a low-level spinal lesion. Paraplegic subjects with a high-level lesion, above the fifth thoracic vertebra, T5, have an impaired sympathetic outflow that also affects the neural pathways directly innervating the heart: this condition might influence both the long-term correlations of the heart rate (thus, the heart rate irregularity) [18] and the overall system complexity. Lesions between T5 and T11 do not directly affect the cardiac autonomic outflows but influence importantly the central autonomic regulation of several vascular districts. These innervations are responsible for vasomotor components that generate the Mayer waves in blood pressure, reflected in the low-frequency spectral power of HRV through the baroreflex [17]. Thus, also in the case of spinal lesions between T5 and T11, we may expect both an altered cardiovascular complexity due to the lesion and an altered heart-rate randomness due to the loss of predictable oscillatory components (the Mayer waves). By contrast, paraplegic individuals with lesions below T11 preserve the oscillatory components of HRV that characterize the power spectrum of AB controls [17]. Nevertheless, their spinal lesion alters the cardiovascular complexity as previously demonstrated by assessing the HRV multifractality [14]. Therefore, paraplegic individuals with spinal lesions below T11 represent a model of human cardiovascular regulation that allows testing the information provided by different entropy metrics, having an intact HRV power spectrum (i.e., second-order statistics) but an altered multifractal spectrum (higher-order statistics). In the following, we will discuss the results we obtained by assessing our dataset of heart rate recordings with Sample, Fuzzy, and Distribution entropy methods.

4.1. SampEn, FuzzyEn, and DistEn

The significances of the factors "Case" and "Posture" highlighted the different nature of DistEn compared to SampEn or FuzzyEn (Table 1) and support the hypothesis that DistEn may identify structural alterations in the cardiovascular system rather than measuring changes in the heart rate randomness, like SampEn or FuzzyEn. In line with the starting hypothesis that a postural shift is not expected to be associated with structural changes affecting the system complexity, DistEn fails to detect the increase in cardiac sympathetic tone and decrease in vagal tone that characterizes the shift from supine to sitting. These autonomic changes are detected by SampEn and FuzzyEn similarly, both describing a decrease in heart rate irregularity in sitting. In fact, the higher sympatho/vagal balance in the sitting position should increase the relative amplitude of the slower HRV components driven by the cardiac sympathetic outflow, increasing the long-range correlation of heart rate values and consequently their beat-to-beat predictability. Our results also show that both SampEn and FuzzyEn quantify a more marked postural decrease of entropy in the AB group. The similarity of the results provided by SampEn and FuzzyEn is not surprising because both count the number of similar segments of length m that remain similar when the segment length increases to $m + 1$, even if the way this number is obtained and processed differs between the methods.

Also in line with the starting hypothesis is our finding that DistEn, unlike SampEn or FuzzyEn, distinguishes the altered integrative autonomic control in SCI participants (the "Case" factor is significant for DistEn only, Table 1). The effect is a greater DistEn in the SCI group. We cannot exclude that SampEn and FuzzyEn may also partially reflect the effect of the spinal lesion on the system complexity and Figure 4 might suggest a tendency for SampEn/FuzzyEn to separate the AB and SCI groups in Supine. However, the difference is far from being significant, supporting the hypothesis that SampEn and FuzzyEn are more sensitive to changes in the signal randomness, and DistEn to alterations in the system complexity, as formulated in [9,10].

Our results, therefore, indicate that the HRV complexity, as quantified by DistEn, is greater in the SCI group. We cannot indicate how the impairment of the autonomic integrative control due to the spinal cord lesion may have produced an increased complexity. However, our results on the synthesized series might help to formulate a tentative hypothesis. The trends we observed in our AB and SCI volunteers have the same sign of the differences between white noise and chaotic series: greater DistEn and lower SampEn/FuzzyEn for the chaotic series compared to white noise, as in SCI compared to AB participants. Looking at the ePDF functions of white noise and chaotic series (e.g., see Figure 7) we found that the chaotic dynamics produced distances equally distributed at all the amplitudes; by contrast, even if white noise is totally unpredictable the distances between its vectors have certain amplitudes more likely than others. This might suggest that the presence of an effective integrative autonomic control allows the higher centers of the nervous system to coordinate the local vascular regulations orchestrating harmonically the cardiovascular dynamics and shaping the ePDF in a sort of bell curve. This would avoid that independent regulations of the local vascular districts reverberate into less structured, and flatter, ePDF, resulting in a higher DistEn.

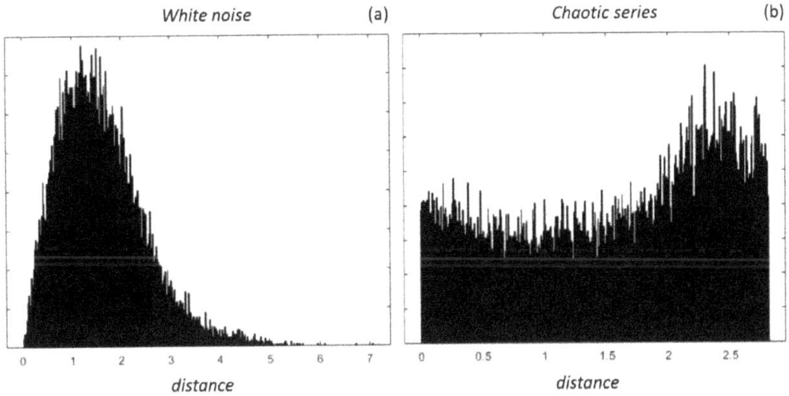

Figure 7. Empirical probability distribution functions of distances between vectors at $m = 2$ for white noise series (panel (**a**)) and chaotic series (panel (**b**)).

4.2. mSE, mFE and mDE

The multiscale entropy approach was originally proposed to distinguish the sample entropy of the uncorrelated white noise process from the entropy of series with long-term correlation (like pink noise), as a tool for distinguishing time-series randomness from system complexity. Our results on the synthesized series confirm the ability of the multiscale approach to distinguish between white and pink noises not only for SampEn but also for FuzzyEn. By contrast, the multiscale approach appears much less efficient in separating white from pink noise when applied to DistEn.

Furthermore, the multiscale approach suggests complementary roles of Sample, Fuzzy, and Distribution Entropy in describing the HRV dynamics. A result that deserves to be discussed is that at scales $\tau > 1$, mSE and mFE provide substantially different information even if both estimate the K-S entropy by counting points close to each other in m and $m + 1$ dimensions. This contrasts strikingly with the similar behavior of SampEn and FuzzyEn in distinguishing between groups or postures (Figure 4).

In fact, while both SampEn and FuzzyEn distinguish similarly the posture change in the AB group, mSE only and not mFE differs significantly between Supine and Sitting in the AB group. A possible explanation is offered by the analysis of the synthesized series (Figure 3). The way mSE and mFE change with τ is similar for all the signals. However, the relative differences between signals are not the same for mSE and mFE. For instance, let us consider white and Brown noises: for both mSE and mFE, the white noise decreases with τ from its highest value at $\tau = 1$ and the Brown noise increases with τ from a much lower value at $\tau = 1$. However, the slope of Brown noise as τ increases differs between mSE and mFE and at around $\tau = 4$ mSE is much higher for the white than Brown noise while the mFE of white and Brown noises are similar. We might hypothesize that the fuzzy and dichotomous thresholds used for classifying segments as similar or dissimilar weight differently in strongly correlated series, such as the Brown noise. The consequence is that the mSE and mFE profiles with τ may differ depending on the level of correlation among samples, justifying why the Sample and Fuzzy methods may provide different information on the HRV entropy at scales greater than one.

The same reasoning may also explain a second interesting difference between mSE and mFE, i.e., the ability of these multiscale methods to distinguish between AB and SCI groups. In fact, at $\tau \geq 10$ (scale likely influenced by vasomotor components such as the Mayer waves) the mFEs of AB and SCI groups in sitting position diverge while the mSEs almost overlap each other. The greater mFE in sitting SCI participants suggests a higher system complexity, in line with the result provided by DistEn. Furthermore, differences between groups seem to appear in the supine position too, in this case for mSE around $\tau = 7$ (scales more likely influenced by respiration). Our results, therefore, suggest that the Sample and Fuzzy entropy methods provide complementary information on HRV when considered over multiple scales.

By contrast, no significant differences between groups or postures are identified by mDE at any $\tau > 1$. In particular, the difference between groups that characterize DistEn (i.e., mDE at $\tau = 1$) vanishes as τ increases and the mDEs of the AB and SCI groups practically overlap each other at $\tau > 7$. This is not completely unexpected because DistEn has been proposed to detect alterations in the system complexity even at scale 1, without requiring a multiscale analysis, as for Sample and Fuzzy entropies; and because the way mSE and mFE detect the entropic temporal structure depends critically on the definition of the threshold r [26], a parameter that is absent in the calculation of DistEn. However, a deterioration of the Distribution Entropy method in revealing structural alterations in the system complexity at larger scales is possible: actually, the mDE inability to detect differences between white noise and the chaotic signals at $\tau > 2$ was reported previously [25] and confirmed by our results (Figure 3). Even if mDE proved to be valid in distinguishing multiscale entropy alterations with aging and congestive heart failure [25], likely thanks to its capacity to provide statistically consistent estimates even on very short series, the low performance in distinguishing purely random from deterministic chaotic signals at $\tau > 1$ may be responsible for the lack of significant differences between our AB and SCI groups. Interestingly, a recent study suggested that the relatively poor performance of mDE in distinguishing between chaotic and random series at larger scales is due to neglecting more complete quantifications of the distance between vectors in the embedding space [24]. These authors introduced an improved distribution entropy method taking into account also the orientation-based angular distance and the rank-based Spearman distance between vectors in the phase space, with promising results [24].

4.3. Limitations

Some limitations should be considered in interpreting our results. The first regards the relatively small number of enrolled SCI participants, in particular, female participants. A larger number might have identified as significant some reported statistical tendencies, or allowed identifying sex-related differences. However, it should be considered that the enrollment criteria in the SCI group were particularly stringent, requiring a well-delimited lesion level, the completeness of the lesion, and an active lifestyle to reduce the effect of sedentariness as a confounding factor. Furthermore, some medications could not be suspended (3 SCI participants were taking antibiotics for urinary infection, and another 3 were taking drugs against muscle spasms) and a possible influence of these drugs on the results cannot be completely excluded.

5. Conclusions

We may conclude that Sample, Fuzzy, and Distribution Entropy methods are not alternatives one to the other, but they complement each other providing different information on HRV. Regarding the beat-to-beat scale, DistEn cannot be considered an index of sympathovagal balance, because, unlike SampEn and FuzzyEn, it does not distinguish between supine and sitting postures either in healthy controls or in paraplegic volunteers. However, it demonstrates the ability to distinguish alterations in the integrative autonomic control that SampEn and FuzzyEn fail to detect.

At larger scales, mSE and mFE provide different pieces of information that cannot be derived from mDE and that, taken together, allow obtaining a more complete picture of the alterations induced by the posture shift and the spinal lesion. Therefore, Sample, Fuzzy, and Distribution entropies are not alternative tools but they represent different aspects of a more complex picture.

Author Contributions: Conceptualization, P.C. and G.M.; methodology, A.F., P.C. and G.M.; software, A.F.; validation, A.F.; formal analysis, A.F. and P.C.; investigation, G.M.; resources, G.M.; data curation, A.F.; writing—original draft preparation, P.C.; writing—review and editing, P.C., A.F., G.M. and G.P.; visualization, A.F.; supervision, P.C.; project administration, G.M.; funding acquisition, G.M. and G.P. All authors have read and agreed to the published version of the manuscript.

Funding: This research was funded by the Italian Ministry of Health (Ricerca Corrente RC2022).

Institutional Review Board Statement: The study was conducted in accordance with the Declaration of Helsinki, and approved by the Ethics Committee of Fondazione Don Carlo Gnocchi in 1999.

Data Availability Statement: The data supporting the main findings of this study are available at URL https://doi.org/10.5281/zenodo.7491970 (accessed on 28 January 2023) with access granted on justified request to researchers meeting the criteria for access to confidential data due to the hospital research policy and restrictions requested by the ethical committee.

Conflicts of Interest: The authors declare no conflict of interest. The funders had no role in the design of the study; in the collection, analyses, or interpretation of data; in the writing of the manuscript; or in the decision to publish the results.

Appendix A

The multiscale Distribution Entropy has been evaluated for the logistic regression with $\omega = 3.5$ (periodic signal) and $\omega = 4$ (chaotic signal) in two previous papers [24,25]. While mDE estimates for the chaotic signal are similar to our estimates (Figure 3c,f), the estimates of the periodic signal present a systematic oscillation repeating at scales multiple of $\tau = 4$ that is absent in our estimates. Differently from [24,25], we used a Butterworth filter for evaluating scales $\tau > 1$ as suggested in [22] because of its narrower transition band compared to the moving average filter used in previous estimates. To verify that the reported systematic oscillation in multiscale Distribution Entropy of the periodic signal is an artifact generated by the broad transition band of the filter, we estimated its mDE again using the classic coarse-graining approach (which consists in decimating the signal by taking one sample every τ and by substituting the selected sample with the average over τ contiguous samples) and the moving-average approach followed in [24,25] (which consists of filtering the signal with a moving average of order τ and calculating the distribution entropy by introducing the delay $\delta = \tau$ among the filtered samples). Figure A1 shows that mDE estimates after coarse-graining or moving average present a clear periodicity at multiples of $\tau = 4$ and that its values depend on the filter (e.g., for m = 3 mDE shows a relative minimum after coarse graining and a relative maximum after the moving average, at scales $\tau = 2 + 4n$, with $n = 0, 1, 2, \ldots$). This result demonstrates that the periodicity is introduced by the filtering or coarse graining method and suggests that it is an artifact introduced by the broad transition band of the moving average.

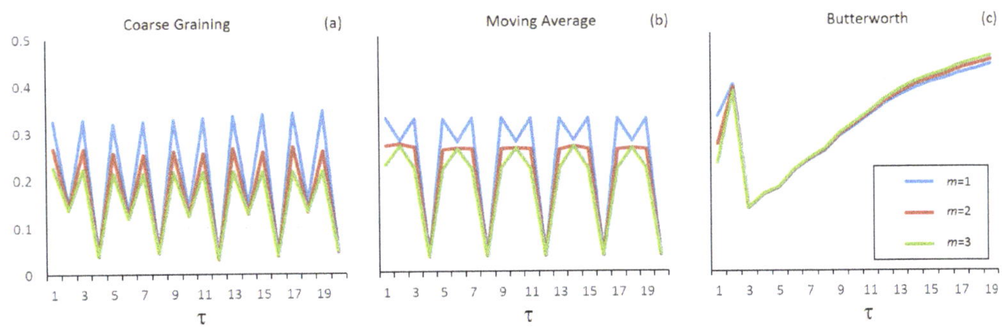

Figure A1. Multiscale Distribution Entropy of the periodic signal estimated with the traditional coarse-graining approach (**a**), with the delay method using the moving average filter (**b**), and the Butterworth filter (**c**).

References

1. Chen, C.; Jin, Y.; Lo, I.L.; Zhao, H.; Sun, B.; Zhao, Q.; Zheng, J.; Zhang, X.D. Complexity Change in Cardiovascular Disease. *Int. J. Biol. Sci.* **2017**, *13*, 1320–1328. [CrossRef] [PubMed]
2. Pincus, S.M. Approximate entropy as a measure of system complexity. *Proc. Natl. Acad. Sci. USA* **1991**, *88*, 2297–2301. [CrossRef] [PubMed]
3. Pincus, S.M.; Gladstone, I.M.; Ehrenkranz, R.A. A regularity statistic for medical data analysis. *J. Clin. Monit. Comput.* **1991**, *7*, 335–345. [CrossRef] [PubMed]
4. Grassberger, P.; Procaccia, I. Estimation of the Kolmogorov entropy from a chaotic signal. *Phys. Rev. A* **1983**, *28*, 2591–2593. [CrossRef]
5. Takens, F. On the Numerical Determination of the Dimension of an Attractor. In *Dynamical Systems and Bifurcations*; Braaksma, B.L.J., Broer, H.W., Takens, F., Eds.; Lecture Notes in Mathematics: Berlin/Heidelberg, Germany; Springer: Berlin/Heidelberg, Germany, 1985; Volume 1125, pp. 99–106. ISBN 978-3-540-15233-0.
6. Richman, J.S.; Moorman, J.R. Physiological time-series analysis using approximate entropy and sample entropy. *Am. J. Physiol. Heart Circ. Physiol.* **2000**, *278*, H2039–H2049. [CrossRef]
7. Chen, W.; Wang, Z.; Xie, H.; Yu, W. Characterization of Surface EMG Signal Based on Fuzzy Entropy. *IEEE Trans. Neural Syst. Rehabil. Eng.* **2007**, *15*, 266–272. [CrossRef]
8. Azami, H.; Li, P.; Arnold, S.E.; Escudero, J.; Humeau-Heurtier, A. Fuzzy Entropy Metrics for the Analysis of Biomedical Signals: Assessment and Comparison. *IEEE Access* **2019**, *7*, 104833–104847. [CrossRef]
9. Li, P.; Liu, C.; Li, K.; Zheng, D.; Liu, C.; Hou, Y. Assessing the Complexity of Short-Term Heartbeat Interval Series by Distribution Entropy. *Med. Biol. Eng. Comput.* **2015**, *53*, 77–87. [CrossRef]
10. Gao, L.; Gaba, A.; Cui, L.; Yang, H.; Saxena, R.; Scheer, F.A.J.L.; Akeju, O.; Rutter, M.K.; Lo, M.; Hu, K.; et al. Resting Heartbeat Complexity Predicts All-Cause and Cardiorespiratory Mortality in Middle- to Older-Aged Adults From the UK Biobank. *JAHA* **2021**, *10*, e018483. [CrossRef]
11. Costa, M.; Goldberger, A.L.; Peng, C.-K. Multiscale Entropy Analysis of Complex Physiologic Time Series. *Phys. Rev. Lett.* **2002**, *89*, 068102. [CrossRef]
12. Karmakar, C.; Udhayakumar, R.K.; Li, P.; Venkatesh, S.; Palaniswami, M.S. Stability, Consistency and Performance of Distribution Entropy in Analysing Short Length Heart Rate Variability (HRV) Signal. *Front. Physiol.* **2017**, *8*, 720. [CrossRef]
13. Brooks, C.M.C. Introduction to the Integrative Role of the Autonomic Nervous System in the Regulation of Cardiovascular Function. In *Cardiovascular Physiology Neural Control Mechanisms*; Elsevier: Amsterdam, The Netherlands, 1981; pp. 21–22. ISBN 978-0-08-026821-7.
14. Castiglioni, P.; Merati, G.; Parati, G.; Faini, A. Decomposing the Complexity of Heart-Rate Variability by the Multifractal–Multiscale Approach to Detrended Fluctuation Analysis: An Application to Low-Level Spinal Cord Injury. *Physiol. Meas.* **2019**, *40*, 084003. [CrossRef] [PubMed]
15. Castiglioni, P.; Parati, G.; Di Rienzo, M.; Carabalona, R.; Cividjian, A.; Quintin, L. Scale Exponents of Blood Pressure and Heart Rate during Autonomic Blockade as Assessed by Detrended Fluctuation Analysis. *J. Physiol.* **2011**, *589*, 355–369. [CrossRef] [PubMed]
16. Castiglioni, P.; Merati, G.; Parati, G.; Faini, A. Complexity in Heart Rate Variability after Postural Sympathovagal Change by Sample, Fuzzy, and Distribution Entropy. In Proceedings of the 2022 12th Conference of the European Study Group on Cardiovascular Oscillations (ESGCO), Štrbské Pleso, Slovakia, 9 October 2022; IEEE: Štrbské Pleso, Slovakia; pp. 1–2.
17. Castiglioni, P.; Merati, G. Fractal Analysis of Heart Rate Variability Reveals Alterations of the Integrative Autonomic Control of Circulation in Paraplegic Individuals. *Physiol. Meas.* **2017**, *38*, 774–786. [CrossRef]
18. Merati, G.; DiRienzo, M.; Parati, G.; Veicsteinas, A.; Castiglioni, P. Assessment of the Autonomic Control of Heart Rate Variability in Healthy and Spinal-Cord Injured Subjects: Contribution of Different Complexity-Based Estimators. *IEEE Trans. Biomed. Eng.* **2006**, *53*, 43–52. [CrossRef] [PubMed]
19. Borin, A.M.S.; Silva, L.E.V.; Murta, L.O. Modified Multiscale Fuzzy Entropy: A Robust Method for Short-Term Physiologic Signals. *Chaos* **2020**, *30*, 083135. [CrossRef] [PubMed]
20. Faini, A.; Castiglioni, P. Comment on "Modified Multiscale Fuzzy Entropy: A Robust Method for Short-Term Physiologic Signals" [Chaos 30, 083135 (2020)]. *Chaos* **2021**, *31*, 018103. [CrossRef]
21. Castiglioni, P.; Parati, G.; Faini, A. Information-Domain Analysis of Cardiovascular Complexity: Night and Day Modultions of Entropy and the Effects of Hypertension. *Entropy* **2019**, *21*, 550. [CrossRef]
22. Valencia, J.F.; Porta, A.; Vallverdu, M.; Claria, F.; Baranowski, R.; Orlowska-Baranowska, E.; Caminal, P. Refined Multiscale Entropy: Application to 24-h Holter Recordings of Heart Period Variability in Healthy and Aortic Stenosis Subjects. *IEEE Trans. Biomed. Eng.* **2009**, *56*, 2202–2213. [CrossRef]
23. Govindan, R.B.; Wilson, J.D.; Eswaran, H.; Lowery, C.L.; Preißl, H. Revisiting Sample Entropy Analysis. *Phys. A Stat. Mech. Its Appl.* **2007**, *376*, 158–164. [CrossRef]
24. Deka, B.; Deka, D. An Improved Multiscale Distribution Entropy for Analyzing Complexity of Real-World Signals. *Chaos Solitons Fractals* **2022**, *158*, 112101. [CrossRef]

25. Lee, D.-Y.; Choi, Y.-S. Multiscale Distribution Entropy Analysis of Short-Term Heart Rate Variability. *Entropy* **2018**, *20*, 952. [CrossRef] [PubMed]
26. Castiglioni, P.; Coruzzi, P.; Bini, M.; Parati, G.; Faini, A. Multiscale Sample Entropy of Cardiovascular Signals: Does the Choice between Fixed- or Varying-Tolerance among Scales Influence Its Evaluation and Interpretation? *Entropy* **2017**, *19*, 590. [CrossRef]

entropy

MDPI

Article

Visibility Graph Analysis of Heartbeat Time Series: Comparison of Young vs. Old, Healthy vs. Diseased, Rest vs. Exercise, and Sedentary vs. Active

Alejandro Muñoz-Diosdado [1], Éric E. Solís-Montufar [1] and José A. Zamora-Justo [1,2,*]

[1] Instituto Politécnico Nacional, Unidad Profesional Interdisciplinaria de Biotecnología, Mexico City 07340, Mexico; amunozdiosdado@gmail.com (A.M.-D.); eric.montufar@gmail.com (É.E.S.-M.)
[2] Instituto Tecnológico de Santo Domingo (INTEC), Santo Domingo 10602, Dominican Republic
* Correspondence: jzamoraj@ipn.mx

Abstract: Using the visibility graph algorithm (VGA), a complex network can be associated with a time series, such that the properties of the time series can be obtained by studying those of the network. Any value of the time series becomes a node of the network, and the number of other nodes that it is connected to can be quantified. The degree of connectivity of a node is positively correlated with its magnitude. The slope of the regression line is denoted by *k-M*, and, in this work, this parameter was calculated for the cardiac interbeat time series of different contrasting groups, namely: young vs. elderly; healthy subjects vs. patients with congestive heart failure (CHF); young subjects and adults at rest vs. exercising young subjects and adults; and, finally, sedentary young subjects and adults vs. active young subjects and adults. In addition, other network parameters, including the average degree and the average path length, of these time series networks were also analyzed. Significant differences were observed in the *k-M* parameter, average degree, and average path length for all analyzed groups. This methodology based on the analysis of the three mentioned parameters of complex networks has the advantage that such parameters are very easy to calculate, and it is useful to classify heartbeat time series of subjects with CHF vs. healthy subjects, and also for young vs. elderly subjects and sedentary vs. active subjects.

Keywords: visibility; tachograms; *k-M* slope; complex networks; average degree; average path length

1. Introduction

Nowadays, the scientific community is studying new tools for physiological signal analysis that can be successfully used for the classification of different groups by computational algorithms [1–4]. Especially, the analyses of electrocardiograms (ECG) by different methods have shown potential applications for the prediagnosis of several diseases [5]. These tools could be helpful for physicians when giving a diagnosis. Some traditional techniques for the analysis are based on measurements of time and frequency [6,7]. For example, De Abreu et al. [8] analyzed the performance of cyclists with different kinds of training by using time and frequency domain markers on beat-to-beat heart period series. However, these methods have limitations due to the sampling frequency of the equipment used to measure and the sensitivity of the results to the definition of the filter bandwidth [9]. For instance, vagal modulation could be studied in high-frequency heartbeat time series, known as respiratory sinus arrhythmia [10]. Hence, new methodologies have emerged for the analysis of these kinds of signals. Guzzetti et al. [11] showed that signals in a group containing patients with chronic heart failure and in a healthy control group have significant differences using linear and nonlinear methods. Therefore, the measurement of the signals' complexity and the application of nonlinear dynamic techniques have shown relevant results for classifying the signals of healthy subjects from those of diseased subjects [2,4,12]. Moreover, the influence of aging, gender, and the physical condition of subjects has been studied with heart rate variability (HRV) analysis with entropy calculations [9,13–15]. Furthermore, the results of Porta et al. [16] suggest that age is a factor in the decrease in vagal modulation and increase in sympathetic modulation, which reduce the complexity of cardiovascular regulation.

On the other hand, to characterize the irregular behavior of time series data, alternative techniques have recently been applied in different areas of science, such as network analysis, based on the concepts of statistical physics and graph theory [17,18]. The idea of transforming a discrete set of data into a network involves considering a set of nodes and the links between them. The transformation of a time series into a network allows the investigation of the dynamics of complex systems through analysis of the properties of the network [19,20]. Several methods have been proposed to convert time series into networks as well as to characterize these series in terms of the parameters with which the network is described. For example, Zhang and Small [21], starting from time series with periodic properties, obtained complex networks representing each cycle as a node. The work of Xu et al. [22] is also important; they analyzed the properties of complex networks and the distribution of subsets of the networks' graphs. Gao et al. [23,24] proposed a methodology to study chaotic time series and developed a new network, which they called a multivariate recurrence network.

One of the most important methodologies in this field of research is the visibility graph algorithm (VGA) [25], which consists of transforming a time series into a complex network while maintaining its basic characteristics and making use of network theory to characterize the original data. Through this mapping, the dynamic characteristics of the time series become the characteristics of the network, and conversely, the features of the time series can be deduced from those of the network. This algorithm has recently been applied in various areas, such as finance [26–30], weather forecasting [31–33], medical science [34–37], and especially in real [20,38–42], experimental [19], and synthetic seismicity analyses [43].

In the VGA approach, a segment connects two time series values that can see each other. In terms of graph theory, each time series value is assigned a node, and there exists a connection (or edge) between two nodes if there is visibility between them. The degree of connectivity is the number of connections per node. Each node is visible at least to its nearest neighbors (left and right), and there is invariance under related transformations of the time series [25]. The graphs developed using the VGA method have been shown to transform periodic, random, and fractal time series into regular, random, and unscaled networks, respectively [25,44,45].

In the seismicity field, Telesca et al. [38] have used the VGA method to study sequences of seismic magnitude, and they observed power law behavior in the distribution of connectivity degrees. This unscaled signature on connectivities is concomitant with other scale-invariant properties observed in earthquake research. Earthquakes are complex phenomena, both in terms of their spatial and temporal characteristics, and many of the laws of seismology exhibit fractal properties, as reflected by scale exponents [46,47].

In the visibility graph methodology, the degree of connectivity k is defined for each event according to its magnitude and position in the time series. The k-M parameter is the slope of the straight regression line, with the magnitude on the x-axis and the degree of connectivity on the ordinate axis. Telesca et al. [19] conducted a similar analysis on regional seismicity in Pannonia and proposed that the correlation between these two parameters is almost universal. Similar results have been obtained when analyzing seismicity time series obtained experimentally [19]. Additional studies in this direction have been performed by Azizzadeh-Roodpish et al. [42,48] for Southern California and the Alaskan Crustal and Aleutian subduction zones; by Khoshnevis et al. [41] on the seismicity of northern Iran; and by Pérez-Oregon et al. [43] in synthetic earthquake time series. All of these studies have strengthened this universal linear correlation.

The variability in the human heart rate is the result of the combination of different physiological control systems that allow the body's functioning to adapt to physical, environmental, or other changes. Time series of long-term R–R intervals (or tachograms) are nonlinear time series that present fractal properties. The behavior of heart rate variability is altered with age, when subjects exercise, and when they are ill [9,49–53]. When HRV is reduced, there is an increased risk of cardiovascular disease [54,55], and higher mortality has also been reported in patients with diseases of the circulatory system [56,57]. As tachograms are nonlinear, inhomogeneous, and non-stationary time series, nonlinear methods can better capture changes in HRV than linear methods [12,58–60]. To analyze these time series, many nonlinear methodologies have been used, such as detrended fluctuation analysis (DFA) [61–63], fractal dimension [64,65], and multifractal analysis [4,66].

Changes in HRV induced by low or intense physical activity have been widely studied [1,13,15,67–74]; when the subject is exercising, cardiovascular disorders can be detected that are not observed when the subject is inactive [70]. Therefore, by introducing extra work for the heart while exercising and monitoring the subject's heart activity with an electrocardiograph, it may be possible to discern heart problems that are not seen when the subject is at rest [68]. Heart rate has been shown to increase during physical activity due to the withdrawal of the parasympathetic system and increased sympathetic activity (of course, this depends on the intensity of the exercise) [69]. Furthermore, it has been reported in animal models that improving physical fitness can reduce problems caused by heart disease [75].

Very interesting works analyzing cardiac interbeat series using the visibility algorithm have been published; for example, Jiang et al. [76] applied the VGA to study the dynamics of the human heartbeat during chi meditation and yoga. Their results indicated that VGA analysis can reveal changes caused by meditation training, which manifest as regular heartbeats, and they concluded that this methodology constitutes a reliable tool for the assessment of the dynamic changes in heartbeat under different physiological and pathological conditions, even when using a short series with only 400 beats. In 2018, Bhaduri et al. [3] studied heartbeat time series taken from the Physionet databases, corresponding to two groups of subjects: one healthy and the other with congestive heart failure. As the distribution degree of the nodes was found to satisfy a power law, they proposed the parameter of power of scale-freeness in visibility graph (PSVG), which could significantly differentiate between healthy and diseased subjects and also allow for the determination of different stages of the disease. Choudhary et al. [77] used the horizontal visibility algorithm and proposed the parameter of grouped horizontal visibility entropy (GHVE) to quantify the complexity as a function of the probability distribution of the observations. They concluded that the proposed GHVE measure is more accurate in distinguishing healthy from pathological subjects than its ungrouped HVE counterpart. Therefore, it is better suited for the detection of changes related to aging, disease severity, and activity levels. Other researchers [78] have investigated the possible impact of meditation on the complexity of heart rate signals using the VGA. Using data from Physionet, they analyzed the dynamics of heartbeat time series both before and during meditation by examining the complexity of the VGs using the graphical index complexity (GIC) [79]. In general, the results obtained showed that heart rate signals were more complex during meditation.

On the other hand, there are other parameters characterizing complex networks, such as the average degree, which is the mean number of neighboring nodes for each node, and the average path length, which is the mean distance between all pairs of nodes [17].

The objective of this work is to study the behavior of the *k-M* parameter for a totally different complex system: the heart rate variability time series [50]. To do this, heartbeat time series obtained from different people in various situations are studied: first, we compare healthy young and elderly subjects, then healthy subjects and patients with congestive heart failure (CHF), and, finally, subjects at rest and while exercising. In addition, we analyze the average degree and average path length of the networks obtained for each time series. We show, in this work, that these parameters are useful to determine differences in time series and, remarkably, that the selected parameters are the simplest to obtain from a network due to the low computational costs for time and resources. It is worth mentioning that the main contribution of this work is to identify the network parameters that can be implemented for the classification of the heartbeat time series of patients with CHF vs. healthy subjects, young vs. elderly subjects, and sedentary vs. active subjects. Although VGA is not a new tool, in this paper we explore its use in heartbeat time series, something that has not been widely studied, in order to enlarge the range of tools that can be used for series analysis.

The remainder of the work is organized as follows. Section 2 presents the methods of the VGA performed (Section 2.1) and the databases used for analysis (Section 2.2); this last subsection is divided according to the kinds of series considered. First, we generate white, pink, and Brownian noises to verify the differences between network parameters in these artificial series; then, we obtain series from Physionet [80] with registers of young and elderly subjects and another database with healthy subjects and subjects with CHF; and, finally, the third database was generated by our own records of subjects at rest and during exercise. Section 3 presents the results of the *k*-*M* slope, the average degree and average path length of all databases mentioned above, and the statistical analysis to validate significant differences between the groups. Section 4 opens the discussion and the contribution of the work. Finally, Section 5 shows the concluding remarks.

2. Materials and Methods

2.1. Visibility Algorithm

The visibility graph algorithm is a method that maps a time series into a complex network. We illustrate the method in Figure 1 in terms of a bar graph, where the height of the bars represents the magnitude of the events. Two events are connected if they are visible to each other, and this happens if any event between them intersects the straight line that connects them. This visibility algorithm [25] has attracted a lot of attention as it can reveal the nonlinear characteristics of time series. By mapping the time series into a graph with connected nodes, this complex graph (or network) inherits various properties of the time series; for example, random time series result in random graphs, periodic time series result in random graphs, and time series with fractal properties result in scale-free networks [17,18]. Let $x(i), 1 \leq i \leq N$ be a time series composed of N data, and consider each data point as a node in the VGA. Nodes $x(i)$ and $x(j)$ are connected only when the following condition is satisfied [25]:

$$\forall l \in (i,j); x(l) < x(i) - (j-l)\frac{x(j)-x(i)}{j-1}. \quad (1)$$

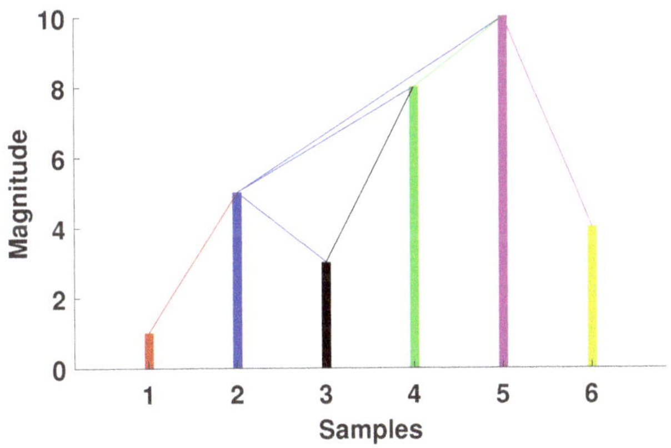

Figure 1. Description of the visibility method. Each bar represents the magnitude of a single value from a time series (called a node) and continuous lines represent connections between nodes. If a node with higher magnitude (taller bar) is between two non-adjacent nodes, these other nodes do not have a connection.

This is the basic definition of an undirected visibility graph. The following properties always hold for a VG graph [25]:

- A node is visible at least to its nearest neighbors (left and right).
- The edges do not have a direction.
- Visibility is not affected by the scaling of either horizontal or vertical axes, nor horizontal or vertical translations.

There are various modified versions of the method, such as the horizontal visibility algorithm and the directed visibility algorithm, in which the connection is defined in only one direction. The direction can be forward (away from the node) or backward (toward the node) [81–83]. In some applications, these variations have been applied with good results; for example, in the analysis of chaotic periodic signals [81,82], as well as in other randomly correlated ones, for which the correlation [81] or the reversibility have been characterized [83].

We can build matrices for the visibility graph of a time series, called the adjacency matrix. The dimension of the adjacency matrix is the number of nodes in the network. If there is a connection between the nodes corresponding to a certain row and column, the element of the matrix is equal to one; otherwise, it is equal to zero. For an undirected visibility graph, this matrix is symmetric, with the elements of the main diagonal equal to zero. There are usually ones around the main diagonal, as neighboring events are more likely to be visible to each other. In the case of a directed visibility graph, the matrix is asymmetric. The degree of connectivity of each node is obtained as the sum of the number of connections in the matrix for each corresponding row [25].

The reliability of this methodology has been confirmed through its application to both artificial fractal series and real series [25]. As has been reported, this method can provide good results with relatively small series; for example, Jiang et al. [76] have claimed that reliable results can be obtained with only 400 data.

In seismicity, the term *k-M* refers to a graph in which the degree of connectivity k is plotted against the magnitude of the earthquake; however, for any time series, the magnitude refers to the time series value that corresponds to that node. It is expected that a small value of the time series will have a small value of k, and a large value of the time series (which, in the visibility diagram, looks like a very high bar) would have a high degree of connectivity as, from this point, many other bars can be seen [84].

2.2. *k-M Slope, Average Degree, and Average Path 0 for White, Pink, and Brownian Noise*

We illustrate the *k-M* calculations for 100 time series generated using white, pink ($1/f$), and Brownian noise. Self-affine series were obtained with $N = 1024$ data and with values of the spectral exponent β as follows: $\beta = 0.5$ (white), $\beta = 1$ (pink), and $\beta = 2$ (Brownian). The methodology of [85–87] was used to generate the time series. In addition, the average degree and average path length of networks generated from the aforementioned series were also calculated. Furthermore, analysis of variance (ANOVA) was used to find significant differences between the networks' parameters calculated; if this was found, the contrast test of Fisher's least significant difference (LSD) was performed to determine exactly which groups showed statistical differences.

2.3. *Heartbeat Time Series*

2.3.1. Young and Elderly Healthy Subjects

In order to evaluate the complexity lost with age [9], the VGA was applied to heartbeat time series obtained from Physionet [49,80], a database containing data from 20 elderly (68–85 years old) and 20 young (21–34 years old) healthy subjects, whose electrocardiography signals (ECG) were recorded while they remained in a resting state watching the movie Fantasia. Each subgroup included equal numbers of women and men. We extracted samples with 1500 events from each time series. Then, we obtained the visibility graph and the corresponding complex network for each time series, and the *k-M* value was obtained by fitting the connectivity degree versus the magnitude (i.e., the duration of the time interval in seconds) of each R–R interval. In addition, the average degree and average path length were also calculated. In order to prove that the differences between the obtained values were statistically significant, we performed Student's t tests with a significance level of 0.05 for comparison.

2.3.2. Healthy Subjects and CHF Patients

In addition, heartbeat time series of healthy subjects and patients with congestive heart failure (CHF) were obtained from Physionet [80], and we considered the following data:

- 54 registers over 24 h from subjects with normal sinus rhythm R–R interval (30 men, aged 28.5–76; 26 women, aged 58–73).
- 29 registers over 24 h from subjects with CHF, New York Heart Association (NYHA) classifications I, II, and III.
- 15 registers over 24 h from subjects with severe CHF, NYHA classifications III and IV (11 men, aged 22–71; 4 women, aged 54–63).

We obtained six-hour sub-series when the subjects were sleeping and also six-hour sub-series when patients were awake. However, for comparison purposes, we only separated sub-series with 1500 data, as with the data on healthy old and young people from the previous section. To obtain the *k-M*, average degree, and average path length, the same procedure was followed as in the previous section. A Student's *t* test was also performed with a significance level of 0.05 for the comparison of these values calculated for both groups. With this analysis, we intended to verify if there is also a loss of complexity in the signals of subjects with CHF, as occurs in the aging process, which is associated with an increase in sympathetic modulation due to the constant cardiac stress generated by the disease.

2.3.3. Cardiac Interbeat Series Obtained at Rest and While Exercising

In total, tachograms of 152 subjects at rest and exercising (walking on a treadmill) were analyzed. Of these 152, 84 were young men and women with an average age of 20.8 years, while the other 68 were middle-aged adults with a mean age of 51.9 years. The measurements were repeated for all subjects. For each participating subject, personal information was collected, including age, gender, weight, height, and so on. Subjects with any disease related to the cardiovascular system were discarded, and it was ensured that they were not taking medication of any kind. The tests were carried out in the morning, at a suitable temperature. The average body mass indices of the subjects were as follows: for young people, it was 25.24 ± 3.53; for middle-aged adults, it was 28.08 ± 4.54. For each record, there were two measurements: a 60 min rest period and a 30 min walking period on a treadmill. A total of 4096 and 2048 data were analyzed for resting and exercise series, respectively. At the beginning of the test, seven electrodes were placed on the subjects to take their ECG recordings, using a Fukuda Denshi Holter monitor model FM-180 with a sampling frequency of 125 Hz. Once the 60 min rest period had ended, the young subjects (under 30 years of age) were put on a commercial electric treadmill to walk at 1.56 m/s for 30 min. The speed for the older subjects was 1.34 m/s. The beat-by-beat signal, or tachogram, was obtained for each digitized ECG using a peak detection algorithm.

The participants were volunteers and signed the informed consent form. The protocol for this research was approved by the Ethics Committee of the FES-UNAM-Mexico under document number CE/FESI/022020/1348. Before the test, all subjects answered the International Physical Activity Questionnaire (IPAQ) in order to classify their physical activity [88]. With the help of this questionnaire, the subjects were classified into three levels of physical activity (high, moderate, and low); all comparisons were made with subjects with high and low physical activity because time series of subjects with moderate results were discarded, so comparisons were made with 31 series of active middle-aged adults, 24 series of sedentary middle-aged adults, 51 series of active young subjects, and 30 series of sedentary young subjects.

In addition, all visibility graphs were generated from the R–R series of each person in the previously described databases, allowing us to compare the *k-M*, the average degree, and the average path length for the different populations. In order to prove that the differences between the obtained values were statistically significant, we performed a hypothesis test by ANOVA followed by Fisher's LSD contrast test to determine which groups showed significant differences. The objective of this analysis was to identify changes in series through VGA. Subjects who performed regular exercise achieved the inhibition of the sympathetic system during physical activity, causing an improvement in the vagal response [8], which could be seen in minimal changes between series taken at rest and those during exercise. This fact was evaluated in both young and middle-aged adult subjects.

3. Results

3.1. Visibility Parameters from White, Pink, and Brownian Noise

Figure 2a shows the visibility graph obtained for white noise, where the node color depends on its degree (k), with darker colors representing higher degrees. The series that generated this graph is shown in Figure 2b. Figure 2c shows the plot of connectivity degree (k) vs. magnitude (M) for each value from the series, and the linear fit of the data by mean least squares, which was used to obtain the *k-M* slope, can also be observed. Figure 3 corresponds to the *1/f* noise, while Figure 4 corresponds to the Brownian noise. It can be seen that the appearance of the networks, as these types of noise are different, presented the following characteristics:

- White noise has more dispersion between nodes, providing a lower average degree in the network and a greater average path length.
- With Brownian noise, the nodes form many large clusters with high connectivity, leading to a greater average degree and a lower average path length.
- For pink noise, the average path length and average degree values were between those obtained for white and Brownian noise.

From the plots of k (connectivity degree) vs. M (magnitude), we obtained the linear fits: for white noise, $k = 4.8787\ M + 10.4871$; for $1/f$ noise, $k = 13.5515\ M + 0.0853$; and, for Brownian noise, $k = 19.4626\ M - 5.0672$.

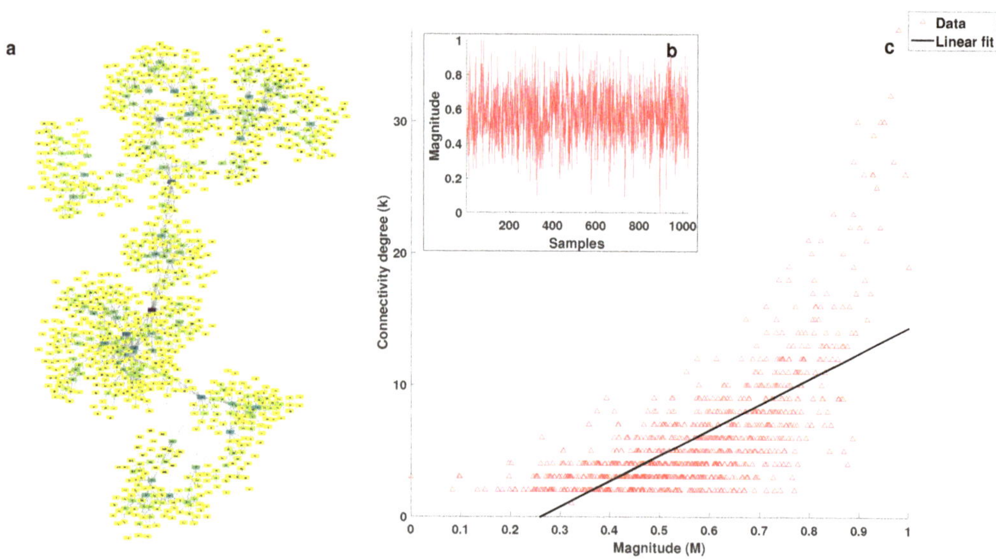

Figure 2. (**a**) Visibility graph generated for the white noise shown in (**b**); and (**c**) k vs. M plot with linear fit.

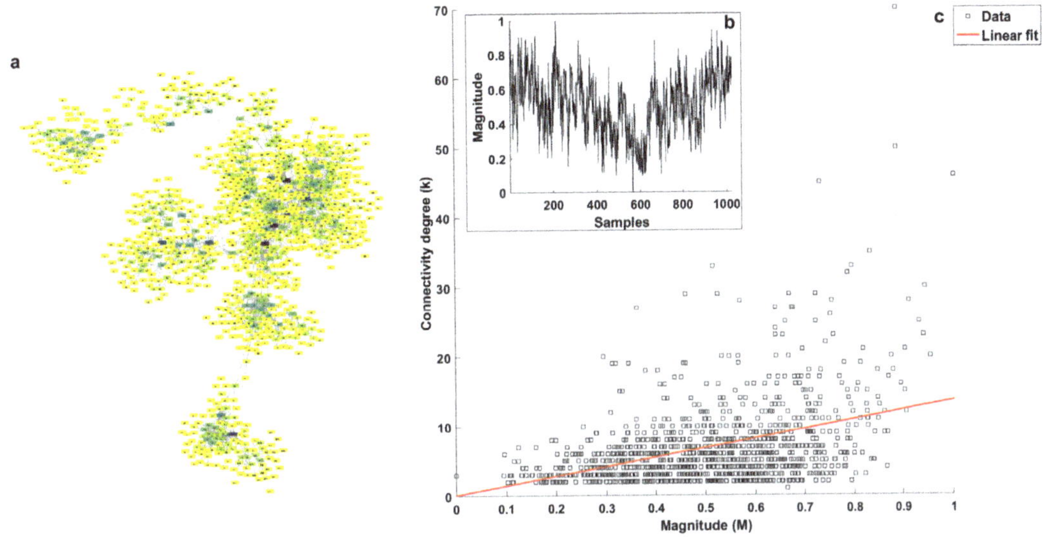

Figure 3. (**a**) Visibility graph generated for the pink noise shown in (**b**); and (**c**) k vs. M plot with linear fit.

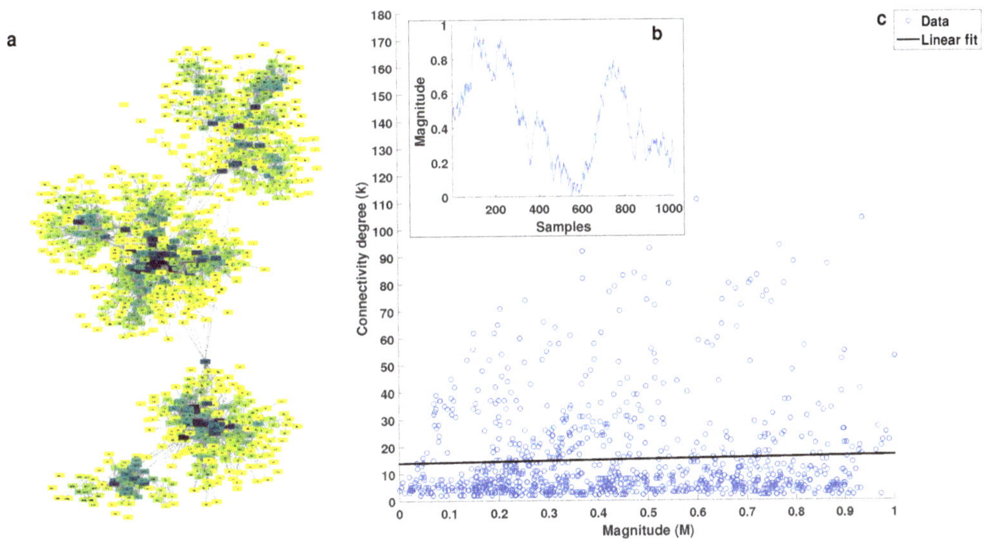

Figure 4. (**a**) Visibility graph generated for the Brownian noise shown in (**b**); and (**c**) k vs. M plot with linear fit.

In the graph of k vs. M, despite the fact that the distribution of points was not close to a straight line, we proved that the correlation was significant with the corresponding hypothesis test for the white, *1/f*, and Brownian noises. Considering a level of significance of 0.05, we obtained p-values less than 3.5×10^{-3} in all cases; therefore, the correlation was significant despite the large spread of the data in the k versus M plots, and other mathematical models were not considered to fit the data since the linear fit was statistically validated, as indicated above.

On the other hand, 100 different series of white noise with 1024 data as well as 100 series of $1/f$ noise and Brownian noise were generated. For each of the three hundred series, the value of the slope *k-M* was obtained. The averages for each type were as follows: for white noise, 21.6203 ± 2.1061; for $1/f$ noise, 15.4359 ± 2.4233; and, for Brownian noise, 5.8246 ± 6.9253.

Figure 5a shows the average values of k and M from the 100 series for each kind of noise. It is worth mentioning, in this context, that the hypothesis tests to validate the correlation coefficients of the linear fit to calculate the *k-M* slope were also performed. In addition, Figure 5b shows the distribution of the *k-M* slopes obtained from the series, from which the differences between the *k-M* values of these three kinds of series can be observed. The average values of the means are clearly separated, although the values of the standard deviations are large, especially for Brownian noise. However, the ANOVA results indicated significant differences between the *k-M* of the three groups. Since differences were found by ANOVA, posthoc tests were necessary for evaluating pair-wise differences; hence, Fisher's least significant difference (LSD) procedure was performed, and the results showed that there were significant differences between all the pairs of noise types ($p < 1.0 \times 10^{-5}$ in all cases, $\alpha = 0.05$).

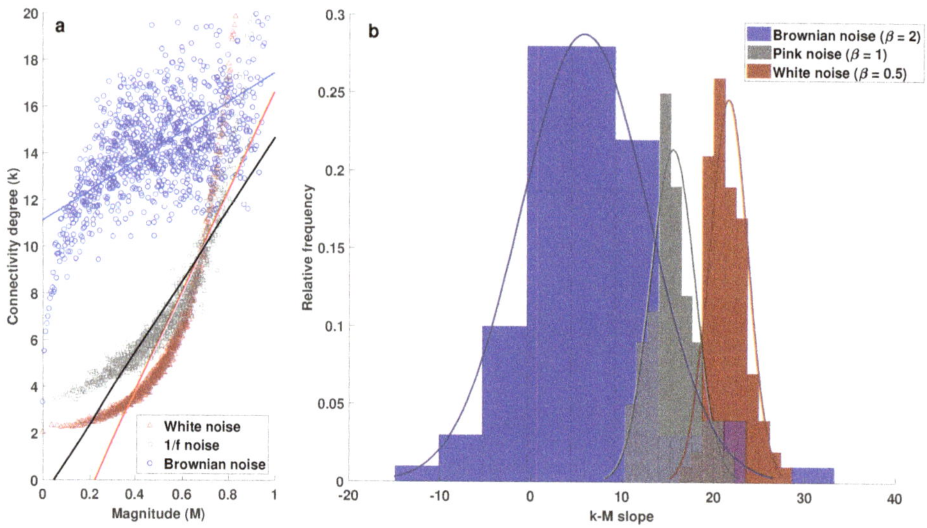

Figure 5. (**a**) Average connectivity degree vs. average magnitude plot for white, pink, and Brownian noise; and (**b**) histogram and distribution of *k-M* slopes from 100 series of the above-mentioned noises.

It is precisely the very different values obtained for the average of the *k-M* slopes that motivated us to think that this parameter may distinguish between different time series, especially those of cardiac interbeat. In terms of the β spectral exponent, it has already been reported that the R–R series has a β value of around one or a DFA exponent value of one [60–62], which corresponds to the *1/f* noise; however, deviations from this value may occur in the elderly or in people with heart disease.

In addition, we compared the average degree and average path length in the white, pink, and Brownian noise networks. Figure 6 shows the histogram of the average degrees, from which we observed similar differences to those mentioned above. The values obtained were as follows: 5.8249 ± 0.0855 for white noise, 7.0797 ± 0.1734 for pink noise, and 14.2925 ± 2.0838 for Brownian noise. Furthermore, an ANOVA was performed, and the results indicated significant differences between the average degrees for the three series.

Thus, a contrast pair-wise test was necessary, and in all cases, there were significant differences when the corresponding Fisher's LSD test was performed ($p < 1.0 \times 10^{-5}$ in all comparisons). On the contrary, the average path length values did not present significant differences.

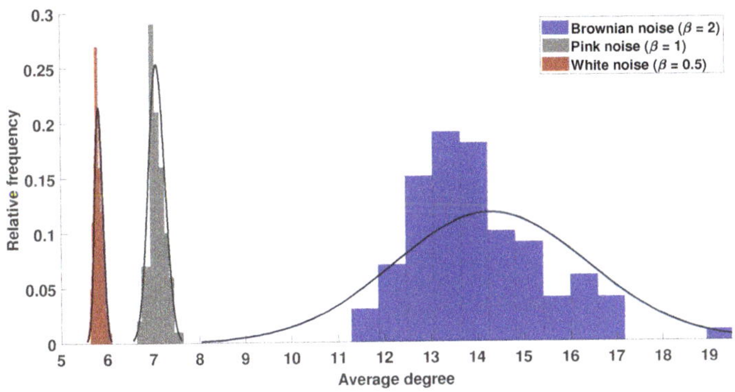

Figure 6. Histogram of average degree for white, pink, and Brownian noises.

3.2. *Analysis of Physionet Database*

By analyzing the Fantasia database, it was possible to observe a difference between the *k-M* plots of the two populations. While the mean and standard deviation of the slopes of the *k-M* plots for young people had lower values, these values were higher for the older subjects. The obtained values for the mean and standard deviation for both populations can be observed in Table 1. These two populations were statistically different, with a *p*-value of $p = 0.004$. It is worth mentioning that the average degree and average path length calculated for these series can also be observed in Table 1; however, these results did not indicate significant differences between populations. These results suggest that the *k-M* value could be a network parameter that is useful to evaluate the aging process and the complexity loss presented in the series of the elderly subjects, which also indicates a reduction in regulatory mechanisms.

Table 1. Values of mean and standard deviation for the slope of the *k-M* plots, average degree, and average path length for the Physionet database.

	k-M Slope	Average Degree	Average Path Length
Young (Fantasia)	37.00 ± 4.05	7.48 ± 0.88	5.09 ± 0.26
Elderly (Fantasia)	65.00 ± 6.40	7.90 ± 1.28	5.02 ± 0.43
Healthy (awake)	47.38 ± 6.12	9.15 ± 1.35	4.62 ± 0.63
CHF (awake)	137.73 ± 11.58	8.15 ± 1.23	4.37 ± 0.81
Healthy (asleep)	48.09 ± 4.88	8.41 ± 1.26	4.79 ± 0.55
CHF (asleep)	120.97 ± 10.07	7.68 ± 0.97	4.37 ± 0.46

Similar results were observed when comparing the healthy subjects with the congestive heart failure (CHF) patients from the Physionet database. The series were split into asleep and awake periods, which were compared separately for the two populations. In both periods, the healthy subjects reported lower values for the slope in the *k-M* plot than the CHF patients, and these differences resulted in significant *p*-values ($p_{awake} = 9.0 \times 10^{-5}$, $p_{sleep} = 9.0 \times 10^{-5}$). The results of this analysis are provided in Table 1. Furthermore, Table 1 also presents the average degree and average path length calculated for these series, from which it was found that there were significant differences between the average degree of healthy and diseased subjects in awake and asleep period ($p_{awake} = 2.7 \times 10^{-3}$, $p_{asleep} = 2.4 \times 10^{-4}$); meanwhile, for the average path length, only significant differences were observed for the asleep series ($p_{asleep} = 3.73 \times 10^{-4}$). Hence, these results indicate that these network parameters can be useful tools for the classification of tachograms from healthy subjects and those with CHF, but apparently it is better to make the comparisons when the patients are asleep.

3.3. Heartbeat Time Series of Exercising Subjects

Figure 7 shows the recorded ECG signal from a young subject during rest and exercise testing. From the Figure, it can be easily identified when the person is at rest and when they are on the treadmill due to the time intervals between the R peaks decreasing in the exercise period, caused by the blood and oxygen demand of the skeletal muscle cells to perform the movement. Therefore, it is easy to differentiate the tachograms at rest and those of physical activity once the artifacts are removed. When the Holter Fukuda Denshi FM 180 was used to measure the ECG signal of subjects at rest and walking at low speeds, the signal obtained was very stable and has almost no artifacts, so our results were not affected by the presence of many non-stationarities [89]. The time series for physical activity were not taken from the time the subject got on the treadmill, as the signal takes time to stabilize; that is, data in the rest–exercise transition region were not considered.

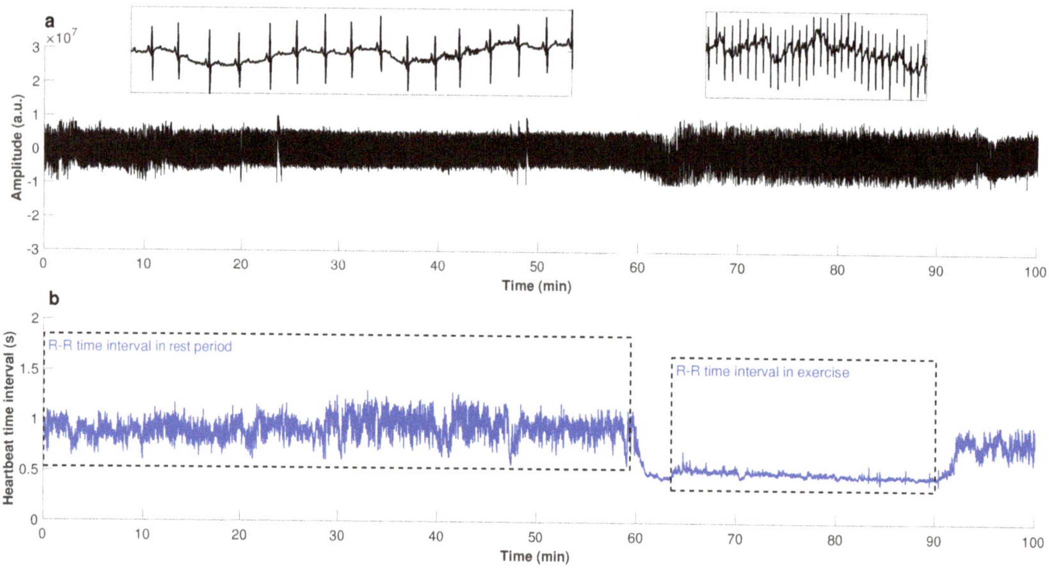

Figure 7. (**a**) Electrocardiography signal; and (**b**) R–R time interval series in rest period and during physical activity.

A sample of networks for rest and activity for a young subject are shown in Figure 8, from which it can be observed that the rest series graph generated more clusters and the nodes have a greater connectivity degree compared with the graph derived from the exercise series. We compared the *k-M* values for young and middle-aged subjects' time series at rest and during physical activity, from which it was found that differences were higher than those in the previous results. The *k-M* mean and standard deviation for this test are displayed in Table 2. In this case, an ANOVA was used to compare the calculations of the *k-M* slope for the four groups. Using this test, significant differences were found between the *k-M* means. Hence, we proceeded to performed a contrast test by using Fisher's LSD method, obtaining the following results:

- Series recorded of young subjects and middle aged adults at rest did not show significant differences ($p = 0.26$).

- Series of young subjects at rest comparing with those during exercise showed significant differences ($p < 1 \times 10^{-5}$).
- Series of middle-aged adults at rest comparing with those recorded while exercising showed significant differences ($p < 1 \times 10^{-5}$).
- Series of young subjects and middle-aged adults during exercise showed significant difference ($p = 5.7 \times 10^{-3}$).

These results showed that the *k-M* slope of the two populations at rest are similar and this is attributed to the classification of all subjects as healthy, since it has been reported [13] that healthy subjects, during rest periods, conserve their heart rate variability and tachogram complexity. However, during physical efforts, the HRV and heart interbeat time series' complexity decreased, producing *k-M* slopes that were significantly higher in the exercise series than those during rest in both populations.

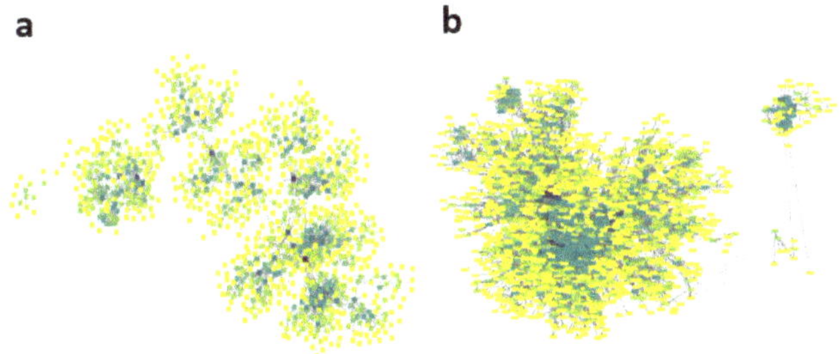

Figure 8. (**a**) Visibility graph of a young subject at rest; (**b**) visibility graph of same subject during exercise. The node color depends on its connectivity degree, where darker colors represent higher degrees.

Table 2. Mean and standard deviation of the slope of *k-M* plots, average degree, and average path length for young and middle-aged subjects at rest and during exercise.

	k-M Slope	Average Degree	Average Path Length
Young (rest)	27.24 ± 14.14	8.32 ± 1.07	5.80 ± 0.89
Adults (rest)	50.15 ± 20.04	7.87 ± 0.98	6.41 ± 3.47
Young (exercise)	196.97 ± 118.60	6.99 ± 1.10	5.18 ± 0.68
Adults (exercise)	214.27 ± 50.34	6.28 ± 0.75	5.36 ± 0.87

In order to ensure that the observed results were trustworthy and reproducible, we compared the *k-M* values of young and adult subjects from the Fantasia database with those we measured using a Holter monitor. As the Fantasia series were recorded while the subjects were watching a movie, we compared these series with our subjects at rest. The means and standard deviations can be observed in Tables 1 and 2 for the Fantasia database and the subjects at rest, respectively. The Student's t tests indicated that there were no significant differences for these populations; that is, young subjects from the Fantasia database were comparable to those measured at rest (and the same for the adults). The *p*-values obtained in these statistical tests were $p_{young} = 0.56$ and $p_{adult} = 0.81$.

The IPAQ results were used to compare the *k-M* slope differences between subjects with high physical activity levels and those who preferred a sedentary lifestyle. The *k-M* mean and standard deviation of these groups at rest and during activity can be observed in Table 3. In addition, Figure 9 depicts the *k-M* values obtained from series for the two groups, with the continuous line representing the average. The ANOVA was performed and the results showed significant differences. Furthermore, the following results were obtained by the Fisher's LSD test:

- There were no significant differences between series of subjects with high physical activity levels and sedentary subjects in both populations at rest ($p > 0.36$ in all cases).

- There were significant differences found between series of subjects with high physical activity levels and sedentary subjects in both populations at rest and those recorded during exercise ($p < 1.3 \times 10^{-5}$ in all cases).
- A significant difference was found in series recorded during exercise of young subjects with high physical activity levels and sedentary subjects ($p = 9.7 \times 10^{-6}$).
- A significant difference was found in series recorded during exercise of middle-aged adults with high physical activity levels and sedentary subjects ($p = 9.7 \times 10^{-4}$).
- Series of young subjects who exercise regularly and sedentary middle-aged adults during exercise did not show a significant difference ($p = 0.12$).

Table 3. Values of mean and standard deviation for the slope of the *k-M* plots, average degree, and average path length of young and middle-aged subjects classified as subjects with high physical activity levels and subjects with sedentary lifestyle, in resting period and during physical activity.

Condition	Subjects	*k-M* Slope	Average Degree	Average Path Length
Rest	Young subjects with high physical activity levels	27.83 ± 13.53	8.33 ± 1.09	5.72 ± 0.34
	Young subjects with sedentary lifestyle	26.23 ± 17.99	8.28 ± 1.043	5.93 ± 1.40
Exercise	Young subjects with high physical activity levels	169.23 ± 81.84	7.20 ± 1.04	5.08 ± 0.59
	Young subjects with sedentary lifestyle	244.13 ± 151.22	6.63 ± 1.13	5.35 ± 0.80
Rest	Adults with high physical activity levels	40.82 ± 33.59	7.76 ± 0.89	6.48 ± 3.62
	Adults with sedentary lifestyle	44.07 ± 39.35	8.05 ± 0.92	5.61 ± 0.55
Exercise	Adults with high physical activity levels	131.53 ± 71.64	7.04 ± 1.12	5.00 ± 0.43
	Adults with sedentary lifestyle	196.96 ± 65.71	6.33 ± 0.75	5.423 ± 0.85

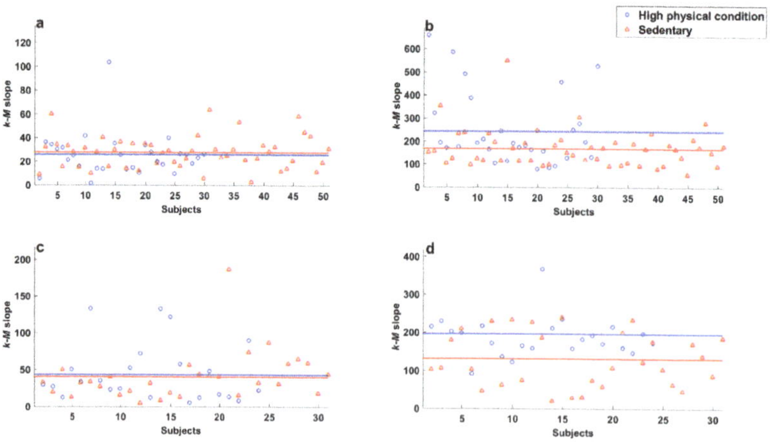

Figure 9. Comparison of *k-M* slope from sedentary young subjects and high-physical activity young subjects in (**a**) resting period and (**b**) during exercise; and those corresponding to middle-aged subjects in (**c**) resting period and (**d**) during physical activity. Markers represent every value of *k-M*, and continuous lines indicate the means.

It can be observed in Table 3, that the *k-M* slopes obtained for the series of subjects increase during exercise, above all in series of young subjects the increase is higher compared with that of middle-aged adults. In addition, the *k-M* is greater in subjects with a sedentary lifestyle in both groups. This fact produces an overlap in the series of sedentary adults and young subjects with high physical activity levels.

It is worth mentioning that significant differences in this parameter were not obtained for the resting series. We also attributed this to all subjects having been declared as healthy; during the rest period, heart rate variability (HRV) is conserved, while during heart effort, the variability decreases. These results indicate that the hearts of subjects with high physical activity levels manage activity conditions better and maintain HRV, while the opposite is observed in sedentary individuals.

Furthermore, the average degree of networks obtained from rest and exercise time series were analyzed. Table 2 shows the results of the means and standard deviations for young subjects and adults. In this case, the average degree of exercise series decreased in physical activity. Significant differences between rest and physical activity periods were obtained by ANOVA test and the contrast Fisher's test showed the following:

- Average degree values of series recorded of subjects of both groups at rest showed significant differences ($p = 0.01$).
- Average degree values of young subjects' series at rest compared with those during exercise showed significant differences ($p < 1 \times 10^{-5}$).
- Average degree values of middle-aged adults' series at rest compared with those recorded while exercising also showed significant differences ($p < 1 \times 10^{-5}$).
- Series of young subjects and middle aged adults during exercise did not show a significant difference ($p = 0.16$).

Similar to the *k-M* slope, comparisons of the average degree calculated from subjects with high physical activity levels and sedentary subjects were performed, with the averages and standard deviations shown in Table 3 and a graphic representation in Figure 10. The ANOVA results indicated that there were significant differences between the averages of the eight groups, and the following results were found by Fisher's LSD test:

- Series of young subjects with high physical activity levels and sedentary subjects at rest did not show a significant difference ($p = 0.85$).
- Series of middle-aged adults with high physical activity levels and sedentary subjects at rest did not show a significant difference ($p = 0.30$).
- Significant differences were found between the average degree values of series of subjects with high physical activity levels and sedentary subjects from both populations at rest and those recorded during exercise ($p < 6.0 \times 10^{-3}$ in all cases).
- A significant difference was found in exercise series of young subjects with high and low physical activity levels ($p = 1.5 \times 10^{-2}$).
- A significant difference was found in exercise series of middle-aged subjects with high and low physical activity levels ($p = 1.0 \times 10^{-2}$).
- Series of young subjects and middle-aged adults classified with high physical activity levels while exercising did not show a significant difference ($p = 0.50$).
- Series of sedentary young subjects and middle-aged adults in exercise period did not show a significant difference ($p = 0.28$).

It was also found that there were no significant differences in the rest series; again, this was attributed to all subjects in database being healthy, such that the real difference in HRV was observed only during physical activity. In addition, the results suggest that the average degree calculated in series of subjects who exercise regularly are similar regardless of age, and the same occurs for sedentary subjects.

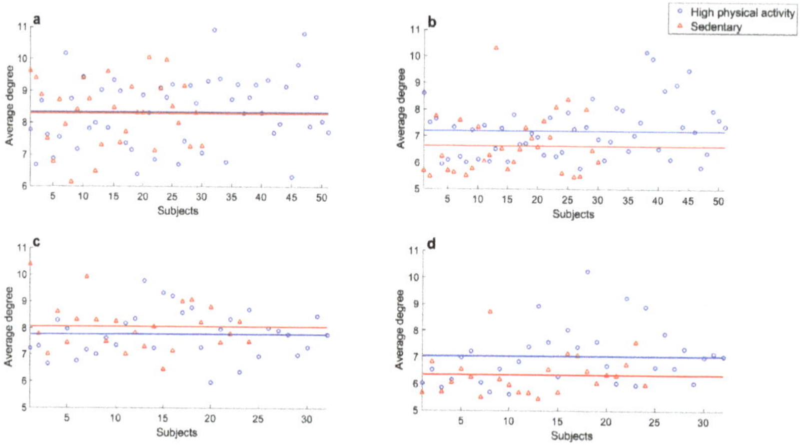

Figure 10. Comparison of average degree from visibility graph of sedentary young subjects and high-physical activity young subjects in (**a**) resting period and (**b**) during physical activity; and those corresponding to middle-aged subjects in (**c**) resting period and (**d**) during physical activity. Markers represent every value of average degree and continuous lines indicate the means.

The average path length of networks obtained from resting and exercise series were also analyzed. Table 2 shows the results in terms of the averages and standard deviations of young subjects and adults in both conditions. In this case, the average path length is lower during exercise and this difference is greater in middle-aged adults. The ANOVA was performed and the Fisher's LSD test showed the following:

- Average path length values of series of both groups at rest did not show significant differences ($p = 0.22$).
- Average path length values of series during rest period and exercise of both groups showed significant differences ($p = 6.0 \times 10^{-3}$ in all cases).
- Average path length values of series of both groups recorded during exercise did not show significant differences ($p = 0.99$).

Finally, the comparisons of the average path length of networks from subjects with high and low physical activity were performed. The results are provided in Table 3 and depicted in Figure 11, and we obtained the following significant differences:

- Average path length values of middle-aged adults with high physical activity levels at rest and during exercise showed a significant difference ($p = 4.5 \times 10^{-5}$).
- Average path length values of young subjects with high physical activity levels at rest and during exercise showed a significant difference ($p = 0.021$).

In these comparisons, significant differences were only obtained in the average path length of series of active subjects at rest and during physical activity. On the contrary, this parameter for series of sedentary subjects in both populations was similar.

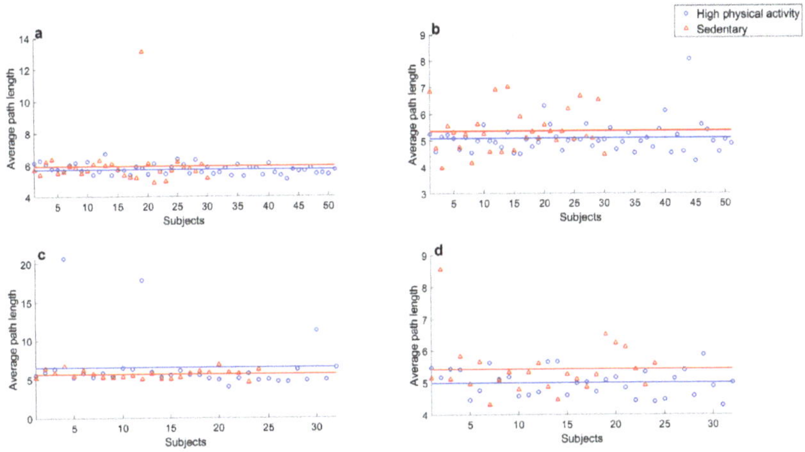

Figure 11. Comparison of average path length from visibility graph of sedentary young subjects and high-physical activity young subjects in (**a**) resting period and (**b**) during exercise; and those corresponding to middle-aged subjects in (**c**) resting period and (**d**) during physical activity. Markers represent every value of average path length and continuous lines indicate means.

4. Discussion

Once we had analyzed the selected visibility graph parameters for white, pink, and Brownian noises and we had verified the differences, we decided to applied these calculations in heartbeat time series, since it has been reported that series of young and healthy subjects present long-term correlations [1], such as in the pink noise. However, some conditions (age or illness) modify this behavior, causing decorrelations or short-term correlations in series, such as in the white and Brownian noises, respectively [60–62,90].

Generally the work was divided in three stages. In the first stage, we analyzed the series of young and elderly subjects watching the movie Fantasia, and significant differences were obtained in the *k*-*M* values between both groups, which were attributed to a complexity loss due to age [14,66]. This reduction also indicated alterations in the homeostatic regulatory mechanisms, which may compromise the capacity of physiological adaptations before several external conditions [9,10]. The second stage showed significant differences in three parameters calculated in series of healthy and CHF subjects, especially during sleeping periods. This was attributed to the fact that the complexity changes caused by the aging process are also present in the disease process [9]. In addition, Guzzetii et al. [11] reported that neuroendrocrine activation present in chronic heart failure can also be associated with the parameters of the time and frequency domain analysis of heartbeat time series. In this work, we confirmed that these changes also could be identified with VG parameters. Finally, the third stage consisted of the comparison between series at rest and during exercise of both sedentary and high physical activity subjects. The analysis concluded that there were significant differences in three parameters mentioned above between groups in both conditions, which is associated to the fact that the subjects who regularly performed aerobic exercise promote the inhibition of sympathetic regulation and the enhancement of vagal regulation [8], which causes a reduction in the loss of complexity of the signal in athletes compared to sedentary subjects. This was also identified by the VGA realized in this work.

In this study, we showed that the value of the slope for the *k*-*M* plots of the visibility graph, the average degree, and the average path length created from the R–R time series vary significantly for subjects under different conditions. It was confirmed that *k*-*M* values are useful for classifying heartbeat time series of young and elderly subjects due to these shown significant differences. In addition, with these parameters it can be distinguished between series of healthy subjects and patients with CHF, series of subjects at rest and during exercise, and series of sedentary subjects and subjects with high physical activity levels. Furthermore, the networks parameters selected are some of the simplest and fastest to calculate and, even so, the results were effective in identifying between different groups. This methodology has the advantage over traditional techniques of not being sensitive to the definition of the filter bandwidth to analyze the series.

Several studies [1,2,4,8,12,13,51–53] have showed the correlation between the results of nonlinear dynamic techniques applied to heartbeat time interval series (or tachograms) and the health status or age of a subject. When we observed the graphical representation of the networks obtained (Figures 2a–4a and 8), noticeable differences could be observed with the naked eye. However, the task of distinguishing between different types of time series should not be based solely on visual inspection. For this reason, several parameters were considered; in particular, in this work, we emphasized the parameter *k-M*, which has been used with great success in the analysis of seismic time series [38]. As we have attempted to show, the three network parameters used can distinguish between cardiac interbeat time series of young subjects and the elderly, between healthy subjects and CHF patients, people at rest and people exercising, as well as sedentary people and people who exercise regularly. The latter allowed us to glimpse some possible applications to help characterize the physical state of people, as the complex networks obtained using the VGA algorithm from the cardiac interbeat time series of sedentary people are very different to those of people who exercise regularly, as has already been reported according to other types of parameters [13,72].

Therefore, our work adds complementary information to the results that other authors have reported, with the advantage that they were applied to a wider variety of groups of subjects. As mentioned above, the VG algorithm has been used in several previously reported analyses [3,76–78], in which the time series of heartbeats were studied. However, the parameters that have been used to study them previously differed from those used in this work; in fact, in no previous work has the *k-M*, the average degree, and the average length of the path been used together to distinguish heartbeat time series from subjects under different conditions. Hence, this is the first time that the VGA algorithm has been used to analyze heartbeat time series of subjects who exercise in an attempt to distinguish between sedentary and active individuals. This means that these parameters can be used to characterize the physical condition of a person, and may be used to measure the progress of people who undergo an exercise routine to improve their physical condition.

As has been reported by Bhaduri et al. [3], this methodology can distinguish between healthy individuals and individuals with congestive heart failure. We also considered this aspect, but using the three parameters mentioned above, we found that we can also distinguish between the series of heartbeats of young and old people. Of course, such separations are made with respect to the mean values, namely, the means of the two populations are different and their difference is statistically significant, but this does not mean that there are no overlaps; that is, at the time of performing the separation, there will always be some cases of healthy people who appear ill and some ill people who appear healthy. This overlap is not exclusive to our work, but also occurs in other works already published. For this reason, the joint use of several parameters is necessary to make classifications; in addition to the list of already existing ones, we propose these three relatively new parameters. In this sense, what we propose is that, if we wish for these types of tools to be used, for example, to provide a diagnosis, we should use many of these techniques together (even considering traditional analysis techniques), including the ranges of values for these and other parameters with respect to the control groups of healthy people, in such a way that deviations from these values can be quantified. Then, when several deviations are observed in different parameters, it can be affirmed that the person has a health concern, which could help the physician to provide a final diagnosis.

5. Conclusions

In this work we used the visibility algorithm to map time series in complex networks and calculate for these networks the parameters *k-M*, average degree, and average path length, initially for white, pink, and Brownian noise time series. We used a broad set of parameters that are used for complex networks, but we decided to only focus on those that are easier to evaluate and that also gave us better results. Since the *k-M* parameter and the average degree of the aforementioned noise networks showed significant differences, we concluded that such parameters can be used to distinguish between heartbeat time series from different groups, for which comparisons were made, for example, in young people vs. elderly people, healthy subjects vs. CHF patients, resting subjects vs. exercising subjects, and sedentary youth or adults vs. regularly exercising youth or adults.

Using time series obtained from Physionet [80], we found that there are significant differences in *k-M* values between the heartbeat times series of young people and those of the elderly, which is attributed to the fact that heart rate variability decreases with age [66]. In addition, with Physionet data, we found statistically significant differences in these parameters when comparing series of healthy subjects and patients with congestive heart failure. In our research group, we have made electrocardiography (ECG) recordings using Holter monitors of people who are first at rest and then performing physical activity. The physical condition of the participating subjects was characterized using the IPAQ questionnaire [88], so we were able to compare heartbeat time series of people with a sedentary lifestyle (low physical activity) and people who exercise regularly (high physical activity). There are significant differences in both sedentary and active young subjects and middle-aged adults. In addition, it was found that there are no significant differences in the series of the rest period, neither in young people nor in middle-aged adults. However, during physical effort, the *k-M* slope, the average degree, and the average path length of the series from sedentary and highly physically active young and middle-aged adults showed statistically significant differences.

The *k-M* parameter has given excellent results applied to real and synthetic seismicity time series. In this case, the results showed that the calculation of this parameter with the average degree and average path length, despite them having been described several years earlier for different applications, could be useful as a new tool for ECG analysis and the identification of subjects with CHF and sedentary subjects and for evaluation of aging process. We are continuing to work on increasing the size of our database with the aim that in the future other complex network parameters can be incorporated into this methodology.

Author Contributions: Conceptualization, methodology, writing—original draft preparation, writing—review and editing, and supervision, A.M.-D.; software and writing—original draft preparation, É.E.S.-M.; software, writing—review and editing, validation, and formal analysis, J.A.Z.-J. All authors have read and agreed to the published version of the manuscript.

Funding: This research was funded by Secretaría de Educación, Ciencia, Tecnología e Innovación (SECTEI) of Mexico City grant number SECTEI/271/2019 and the Consejo Nacional de Ciencia y Tecnología of Mexico grant number 4647203.

Institutional Review Board Statement: Not applicable.

Data Availability Statement: All data used in the present work can be requested from the corresponding author.

Acknowledgments: Authors acknowledge all the participants in this project and the contribution of the members of Exercise Physiology's Lab from Facultad de Estudios Superiores Iztacala, UNAM, Mexico.

Conflicts of Interest: The authors declare no conflict of interest.

Abbreviations

The following abbreviations are used in this manuscript:

ANOVA	Analysis of Variance
CHF	Congestive heart failure
DFA	Detrended fluctuation analysis
ECG	Electrocardiogram
GHVE	Grouped horizontal visibility entropy
GIC	Graphical index complexity
HRV	Heart rate variability
HVE	Horizontal visibility entropy
IPAQ	International Physical Activity Questionnaire
LSD	Least significant difference
NYHA	New York Heart Association
PSVG	Power of scale-freeness in visibility graph
VG	Visibility graph
VGA	Visibility graph algorithm

References

1. Salcedo-Martínez, A.; Zamora-Justo, J.; Pérez-López, N.; Gálvez-Coyt, G.; Muñoz-Diosdado, A. The detrended fluctuation analysis of heartbeat intervals in time series during stress tests. *AIP Conf. Proc.* **2019**, *2090*, 050013.
2. Salcedo-Martínez, A.; Zamora-Justo, J.; Muñoz-Diosdado, A. Analysis of the Hurst exponent in RR series of healthy subjects and congestive patients in a state of sleep and wakefulness and in healthy subjects in physical activity. *AIP Conf. Proc.* **2021**, *2348*, 040009.
3. Bhaduri, A.; Bhaduri, S.; Ghosh, D. Visibility graph analysis of heart rate time series and bio-marker of congestive heart failure. *Phys. A Stat. Mech. Appl.* **2017**, *482*, 786–795. [CrossRef]
4. Aguilar-Molina, A.M.; Angulo-Brown, F.; Muñoz-Diosdado, A. Multifractal Spectrum Curvature of RR Tachograms of Healthy People and Patients with Congestive Heart Failure, a New Tool to Assess Health Conditions. *Entropy* **2019**, *21*, 581. [CrossRef]
5. Zamora-Justo, J.; Gutiérrez-Calleja, R.; Muñoz-Diosdado, A. Acquisition software development for monitor Holter prototype signals and its use for pre-diagnosis of cardiac damage based on nonlinear dynamic techniques. *AIP Conf. Proc.* **2016**, *1747*, 090001. [CrossRef]
6. Melo, R.C.; Quitério, R.J.; Takahashi, A.C.M.; Silva, E.; Martins, L.E.B.; Catai, A.M. High eccentric strength training reduces heart rate variability in healthy older men. *Br. J. Sports Med.* **2008**, *42*, 59–63. [CrossRef]
7. Milan-Mattos, J.C.; Porta, A.; Perseguini, N.M.; Minatel, V.; Rehder-Santos, P.; Takahashi, A.C.M.; Mattiello, S.M.; Catai, A.M. Influence of age and gender on the phase and strength of the relation between heart period and systolic blood pressure spontaneous fluctuations. *J. Appl. Physiol.* **2018**, *124*, 791–804. [CrossRef] [PubMed]
8. De Abreu, R.; Porta, A.; Rehder-Santos, P.; Cairo, B.; Da Silva, C.D.; De Favari Signini, E.; Sakaguchi, C.A.; Catai, A.M. Effects of inspiratory muscle-training intensity on cardiovascular control in amateur cyclists. *Am. J. Physiol. Regul. Intgr. Comp. Physiol.* **2019**, *317*, R891–R902. [CrossRef] [PubMed]
9. Takahashi, A.C.M.; Porta, A.; Melo, R.C.; Quitério, R.J.; da Silva, E.; Borghi-Silva, A.; Tobaldini, E.; Montano, N.; Catai, A.M. Aging reduces complexity of heart rate variability assessed by conditional entropy and symbolic analysis. *Intern. Emerg. Med.* **2012**, *7*, 229–235. [CrossRef] [PubMed]
10. La Rovere, M.T.; Porta, A.; Schwartz, P.J. Autonomic Control of the Heart and Its Clinical Impact. A Personal Perspective. *Front. Physiol.* **2020**, *11*, 582. [CrossRef]
11. Guzzetti, S.; Mezzetti, S.; Magatelli, R.; Porta, A.; De Angelis, G.; Rovelli, G.; Malliani, A. Linear and non-linear 24 h heart rate variability in chronic heart failure. *Auton. Neurosci. Basic Clin.* **2000**, *86*, 114–119. [CrossRef] [PubMed]
12. Ivanov, P.C.; Amaral, L.A.N.; Goldberger, A.L.; Havlin, S.; Rosenblum, M.G.; Stanley, H.E.; Struzik, Z.R. From 1/f noise to multifractal cascades in heartbeat dynamics. *Chaos* **2001**, *11*, 641–652. [CrossRef] [PubMed]
13. Solís-Montufar, E.; Gálvez-Coyt, G.; Muñoz-Diosdado, A. Entropy Analysis of RR-Time Series From Stress Tests. *Front. Physiol.* **2020**, *11*, 981. [CrossRef]
14. Catai, A.M.; Takahashi, A.C.M.; Perseguini, N.M.; Milan, J.C.; Minatel, V.; Rehder-Santos, P.; Marchi, A.; Bari, V.; Porta, A. Effect of the postural challenge on the dependence of the cardiovascular control complexity on age. *Entropy* **2014**, *16*, 6686–6704. [CrossRef]
15. Shi, B.; Zhang, Y.; Yuan, C.; Wang, S.; Li, P. Entropy Analysis of Short-Term Heartbeat Interval Time Series during Regular Walking. *Entropy* **2017**, *19*, 568. [CrossRef]
16. Porta, A.; Tobaldini, E.; Guzzetti, S.; Furlan, R.; Montano, N.; Gnecchi-Ruscone, T. Assessment of cardiac autonomic modulation during graded head-up tilt by symbolic analysis of heart rate variability. *Am. J. Physiol. Heart Circ. Physiol.* **2007**, *293*, H702–H708. [CrossRef] [PubMed]
17. Albert, B.; Barabási, A.R. Statistical mechanics of complex networks. *Rev. Mod. Phys.* **2002**, *74*, 47. [CrossRef]

18. Newman, M.E.J. *Networks: An introduction*; Oxford University Press: Oxford, UK, 2010.
19. Telesca, L.; Lovallo, M.; Ramírez-Rojas, A.; Flores-Marquez, L. Relationship between the Frequency Magnitude Distribution and the Visibility Graph in the Synthetic Seismicity Generated by a Simple Stick-Slip System with Asperities. *PLoS ONE* **2014**, *9*, e106233. [CrossRef] [PubMed]
20. Aguilar-SanJuan, B.; Guzmán-Vargas, L. Earthquake magnitude time series: Scaling behavior of visibility networks. *Eur. Phys. J. B* **2013**, *86*, 454. [CrossRef]
21. Zhang, J.; Small, M. Complex network from pseudoperiodic time series: Topology versus dynamics. *Phys. Rev. Lett.* **2006**, *96*, 238701. [CrossRef]
22. Xu, X.; Zhang, J.; Small, M. Superfamily phenomena and motifs of networks induced from time series. *Proc. Natl. Acad. Sci. USA* **2008**, *105*, 19601–19605. [CrossRef] [PubMed]
23. Gao, Z.K.; Jin, N.D.; Wang, W.X.; Lai, Y.C. Motif distributions in phase space networks for characterizing experimental two-phase flow patterns with chaotic features. *Phys. Rev. E* **2010**, *82*, 016210. [CrossRef]
24. Gao, Z.K.; Zhang, X.W.; Jin, N.D.; Donner, R.V.; Marwan, N.; Kurths, J. Recurrence networks from multivariate signals for uncovering dynamic transitions of horizontal oil-water stratified flows. *Europhys. Lett.* **2013**, *103*, 50004. [CrossRef]
25. Lacasa, L.; Luque, B.; Ballesteros, F.; Luque, J.; Nuno, J.C. From time series to complex networks: The visibility graph. *Proc. Natl. Acad. Sci. USA* **2008**, *105*, 4972–4975. [CrossRef]
26. Quian, M.C.; Jiang, Z.Q.; Zhou, W.X. Universal and nonuniversal allometric scaling behaviors in the visibility graphs of world stock market indices. *J. Phys. A Math. Theor.* **2010**, *43*, 335002. [CrossRef]
27. Wang, N.; Li, D.; Wang, Q. Visibility graph analysis on quarterly macroeconomic series of China based on complex network theory. *Phys. A Stat. Mech. Appl.* **2012**, *391*, 6543–6555. [CrossRef]
28. Long, Y. Visibility graph network analysis of gold price time series. *Phys. A Stat. Mech. Appl.* **2013**, *392*, 3374–3384. [CrossRef]
29. Zhang, R.; Ashuri, Y.S.B.; Deng, Y. Forecasting construction cost index based on visibility graph: A network approach. *Phys. A Stat. Mech. Appl.* **2018**, *493*, 239–252. [CrossRef]
30. Ni, X.; Jiang, Z.Q.; Zhou, W.X. Degree distributions of the visibility graphs mapped from fractional Brownian motions and multifractal random walks. *Phys. Lett. A* **2009**, *373*, 3822–3826. [CrossRef]
31. Elsner, J.B.; Jagger, T.H.; Fogarty, E.A. Visibility network of United States hurricanes. *Geophys. Res. Lett.* **2009**, *36*, L16702. [CrossRef]
32. Chen, S.; Hu, Y.; Mahadevan, S.; Deng, Y. A visibility graph averaging aggregation operator. *Phys. A Stat. Mech. Appl.* **2014**, *403*, 1–12. [CrossRef]
33. Jiang, W.; Wei, B.; Zhan, J.; Xie, C.; Zhou, D. A visibility graph power averaging aggregation operator: A methodology based on network analysis. *Comput. Ind. Eng.* **2016**, *101*, 260–268. [CrossRef]
34. Zhu, G.; Li, Y.; Wen, P.P. An efficient visibility graph similarity algorithm and its application on sleep stages classification. In Proceedings of the International Conference on Brain Informatics, Macau, China, 4–7 December 2012; Springer: Berlin/Heidelberg, Germany, 2012; pp. 185–195.
35. Nilanjana, P.; Anirban, B.; Susmita, B.; Dipak, G. Noninvasive alarm generation for sudden cardiac arrest: A pilot study with visibility graph technique. *Transl. Biomed.* **2016**, *7*, 79.
36. Yu, M.; Hillebrand, A.; Gouw, A.A.; Stam, C.J. Horizontal visibility graph transfer entropy (HVG-TE): A novel metric to characterize directed connectivity in large-scale brain networks. *NeuroImage* **2017**, *156*, 249–264. [CrossRef] [PubMed]
37. Bhaduri, A.; Ghosh, D. Quantitative Assessment of Heart Rate Dynamics during Meditation: An ECG Based Study with Multi-Fractality and Visibility Graph. *Front. Physiol.* **2016**, *7*, 44. [CrossRef] [PubMed]
38. Telesca, L.; Lovallo, M. Analysis of seismic sequences by using the method of visibility graph. *Europhys. Lett.* **2012**, *97*, 50002. [CrossRef]
39. Telesca, L.; Lovallo, M.; Ramírez-Rojas, A.; Flores-Marquez, L. Investigating the time dynamics of seismicity by using the visibility graph approach: Application to seismicity of Mexican subduction zone. *Phys. A Stat. Mech. Appl.* **2013**, *392*, 6571–6577. [CrossRef]
40. Telesca, L.; Lovallo, M.; Toth, L. Visibility graph analysis of 2002–2011 Pannonian seismicity. *Phys. A Stat. Mech. Appl.* **2014**, *416*, 219–224. [CrossRef]
41. Khoshnevis, N.; Taborda, R.; Azizzadeh-Roodpish, S.; Telesca, L. Analysis of the 2005–2016 Earthquake Sequence in Northern Iran Using the Visibility Graph Method. *Pure Appl. Geophys.* **2017**, *174*, 4003–4019. [CrossRef]
42. Azizzadeh-Roodpish, S.; Cramer, C.H. Visibility Graph Analysis of Alaska Crustal and Aleutian Subduction Zone Seismicity: An Investigation of the Correlation between b Value and k–M Slope. *Pure Appl. Geophys.* **2018**, *175*, 4241–4252. [CrossRef]
43. Perez-Oregon, J.; Muñoz-Diosdado, A.; Rudolf-Navarro, A.H.; Angulo-Brown, F. Some Common Features Between a Spring-Block Self-Organized Critical Model, Stick–Slip Experiments with Sandpapers and Actual Seismicity. *Pure Appl. Geophys.* **2019**, *177*, 889–903. [CrossRef]
44. Donner, R.V.; Donges, J.F. Visibility graph analysis of geophysical time series: Potentials and possible pitfalls. *Acta Geophys.* **2011**, *60*, 589–623. [CrossRef]
45. Campanharo, A.S.L.O.; Sirer, M.I.; Malmgren, R.D.; Ramos, F.M.; Amaral, L.A.N. Duality between time series and networks. *PLoS ONE* **2011**, *6*, e23378. [CrossRef] [PubMed]
46. Bak, P.; Christensen, K.; Danon, L.; Scanlon, T. Unified Scaling Law for Earthquakes. *Phys. Rev. Lett.* **2002**, *88*, 178501. [CrossRef] [PubMed]

47. Davidsen, J.; Goltz, C. Are seismic waiting time distributions universal? *Geophys. Res. Lett.* **2004**, *31*, 5. [CrossRef]
48. Azizzadeh-Roodpish, S.; Khoshnevis, N.; Cramer, C.H. Visibility graph analysis of southern California. In Proceedings of the Annual Meeting of the Seismological Society of America, Denver, CO, USA, 18–20 April 2017; Volume 10.
49. Iyengar, N.; Peng, C.K.; Morin, R.; Goldberger, A.L.; Lipsitz, L.A. Age-related alterations in the fractal scaling of cardiac interbeat interval dynamics. *Am. J. Physiol. Regul. Integr. Comp. Physiol.* **1996**, *271*, R1078–R1084. [CrossRef]
50. Ivanov, P.C.; Amaral, L.A.N.; Goldberger, A.L.; Havlin, S.; Rosemblum, M.G.; Struzik, Z.R.; Stanley, H.E. Multifractality in human heartbeat dynamics. *Nature* **1999**, *399*, 461–465. [CrossRef]
51. Bernaola-Galván, P.A.; Gómez-Extremera, M.; Romance, A.R.; Carpena, P. Correlations in magnitude series to assess nonlinearities: Application to multifractal models and heartbeat fluctuations. *Phys. Rev. E* **2017**, *96*, 032218. [CrossRef]
52. Gómez-Extremera, M.; Bernaola-Galván, P.A.; Vargas, S.; Benítez-Porres, J.; Carpena, P.; Romance, A.R. Differences in nonlinear heart dynamics during rest and exercise and for different training. *Physiol. Meas.* **2018**, *39*, 084008. [CrossRef]
53. Faes, L.; Gómez-Extremera, M.; Pernice, R.; Carpena, P.; Nollo, G.; Porta, A.; Bernaola-Galván, P. Comparison of methods for the assessment of nonlinearity in short-term heart rate variability under different physiopathological states. *Chaos* **2019**, *29*, 123114. [CrossRef]
54. Tsuji, H.; Venditti, F., Jr.; Manders, E.; Evans, J.; Larson, M.; Feldman, C.; Levy, D. Reduced heart rate variability and mortality risk in an elderly cohort. The Framingham Heart Study. *Circulation* **1994**, *90*, 878–883. [CrossRef] [PubMed]
55. Dietrich, F.D.; Schindler, C.; Schwartz, J.; Barthelemy, J.C.; Tschopp, J.M.; Roche, F.; von Eckardstein, A.; Brändli, O.; Leuenberger, P.; Gold, D.R.; et al. Heart rate variability in an ageing population and its association with lifestyle and cardiovascular risk factors: Results of the sapaldia study. *Europace* **2006**, *8*, 521–529. [CrossRef] [PubMed]
56. Ziegler, D.; Zentai, C.P.; Perz, S.; Rathmann, W.; Haastert, B.; Döring, A.; Meisinger, C. Prediction of mortality using measures of cardiac autonomic dysfunction in the diabetic and nondiabetic population: The monica/kora augsburg cohort study. *Diabetes Care* **2008**, *31*, 556–561. [CrossRef]
57. Drawz, P.; Babineau, D.; Brecklin, C.; He, J.; Kallem, R.R.; Soliman, E.Z.; Xie, D.; Appleby, D.; Anderson, A.H.; Rahman, M.; et al. Heart rate variability is a predictor of mortality in chronic kidney disease: A report from the cric study. *Am. J. Nephrol.* **2013**, *38*, 517–528. [CrossRef]
58. Sugihara, G.; Allan, W.; Sobel, D.; Allan, K.D. Nonlinear control of heart rate variability in human infants. *Proc. Natl. Acad. Sci. USA* **1996**, *93*, 2608–2613. [CrossRef] [PubMed]
59. Voss, B.A.; Schulz, S.; Schroeder, R.; Baumert, M.; Caminal, P. Methods derived from nonlinear dynamics for analyzing heart rate variability. *Phil. Trans. R. Soc. A* **2009**, *367*, 277–296. [CrossRef]
60. Guzmán-Vargas, L.; Muñoz-Diosdado, A.; Angulo-Brown, F. Influence of the loss of time-constants repertoire in pathologic heartbeat dynamics. *Phys. A Stat. Mech. Appl.* **2005**, *348*, 304–316. [CrossRef]
61. Peng, C.K.; Havlin, S.; Stanley, H.E.; Goldberger, A.L. Quantification of scaling exponents and crossover phenomena in nonstationary heartbeat time series. *Chaos* **1995**, *5*, 82–87. [CrossRef]
62. Hu, K.; Ivanov, P.C.; Chen, Z.; Carpena, P.; Stanley, H.E. Effect of trends on detrended fluctuation analysis. *Phys. Rev. E* **2001**, *64*, 011114/1–011114/19. [CrossRef]
63. Muñoz-Diosdado, A.; Vargas, L.G.; Rojas, A.R.; Río-Correa, J.L.D.; Brown, F.A. Some cases of crossover behavior in heart interbeat and electroseismic series. *Fractals* **2005**, *13*, 253–263. [CrossRef]
64. Higuchi, T. Approach to an irregular time series on the basis of the fractal theory. *Phys. D Nonlinear Phenom.* **1988**, *31*, 277–283. [CrossRef]
65. Eke, A.; Herman, P.; Kocsis, L.; Kozak, L.R. Fractal characterization of complexity in temporal physiological signals (Review). *Physiol. Meas.* **2002**, *23*, R1–R38. [CrossRef] [PubMed]
66. Goldberger, A.L.; Amaral, L.A.N.; Hausdorff, J.M.; Ivanov, P.C.; Peng, C.K.; Stanley, H.E. Fractal dynamics in physiology: Alterations with disease and aging. *Proc. Natl. Acad. Sci. USA* **2002**, *99*, 2466–2472. [CrossRef] [PubMed]
67. Tulppo, M.; Makikallio, T.; Takala, T.; Seppanen, T.; Huikuri, H. Quantitative beat-to-beat analysis of heart rate dynamics during exercise. *Am. J. Physiol. Heart Circ. Physiol.* **1996**, *271*, H244–H252. [CrossRef] [PubMed]
68. Karasik, R.; Sapir, N.; Ashkenazy, Y.; Ivanov, P.C.; Dvir, I.; Lavie, P.; Havlin, S. Correlation differences in heartbeat fluctuations during rest and exercise. *Phys. Rev. E* **2002**, *66*, 062902. [CrossRef]
69. Aubert, A.E.; Seps, B.; Beckers, F. Heart rate variability in athletes. *Sports Med.* **2003**, *33*, 889–919. [CrossRef]
70. Leicht, A.S.; Sinclair, W.H.; Spinks, W.L. Effect of exercise mode on heart rate variability during steady state exercise. *Eur. J. Appl. Physiol.* **2008**, *102*, 195–204. [CrossRef]
71. Goya-Esteban, R.; Barquero-Perez, O.; Sarabia-Cachadina, E.; de la Cruz-Torres, B.; Naranjo-Orellana, J.; Rojo-Alvarez, J.L. A modified Hilbert-Huang algorithm to assess spectral parameters in intense exercise. In Proceedings of the Computing in Cardiology Conference (CinC), Zaragoza, Spain, 22–25 September 2013; IEEE: Krakow, Poland, 2013; pp. 409–412.
72. Weippert, M.; Behrens, M.; Rieger, A.; Behrens, K. Sample entropy and traditional measures of heart rate dynamics reveal different modes of cardiovascular control during low intensity exercise. *Entropy* **2014**, *16*, 5698–5711. [CrossRef]
73. Taylor, K.A.; Wiles, J.D.; Coleman, D.D.; Sharma, R.; O'Driscoll, J.M. Continuous cardiac autonomic and haemodynamic responses to isometric exercise. *Med. Sci. Sports Exerc.* **2017**, *49*, 1511–1519. [CrossRef]
74. Richman, J.; Moorman, J. Physiological time-series analysis using approximate entropy and sample entropy. *Am. J. Physiol. Heart Circ. Physiol.* **2000**, *278*, H2039–H2049. [CrossRef]

75. Amaral, L.; Silva, F.A.; Correia, V.B.; Andrade, C.E.F.; Dutra, B.A.; Oliveira, M.V.; de Magalhães, A.C.M.; Volpini, R.A.; Seguro, A.C.; Coimbra, T.M.; et al. Beneficial effects of previous exercise training on renal changes in streptozotocin-induced diabetic female rats. *Exp. Biol. Med.* **2016**, *241*, 437–445. [CrossRef] [PubMed]
76. Jiang, S.; Bian, C.; Ning, X.; Ma, Q.D.Y. Visibility graph analysis on heartbeat dynamics of meditation training. *Appl. Phys. Lett.* **2013**, *102*, 253702. [CrossRef]
77. Choudhary, G.I.; Aziz, W.; Khan, I.R.; Rahardja, S.; Fränti, P. Analysing the Dynamics of Interbeat Interval Time Series Using Grouped Horizontal Visibility Graph. *IEEE Access* **2019**, *7*, 9926–9934. [CrossRef]
78. Nasrolahzadeh, M.; Mohammadpoory, Z.; Haddadnia, J. Analysis of heart rate signals during meditation using visibility graph complexity. *Cogn. Neurodyn.* **2019**, *13*, 45–52. [CrossRef] [PubMed]
79. Kim, J.; Wilhelm, T. What is a complex graph? *Phys. A Stat. Mech. Appl.* **2008**, *387*, 2637–2652. [CrossRef]
80. Goldberger, A.; Amaral, L.; Glass, L.; Hausdorff, J.; Ivanov, P.C.; Mark, R.; Stanley, H.E. PhysioBank, PhysioToolkit, and PhysioNet: Components of a new research resource for complex physiologic signals. *Circulation* **2000**, *101*, e215–e220. [CrossRef] [PubMed]
81. Luque, B.; Lacasa, L.; Ballesteros, F.; Luque, J. Horizontal visibility graphs: Exact results for random time series. *Phys. Rev. E* **2009**, *80*, 046103. [CrossRef]
82. Nuñez, A.; Lacasa, L.; Valero, E.; Gómez, J.; Luque, B. Detecting series periodicity with horizontal visibility graphs. *Int. J. Bifurc. Chaos* **2012**, *22*, 1250160. [CrossRef]
83. Lacasa, L.; Nuñez, A.; Roldán, E.; Parrondo, J.M.R.; Luque, B. Time series irreversibility: A visibility graph approach. *Eur. Phys. J. B* **2012**, *85*, 217. [CrossRef]
84. Pierini, J.O.; Lovallo, M.; Telesca, L. Visibility graph analysis of wind speed records measured in central Argentina. *Phys. A Stat. Mech. Appl.* **2012**, *391*, 5041–5048. [CrossRef]
85. Malamud, B.D.; Turcotte, D.L. Self-Affine Time series: Generation and Analyses. *Adv. Geophys.* **1999**, *40*, 1–90.
86. Mandelbrot, B.B.; Ness, J.W.V. Fractional Brownian motions, fractional noises and applications. *SIAM Rev.* **1968**, *10*, 422–437. [CrossRef]
87. Gálvez-Coyt, G.; Muñoz-Diosdado, A.; Del-Río-Correa, J.L.; Angulo-Brown, F. A comparative study of validity ranges of some fractal methods. *Fractals* **2010**, *18*, 235–246. [CrossRef]
88. Craig, C.; Marshall, A.; Sjöström, M.; Bauman, A.; Booth, M.; Ainsworth, B.; Pratt, M.; Ekelund, U.; Yngve, A.; Sallis, J.; et al. International physical activity questionnaire: 12-country reliability and validity. *Med. Sci. Sports Exerc.* **2003**, *35*, 1381–1395. [CrossRef]
89. Cohen, M.; Taylor, J. Short-term cardiovascular oscillations in man: Measuring and modelling the physiologies. *J. Physiol.* **2002**, *542*, 669–683. [CrossRef]
90. Francis, D.; Willson, K.; Georgiadou, P.; Wensel, R.; Davies, L.; Coats, A.; Piepoli, M. Physiological basis of fractal complexity properties of heart rate variability in man. *J. Physiol.* **2002**, *15*, 619–629. [CrossRef]

entropy

MDPI

Article

Autonomic Nervous System Influences on Cardiovascular Self-Organized Criticality

Jacques-Olivier Fortrat [1,*] and Guillaume Ravé [2]

[1] CHU Angers, Médecine Vasculaire, INSERM, CNRS, MITOVASC, Equipe CarMe, SFR ICAT, Université d'Angers, 49933 Angers, France
[2] Toulouse Football Club, 1 Allée Gabriel Biénès, 31400 Toulouse, France; rave.guillaumme@gmail.com
* Correspondence: jofortrat@chu-angers.fr; Tel.: +33-241-353-689

Abstract: Cardiovascular self-organized criticality has recently been demonstrated. We studied a model of autonomic nervous system changes to better characterize heart rate variability self-organized criticality. The model included short and long-term autonomic changes associated with body position and physical training, respectively. Twelve professional soccer players took part in a 5-week training session divided into "Warm-up", "Intensive", and "Tapering" periods. A stand test was carried out at the beginning and end of each period. Heart rate variability was recorded beat by beat (Polar Team 2). Bradycardias, defined as successive heart rates with a decreasing value, were counted according to their length in number of heartbeat intervals. We checked whether bradycardias were distributed according to Zipf's law, a feature of self-organized criticality. Zipf's law draws a straight line when the rank of occurrence is plotted against the frequency of occurrence in a log–log graph. Bradycardias were distributed according to Zipf's law, regardless of body position or training. Bradycardias were much longer in the standing position than the supine position and Zipf's law was broken after a delay of four heartbeat intervals. Zipf's law could also be broken in some subjects with curved long bradycardia distributions by training. Zipf's law confirms the self-organized nature of heart rate variability and is strongly linked to autonomic standing adjustment. However, Zipf's law could be broken, the significance of which remains unclear.

Keywords: autonomic nervous system; baroreflex; blood pressure regulation; cardiovascular dynamics; heart rate variability; homeostasis; fractal; self-organized criticality; soccer; Zipf's law

1. Introduction

Maintaining internal conditions is a primary physiological challenge for all organisms, and this is achieved through the stability of body constants [1,2]. It is commonly believed that this stability is achieved by sensors involved in regulatory loops with negative feedback, which is known as the homeostasis principle [1,2]. However, the homeostasis principle fails to explain how an organism adjusts to various demands posed by environmental conditions [1,3–5]. Physiological system adjustability requires cybernetic alterations of each regulatory loop, referred to as resetting, which is complex to integrate into the homeostasis principle [6].

An alternative concept for physiological homeostasis has been proposed from the field of physics. In the late 1980s, Per Bak identified a pile of sand as a dynamic system spontaneously poised at equilibrium without the need to tune any control parameter to a precise value [7–9]. The pile of sand was poised at equilibrium but near to criticality, which is referred to as self-organized criticality. This concept was soon applied to many dynamic systems across various fields, including geophysics, astronomy, sociology, and neurobiology [3–5,7–10].

The interesting point is that systems near to criticality can more easily adjust to changes in environmental conditions [4,5,10]. Criticality means that sudden major events, known as catastrophes or avalanches, may spontaneously occur. In the case of the sand pile model, these major events are sand avalanches, whereas in geophysics, they are earthquakes. According to the Gutenberg–Richter law, the distribution of earthquakes by their magnitude and frequency draws a straight line in a log–log graph [7]. Straight lines such as these in log–log graphs or power laws are features of self-organized criticality [7].

In the cardiovascular system, major events include vasovagal syncope, which may occur spontaneously in the standing position, even in healthy individuals [11]. We previously demonstrated that spontaneous vasovagal events observed on any cardiovascular monitoring follow the Gutenberg–Richter law [12]. We also showed that focusing on heart rate can reveal cardiovascular self-organized criticality, as decreasing heart rate is a part of a vasovagal event [13]. We refer to these decreasing heart rates as "spontaneous bradycardia" and demonstrated that they follow another power law, Zipf's law [13]. This law states that the longer the bradycardia, the less frequent its occurrence, and that this relationship presents a perfect straight line on a log–log graph.

Cardiovascular adjustment to the standing position is mainly due to the autonomic nervous system [11,14]. To better understand cardiovascular self-organized criticality, we studied Zipf's law in a model of autonomic nervous system changes. This model includes short-term autonomic changes due to alterations in position, as well as long-term changes due to levels of physical training.

2. Materials and Methods

2.1. Subjects

The study involved twelve professional male soccer players, with an average age of 24.9 $\pm$ 1.5 years (mean $\pm$ SEM). The players had normal characteristics, with an average height of 1.77 $\pm$ 0.03 m and weight of 75 $\pm$ 1.3 kg. Informed consent was obtained from all participants, and the procedures were approved by the Ethics Committee of Angers, France (#2014-36, 30 April 2014) in accordance with the Declaration of Helsinki, Finland.

2.2. Experiment

Heart rate variability monitoring was achieved by using a sports watch on a beat-by-beat basis with a one-millisecond resolution for RR interval measurements (the interval between two consecutive heartbeats, Polar Team 2 Heart Rate Monitor). Continuous heart rate recordings were obtained during stand tests that consisted of a ten-minute supine period followed by a seven-minute period in the standing position. The stand tests were performed at 8:30 am, before breakfast. Four stand tests were conducted for each player during a five-week preparation period for the competition season, after the season break. The preparation period included three phases: a two-week "Warm-Up" period, followed by a two-week "Intensive" training period, and finally a one-week "Tapering" period. The first standing test took place after the season break and before the training period. The three subsequent standing tests were conducted at the end of each phase of the training period. Individual performance was not assessed in this experiment due to the absence of a known objective measurement to evaluate it within the context of a so-called team intermittent sport. In soccer, performance relies primarily on abilities rather than aerobic capacity and is influenced not only by physical fitness but also by situational factors such as tactics, scoring, and refereeing [15].

2.3. Data Analysis

Heart rate recordings were analyzed during stand tests to obtain the mean heart rate in both the supine and standing positions. Additionally, five-minute heart rate recordings were analyzed in both positions to detect bradycardia sequences and determine their distribution according to Zipf's law. A bradycardia sequence was defined as successive heartbeat intervals with a decreasing heart rate value. For each five-minute recording, bradycardia sequences were counted and classified by their length in number of heartbeat intervals. The rank of bradycardia sequences of the same length was determined by classifying them according to their frequency of occurrence. A natural log–log diagram was plotted for each five-minute heart rate recording, with the number of sequences on the x-axis and their rank on the y-axis. Linear regression was performed for each diagram to obtain correlation coefficients and slopes. In cases where diagrams showed a tipping point, its position was determined by the best linear fit.

2.4. Statistics

Data were presented as the mean ± standard error of the mean (SEM). Statistical analyses were conducted using Prism 8 software (GraphPad Software, San Diego, CA, USA). We considered that a Zipf's graph distribution would fit a straight line when the absolute value of the correlation coefficient $|r|$ exceeded 0.95, as this high cut-off is commonly used to test for power laws [7,12,13]. The normality of variable distributions was assessed using D'Agostino–Pearson omnibus K2 tests. Pearson correlations were performed when necessary. Linear mixed models were employed to model the relationship between position and training, followed by Tukey's multiple comparison tests when applicable. Statistical significance was set at $p < 0.05$.

3. Results

The quality of the data collected during this field experiment was good, as demonstrated in Figure 1.

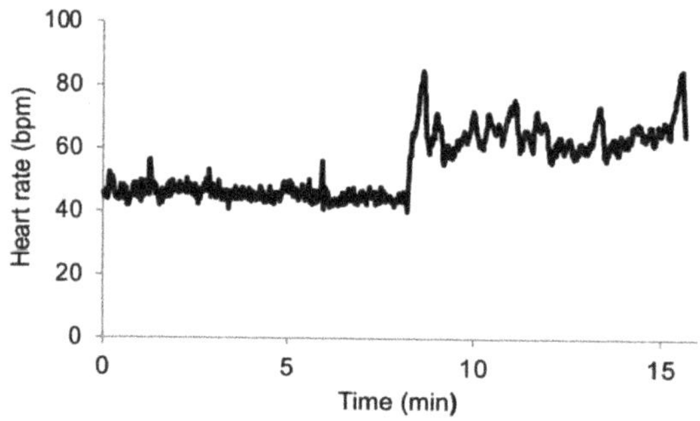

Figure 1. Example of beat-to-beat heart rate recording in a single player during a stand test following ectopic beat removal. The recording starts in the supine position and then transitions to the standing position. bpm: beats per minute.

Heart rate decreased with each stage of training; the decrease was significant from the end of the warm-up period but remained unchanged thereafter (Figure 2). As expected, standing position increased heart rate, but standing heart rate decreased with training and followed the same pattern as supine heart rate (Figure 2). These changes in heart rate indicate changes in the autonomic nervous system resulting from both training and standing position, as previously reported and as demonstrated by spectral analysis during this experiment [16].

The number of heartbeat intervals involved in bradycardia remained unchanged during the training period (Table 1). However, this number was slightly but significantly lower in the standing position (Table 1).

The bradycardia distribution diagrams showed a short range of distribution in the supine position (Figure 3a), extending to 5.2 ± 0.4 heartbeat intervals. A straight line was observed in nine out of twelve players with $|r| > 0.95$. The $|r|$ values of the three remaining players who did not conform to a straight line were 0.93, 0.94, and 0.94.

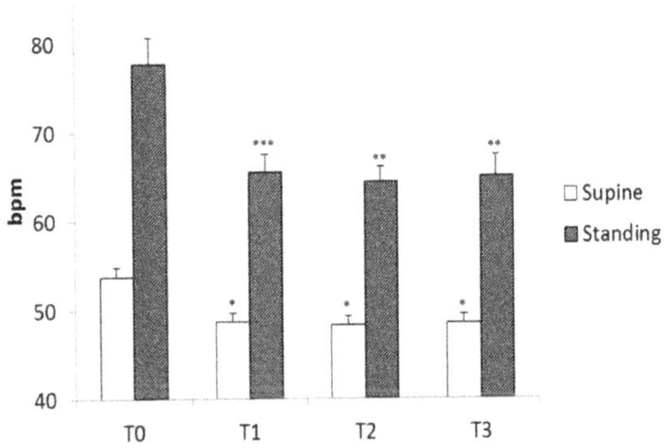

Figure 2. Mean heart rate of the players during four stand tests (T0 to T3) in the supine and standing positions. T0 was the first stand test before the training period; T1 was after the "Warm-Up" period; T2 was after the "Intensive" period; and T3 was after the "Tapering" period. bpm: beats per minute, *: $p < 0.05$; **: $p < 0.01$; ***: $p < 0.001$ vs. T0, linear mixed model followed by Tukey's multiple comparison test.

Table 1. Normalized number of heartbeat intervals involved in bradycardia sequences during five-minute recordings in supine and standing positions. The number of heartbeat intervals is normalized by the total number of heartbeat intervals in the entire recording. Recordings were conducted at each stage of a five-week preparation for a competition season: before the start of preparation, at the end of the warm-up period, at the end of intensive training, and at the end of the tapering period (T0, T1, T2, and T3, respectively).

	T0	T1	T2	T3
Supine	0.76 ± 0.04	0.71 ± 0.06	0.71 ± 0.06	0.71 ± 0.04
Standing	0.61 ± 0.03	0.63 ± 0.02	0.64 ± 0.03	0.68 ± 0.05

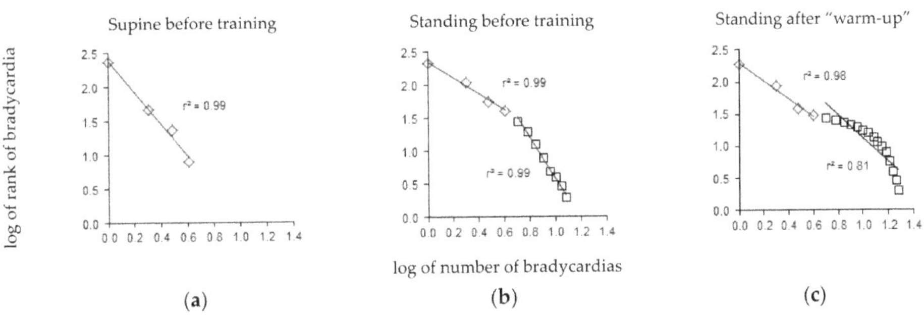

Figure 3. Zipf's distribution of bradycardia sequences. (**a**) In the supine position before the training period (T0), nine out of twelve players displayed a straight line pattern. (**b**) In the standing position before the training period (T0), all twelve players displayed a pattern of two straight lines with a tipping point at four heartbeat intervals. (**c**) In the standing position after the "Warm-Up" period (T1), only two players displayed a pattern of a curved distribution of bradycardia longer than four heartbeat intervals. Each panel displays a representative example from a single player.

The range of bradycardia distribution was much larger in the standing position, extending to 10.9 ± 0.9 heartbeat intervals (Figure 3b). A tipping point was observed in all twelve players in the standing position, with two lines: one included the first four heartbeat intervals and the other included bradycardia longer than four heartbeat intervals ($|r|$ = 0.98 ± 0.00 and 0.97 ± 0.00, respectively, Figure 3b). Training did not lead to significant changes in the distribution of these longer bradycardias (|r| values were T0: 0.98 ± 0.01, T1: 0.97 ± 0.01, T2: 0.98 ± 0.01, T3: 0.98 ± 0.00). However, the distribution of the long bradycardias was clearly curved in two players in the standing position after the warm-up, with |r| values less than 0.95 (0.90 and 0.94, Figure 3c). The short bradycardias remained distributed according to a straight line in these two cases.

The slope of the short bradycardia distribution was not affected by training, while position had a significant impact (Figure 4).

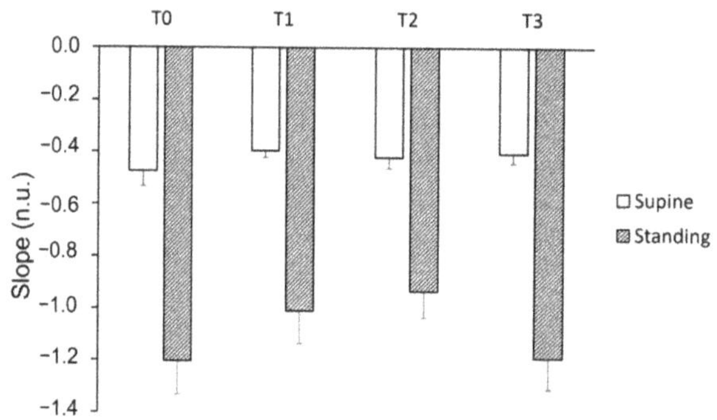

Figure 4. Slope of the distribution of short bradycardia sequences during four stand tests (T0 to T3) in both supine and standing positions. T0 represents the initial stand test conducted before the training period, T1 corresponds to the stand test after the "Warm-Up" period, T2 refers to the stand test following the "Intensive" period, and T3 indicates the stand test conducted after the "Tapering" period. A linear mixed model analysis revealed a lack of effect of training but a strong effect of position ($p < 0.0001$).

Given the similarity in the strong effect of position on this slope and that of heart rate, we sought to establish a correlation between these two variables. We observed a high correlation between position and heart rate (Figure 5). The slope of the long bradycardia distribution could only be studied in the standing position due to the limited range of bradycardia distribution in the supine position (refer to Figure 3a). Throughout the experiment, the slope remained consistent (−0.26 ± 0.02; −0.22 ± 0.06; −0.28 ± 0.04; −0.27 ± 0.02 at T0, T1, T2, and T3, respectively). No correlation was found between the slope of the long bradycardia sequence and heart rate ($r^2 = 0.04$; $p = 0.24$).

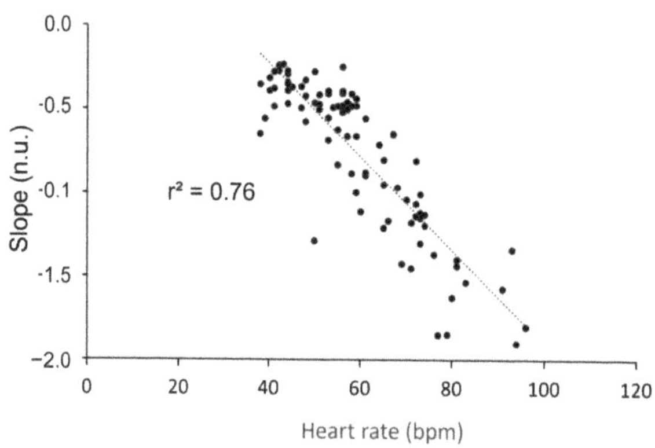

Figure 5. Correlation between heart rate in beats per minute (bpm) and the slope of the distribution of the short bradycardia sequences (no unit, n.u.) during four stand tests performed during a five-week training period. The Pearson's correlation was significant ($p < 0.001$).

4. Discussion

The primary finding of this study is that changes to the autonomic nervous system in response to body position strongly influence cardiovascular self-organized criticality. Specifically, the use of Zipf's law provides evidence of the impact of standing position on heart rate variability recordings.

Professional soccer players are high-performance athletes and cannot be compared to the general population. However, training high-level sport teams provides an exceptional opportunity to conduct long-term experiments on highly motivated and relatively homogeneous groups of healthy individuals. Monitoring heart rate variability in athletes is important for optimizing training, making it an excellent opportunity for studying long-term cardiovascular adjustments [17]. The present study began after a season break when the players were untrained and deconditioned. However, the subsequent training period significantly improved their physical condition, as indicated by the decrease in heart rate. The changes in heart rate also demonstrated the expected autonomic changes resulting from training [18].

The description of heart rate variability in the 1980s challenged the traditional foundation of physiology based on homeostasis, negative feedback regulatory loops, and the stability of bodily constants [18,19]. Homeostasis faced further challenges when Kobayashi and Musha described the fractal patterns of heart rate variability and when further studies revealed multifractality [20,21]. Several alternative views to the principle of homeostasis have been proposed [1,3–5,10], among which self-organized criticality is currently the most compelling [3–5,10]. Self-organized criticality explains the fractal patterns observed in heart rate variability and how a dynamic system operating at criticality can adjust to external changes more effectively than a system relying solely on regulatory loops with negative feedback [1,3–5,10].

There is no single robust way to demonstrate the self-organized criticality of a dynamic system [8,9], as the evidence may be based on several factors. The analysis of transitions between two states, such as the shift from a supine to a standing position, can provide valuable insights, as suggested by Mukli et al. [22]. Unfortunately, the experimental setup required for such analyses was not feasible in the field experiment we are reporting here. However, we have previously demonstrated critical phase transitions during position shifts by replicating, on the cardiovascular system, the seminal study on movement control conducted by Shöner and Kelso [23,24]. Bak et al., who initially described self-organized criticality, focused on the power law observed in earthquakes, known as Gutenberg–Richter's law, which has also been detected in cardiovascular dynamics [12,25]. Others have concentrated on Zipf's law, initially described in language dynamics. Several methods have been employed to demonstrate this law in cardiovascular dynamics [13,26–28]. Lo et al. demonstrated self-organized criticality in sleep–wake rhythms by analyzing not only the distribution of brief awakening episodes but also the sleep duration between these episodes [29]. A similar approach was employed by Racz et al. to argue that connectivity dynamics in the prefrontal cortex reflect the critical state of a brain function [21]. This kind of method could potentially be applied to heart rate variability time series, but would require significantly longer recordings than the 5-min recordings conducted in our study. Conducting longer recordings in future studies would also provide an opportunity to validate the power law relationship using more advanced techniques, as proposed by Clauset et al., instead of relying solely on linear regression [30].

Most studies examining Zipf's law in heart rate variability have used Holter recordings [26–28]. A Holter device records the electrocardiogram over a period of at least 24 h during daily life activities. However, this type of recording includes variability that is a response to daily life events. In our study, we focused on quiet, stationary subjects to eliminate variability due to daily activities. The remaining variability reflects the intrinsic regulatory mechanisms and delays of the cardiovascular system. Holter recordings also do not distinguish between different body positions. In a previous study, we demonstrated the effect of spontaneous movements, fidgeting, and iterative body position shifts on heart rate variability [31]. Amaral et al. confirmed that heart rate variability complexity remains high in subjects who are motionless [32]. In the present study, we aim to account for body position as the standing position leads to well-defined autonomic nervous system adjustments [14,16,18]. Our previous work has also linked criticality and instability in heart rate variability to the pathophysiology of vasovagal syncope, a health problem strongly related to standing position [12,13].

Our study provides clear evidence that Zipf's law is observed in heart rate variability for the standing position, while the results were less conclusive for the supine position. The range of straight lines is limited to a delay of a few beats in the former position. Moreover, alterations of the slope of these bradycardia sequences limited to few beats probably only reflect heart rate alterations since these two variables are strongly correlated. On the contrary, the slope of long bradycardia is not correlated with heart rate and this slope could indicates changes in criticality. This point is supported by the findings of our previous research on vasovagal syncope, which showed that patients with syncope have an increased slope of long bradycardia distribution [13].

The effects of autonomic nervous system adjustments on heart rate due to training are typically noticeable, but their impact on heart rate variability is more challenging to observe, regardless of the analysis tool used [16,18]. In our study, the subtle autonomic changes that occur with training were also not detected by the self-organized criticality assessment tool. However, in two players, Zipf's law was clearly violated after the warm-up period.

We have previously reported that the cardiovascular Zipf's law is violated at a delay of four heartbeat intervals, with the underlying reasons remaining unclear [13]. Our recent study confirms this law is broken, but suggests that the autonomic nervous system is not involved. It is possible that this broken law is simply a finite side effect.

Our study provides robust evidence for the existence of Zipf's law in heart rate variability, supporting the notion of cardiovascular self-organized criticality. However, we also observed instances where Zipf's law was broken, emphasizing the need for further research to elucidate its significance. Moreover, our study introduces a user-friendly tool for characterizing the intricate dynamics of the cardiovascular system, which could potentially aid in the diagnosis of heart rhythm disorders in clinical settings [33].

Author Contributions: Conceptualization, J.-O.F. and G.R.; methodology, J.-O.F. and G.R.; formal analysis, J.-O.F.; investigation, G.R.; data curation, J.-O.F. and G.R.; writing—original draft preparation, J.-O.F.; writing—review and editing, J.-O.F. and G.R.; project administration, G.R. All authors have read and agreed to the published version of the manuscript.

Funding: This research received no external funding.

Institutional Review Board Statement: Informed consent was obtained from all participants, the study was conducted in accordance with the Declaration of Helsinki, and approved by the Ethics Committee of Angers, France (#2014-36, 30 April 2014).

Data Availability Statement: The data presented in this study are available on reasonable request from the corresponding author.

Acknowledgments: We wish to thank the players for taking part in this study. We also wish to thank the Stade Lavallois Mayenne FC, which allowed and facilitated this study.

Conflicts of Interest: The authors declare no conflict of interest.

References

1. Goldstein, D.S. How does homeostasis happen? Integrative physiological, systems biological, and evolutionary perspectives. *Am. J. Physiol.* **2019**, *316*, R301–R317. [CrossRef] [PubMed]
2. Billman, G.E. Homeostasis: The underappreciated and far too often ignored central organizing principle of physiology. *Front. Physiol.* **2020**, *11*, 200. [CrossRef] [PubMed]
3. Struzik, Z.R. Is heart rate variability dynamics poised at criticality? In Proceedings of the 8th Conference of the European Study Group on Cardiovascular Oscillations (ESGCO), Trento, Italy, 25–28 May 2014; IEEE: Piscataway, NJ, USA, 2014.
4. Muñoz, M.A. Colloquium: Criticality and dynamical scaling in living systems. *Rev. Mod. Phys.* **2018**, *90*, 031001. [CrossRef]

5. Leonor Rivera, A.; Toledo-Roy, J.C.; Frank, A. Symmetry and Signs of Self-Organized Criticality in Living Organisms. *J. Phys. Conf. Ser.* **2020**, *1612*, 012024. [CrossRef]
6. Koushanpour, E. Baroreceptor discharge behavior and resetting. In *Baroreceptor Reflexes. Integrative Functions and Clinical Aspect*; Persson, P.B., Kirchheim, H.R., Eds.; Springer: Berlin, Germany, 1991; pp. 9–44.
7. Bak, P. *How Nature Works: The Science of Self-Organised Criticality*; Copernicus Press: New York, NY, USA, 1996; pp. 1–212.
8. Crosby, N.B. Introduction. In *Self-Organized Criticality Systems*; Aschwanden, M.J., Ed.; Open Academic Press: Berlin, Germany; Warsaw, Poland,, 2013; pp. 1–21.
9. Marković, D.; Gros, C. Power laws and self-organized criticality in theory and nature. *Phys. Rep.* **2014**, *536*, 41–74. [CrossRef]
10. Mora, T.; Bialek, W. Are biological systems poised at criticality? *J. Stat. Phys.* **2011**, *144*, 268–302. [CrossRef]
11. Grubb, B.P. Neurocardiogenic syncope and related disorders of orthostatic intolerance. *Circulation* **2005**, *111*, 2997–3006. [CrossRef]
12. Fortrat, J.O.; Gharib, C. Self-organization of blood pressure regulation: Clinical evidence. *Front. Physiol.* **2016**, *7*, 113. [CrossRef]
13. Fortrat, J.O. Zipf's law of vasovagal heart rate variability sequences. *Entropy* **2020**, *22*, 413. [CrossRef]
14. Rowell, L.B. Passive effect of gravity. In *Human Cardiovascular Control*; Oxford University Press: New York, NY, USA, 1993; pp. 3–36.
15. Ravé, G.; Zouhal, H.; Boullosa, D.; Doyle-Baker, P.K.; Saeidi, A.; Abderrahman, A.B.; Fortrat, J.O. Heart rate variability is correlated with perceived physical fitness in elite soccer players. *J. Hum. Kinet.* **2020**, *72*, 141–150. [CrossRef]
16. Ravé, G.; Fortrat, J.O. Heart rate variability in the standing position reflects training adaptation in professional soccer players. *Eur. J. Appl. Physiol.* **2016**, *116*, 1575–1582. [CrossRef] [PubMed]
17. Ravé, G.; Fortrat, J.O.; Dawson, B.; Carre, F.; Dupont, G.; Saeidi, A.; Boullosa, D.; Zouhal, H. Heart rate recovery and heart rate variability: Use and relevance in European professional soccer. *Int. J. Perform. Anal. Sport* **2018**, *18*, 168–183. [CrossRef]
18. Task Force of the European Society of Cardiology and the North American Society of Pacing and Electrophysiology. Heart rate variability: Standards of measurement, physiological interpretation and clinical use. *Circulation* **1996**, *93*, 1043–1065. [CrossRef]
19. Akselrod, S.; Gordon, D.; Ubel, F.A.; Shannon, D.C.; Berger, A.C.; Cohen, R.J. Power spectrum analysis of heart rate fluctuation: A quantitative probe of beat-to-beat cardiovascular control. *Science* **1981**, *213*, 220–222. [CrossRef]
20. Kobayashi, M.; Musha, T. 1/f fluctuation of heartbeat period. *IEEE Trans. Biomed. Eng.* **1982**, *29*, 456–457. [CrossRef]
21. Racz, F.S.; Mukli, P.; Nagy, Z.; Eke, A. Multifractal dynamics of resting-state functional connectivity in the prefrontal cortex. *Physiol. Meas.* **2018**, *39*, 024003. [CrossRef] [PubMed]
22. Mukli, P.; Nagy, Z.; Racz, F.S.; Portoro, I.; Hartmann, A.; Stylianou, O.; Debreczeni, R.; Bereczki, D.; Eke, A. Two-tiered response of cardiorespiratory-cerebrovascular network to orthostatic challenge. *Front. Physiol.* **2021**, *12*, 622569. [CrossRef]
23. Schöner, G.; Kelso, J.A. Dynamic pattern generation in behavioral and neural systems. *Science* **1988**, *239*, 1513–1520. [CrossRef]
24. Fortrat, J.O.; Levrard, T.; Courcinous, S.; Victor, J. Self-organization of blood pressure regulation: Experimental evidence. *Front. Physiol.* **2016**, *7*, 112. [CrossRef]
25. Bak, P.; Tang, C.; Wiesenfeld, K. Self-organized criticality. *Phys. Rev. A* **1988**, *38*, 364–374. [CrossRef]
26. Kalda, J.; Sakki, M.; Vainu, M.; Laan, M. Zipf's law in human heart beat dynamics. *arXiv* **2001**, arXiv:physics/0110075v1.
27. Yang, A.C.; Hseu, S.S.; Yien, H.W.; Goldberger, A.L.; Peng, C.K. Linguistic analysis of the human heartbeat using frequency and rank order statistics. *Phys. Rev. Lett.* **2003**, *90*, 108103. [CrossRef] [PubMed]
28. Rodríguez, J.; Prieto, S.; Correa, C.; Mendoza, F.; Weiz, G.; Soracipa, Y.; Velásquez, N.; Pardo, J.; Martínez, M.; Barrios, F. Physical mathematical evaluation of the cardiac dynamic applying the Zipf-Mandelbrot law. *J. Mod. Phys.* **2015**, *6*, 1881–1888. [CrossRef]
29. Lo, C.C.; Chou, T.; Penzel, T.; Scammell, T.E.; Strecker, R.E.; Stanley, H.E.; Ivanov, P.C. Common scale-invariant patterns of sleep-wake transitions across mammalian species. *Proc. Natl. Acad. Sci. USA* **2004**, *101*, 17545–17548. [CrossRef]
30. Clauset, A.; Shalizi, C.R.; Newman, M.E.J. Power-law distributions in empirical data. *SIAM Rev.* **2009**, *51*, 661–703. [CrossRef]
31. Fortrat, J.O.; Formet, C.; Frutoso, J.; Gharib, C. Even slight movements disturb analysis of cardiovascular dynamics. *Am. J. Physiol.* **1999**, *277*, H261–H267. [CrossRef]
32. Nunes Amaral, L.A.; Ivanov, P.C.; Aoyagi, N.; Hidaka, I.; Tomono, S.; Goldberger, A.L.; Stanley, H.E.; Yamamoto, Y. Behavioral-independent features of complex heartbeat dynamics. *Phys. Rev. Lett.* **2001**, *86*, 6026–6029. [CrossRef]
33. Subramanyan, R. Avalanches in cardiology. *Ann. Pediatr. Cardiol.* **2021**, *14*, 401–407. [CrossRef]

entropy

MDPI

Article

Entropy-Based Multifractal Testing of Heart Rate Variability during Cognitive-Autonomic Interplay

Laurent M. Arsac

Univ. Bordeaux, CNRS, Laboratoire IMS, UMR 5218 Talence, France; laurent.arsac@u-bordeaux.fr

Abstract: Entropy-based and fractal-based metrics derived from heart rate variability (HRV) have enriched the way cardiovascular dynamics can be described in terms of complexity. The most commonly used multifractal testing, a method using q moments to explore a range of fractal scaling in small-sized and large-sized fluctuations, is based on detrended fluctuation analysis, which examines the power–law relationship of standard deviation with the timescale in the measured signal. A more direct testing of a multifractal structure exists based on the Shannon entropy of bin (signal subparts) proportion. This work aims to reanalyze HRV during cognitive tasks to obtain new markers of HRV complexity provided by entropy-based multifractal spectra using the method proposed by Chhabra and Jensen in 1989. Inter-beat interval durations (RR) time series were obtained in 28 students comparatively in baseline (viewing a video) and during three cognitive tasks: Stroop color and word task, stop-signal, and go/no-go. The new HRV estimators were extracted from the f/α singularity spectrum of the RR magnitude increment series, established from q-weighted stable (log–log linear) power laws, namely: (i) the whole spectrum width (MF) calculated as $\alpha_{max} - \alpha_{min}$; the specific width representing large-sized fluctuations (MF_{large}) calculated as $\alpha_0 - \alpha_{q+}$; and small-sized fluctuations (MF_{small}) calculated as $\alpha_{q-} - \alpha_0$. As the main results, cardiovascular dynamics during Stroop had a specific MF signature while MF_{large} was rather specific to go/no-go. The way these new HRV markers could represent different aspects of a complete picture of the cognitive–autonomic interplay is discussed, based on previously used entropy- and fractal-based markers, and the introduction of distribution entropy (DistEn), as a marker recently associated specifically with complexity in the cardiovascular control.

Keywords: cardiovascular; multifractality; DFA; autonomic control; heart–brain; central autonomic network; cognitive task

1. Introduction

Over the years, complexity metrics derived from cardiovascular signals, e.g., heart rate variability (HRV) or blood pressure fluctuations, have enriched how the neurophysiological control of cardiovascular dynamics can be described. Perhaps the most striking observation after decades of methodological development lies in the fact that there exists no universal method and probably no unique kind of complexity in cardiovascular dynamics. Cardiovascular control involves oscillatory components of vagal and sympathetic modulations that combine to form complex dynamics. Cognitive tasking generates interconnections between cortical/subcortical networks and cardiovascular centers in the brainstem, reflected in HRV dynamics, as described by markers of so-called complexity [1,2]. In this domain, metrics obtained from entropy- or fractal-based analyses of HRV have demonstrated obvious interest in informing beyond the autonomic behavior per se on subtle adaptations in brain–heart interplay.

Entropy-based complexity is believed to grasp the information content in the HRV signal by quantifying the degree of predictability or regularity [3]. Somewhat differently, by adopting the point of view of fractal calculus, scaling properties in HRV quantify the way fluctuation sizes depend on observational timescales [4]. To characterize this dependence, analytic methods have envisaged windowing the entire series of collected data into subparts called "bins", and evaluating a given statistical property in these bins (subparts) against the same property in other, larger bins. The most used method, detrended fluctuation analysis (DFA), regresses the variance in bin samples against a variety of bin sizes on a log–log scale to obtain a scaling exponent α-DFA, defined as the slope of this linear relationship [5]. An extension of DFA consists in determining whether α-DFA (the slope) keeps the same value when modifying the fluctuation function using q moments, which emphasizes large-sized ($q > 0$) or small-sized ($q < 0$) fluctuations in the analyzed series [6]. The heterogeneity in q-dependent α-DFA indicates a multifractal rather than a monofractal behavior of the measured series. It has often been stressed that biological signals exhibit multifractal behaviors [7], which deserved methodological efforts for a quantitative approach of multifractality.

To obtain a multifractal geometry, an alternative method to DFA exists that introduces the concept of entropy [8]. In brief, rather than bin variance, the method has its root in bin proportion (P), where $P_i(L)$ is the amount of the measure, i.e., the sum of the samples in the i-th bin of size L, divided by the amount of all the measure across the entire series, i.e., the sum of all samples in the series (see Appendix A for an example). Then, the collected series is deformed using a predetermined set of q exponents to emphasize large-sized ($q > 0$) and small-sized ($q < 0$) fluctuations. The multifractal spectrum is defined thanks to pairs of f(q,L) and α(q,L), obtained after a mass coefficient (μ_i) is calculated for each scale and each q moment as:

$$\mu_i(q, L) = \frac{[P_i(L)]^q}{\sum_i [P_i(L)]^q} \tag{1}$$

The singularity strength α(q) is the singularity for μ(q)-weighted P(L), estimated by

$$\alpha(q) = \lim_{L \to 0} \frac{\sum_i \mu_i(q, L) \log P_i(L)}{\log L} \tag{2}$$

The estimates α(q) belong to the multifractal spectrum if the Shannon entropy of μ(q,L) evolves with L along a dimension f(q) calculated as

$$f(q) = \lim_{L \to 0} \frac{\sum_i \mu_i(q, L) \log \mu_i(q, L)}{\log L} \tag{3}$$

Generally, a set of predetermined values of q is used (see Section 2.3), including extreme q values that provide P-L relationships not compatible with a linear fit. Based on a threshold linear correlation coefficient as a benchmark, only the values of q providing stable (linear) scaling relationships served to obtain the multifractal spectrum [9]. The spectrum is determined by the curve α,f and the main estimator provided via the method is the spectrum width (MF), calculated as $MF = \alpha_{max} - \alpha_{min}$, thus representing the range of singularities α present in the measured signal.

The method has been successively applied in psychology to explore cognition through movement behavior. Establishing the link between multiplicative cascading and nonlinear multifractal behavior in this complex movement system helped conceptualize executive control as an emerging, nonlinear structure, where the activity of the system components can be assembled and disassembled by the interactions unfolding across multiple scales [8,10]. A behavior of similar nature was shown in postural control [11,12]. To date, the nature of the emerging structure in cardiovascular control during brain–heart interactions has not been described in similar terms, and what cardiovascular complexity really means in this context is not a settled matter. In a recent study that specifically addresses this notion, distribution entropy (DistEn) in HRV has been suggested as a reliable estimate of complexity in a human model of spinal cord injury [13]. It is hoped that entropy-based multifractal testing can complement the rather fuzzy picture of cardiovascular complexity and that applications to cognitive tasks can help highlight specific aspects of HRV complexity. For that, the present reanalysis of the HRV series collected during Stroop, stop-signal, and go/no-go tasks could help complete a picture in which multiscale sample entropy and multiscale multifractal estimators showed specific links with cognitive interference (Stroop), action cancellation (stop-signal), and movement restraint (go/no-go) [2]. Perhaps a multifractal testing combining entropy concepts/calculations may help to capture subtle differences in HRV during cognitive tasks.

Here, it is hypothesized that multifractal testing of HRV using the method of Chhabra and Jensen may provide reliable estimators to complete the picture of cardiovascular complexity. To validate this hypothesis, multifractal testing based on Shannon entropy was used for a reanalysis of HRV measures obtained during cognitive tasks. Some links between task and specific HRV metrics have already been shown by applying multiscale sample entropy and multiscale–multifractal DFA to the same dataset so that new information is expected here.

2. Materials and Methods

2.1. Subjects and Data Collection

Among 37 participants in the study of Bouny et al. [2], we considered only the 28 participants having RR series > 513 samples to finally obtain an RR-increment series with a length of 512 samples. The experimental protocol, which is described in detail in [2], consisted in recording RR series from a bipolar electrode transmitter belt Polar H10 (Polar, Finland) while participants watched an emotionally neutral video (baseline), then performed successively in a randomized order, a Stroop color-word task (Stroop), a go/no-go task (go/no-go), and a stop-signal task (stop signal), all displayed on a computer screen. Briefly, the Stroop task consisted in hitting a keyboard key as fast as possible that corresponded to the ink color of a word naming another color. The test assesses the ability to inhibit cognitive interference. Go/no-go assess the ability to restrain an action thanks to the relative amounts of no-go signals, here 30%. For this, a green or a red ellipsis appeared on the screen indicating 'press the space bar' (go) or 'do not press the space bar' respectively. Reaction time in the 'go' condition is presented as critical. The stop-signal task consisted in displaying an arrow pointing to the left or to the right to trigger a fast response by hitting the left or right key on the keyboard. The stop signal was presented in a short unpredictable delay (50 to 400 ms) after the presentation of the go stimulus, so that the response of the participant is already in the process of completion. So, the task assesses the ability to cancel an action as it imposes the suppression of an already initiated motor response. Each task lasted approximately 8 min with a one-minute pause between each. The participants kept a sitting position during the whole experiment (±40 min). As already attested in the initial publication, the study was approved by the IRB of the faculty des STAPS and followed the rules of the Declaration of Helsinki.

2.2. Measured Series, Analyses, and Main Parameters

The RR series that were free of visually removed artifacts and having length > 513 samples were truncated from the center of the series, which corresponded to the middle of the task. Differentiated series were obtained from HRV recordings and transformed to contain non-negative values by taking the absolute value of the successive increments:

$$|\Delta RR_i| = abs(RR_{i+1} - RR_i)$$

Differentiating provides signals resembling multiplicative cascades, and the absolute value avoids negative numbers that could not contribute to the computation of bin proportion and for which the logarithm is not defined [6,9].

A predetermined list of bin length (L) was set up for the need of the analysis, which corresponded to the observational time scales: 4, 8, 16, 32, 64, and 128 samples. So, the minimal observational scale contained four samples, and the maximal scale contained N/4 samples.

A predetermined set of q moments was considered with q values ranging from −20 to 20. The range of q values was established from a first-round analysis showing that a wider range was unnecessary due to the obvious lack of (log–log) linearity in the corresponding relationships between bin proportion probability and the scales in all the series (see below).

2.3. Main Steps in the Employed Method

For readers who would like to go through the method of Chhabra and Jensen [14] in detail, a tutorial is available in a must-read paper published recently [9]. In short, here, for each bin length L (4 to 128) and each q moment (−20 to 20), the sum of the $|\Delta RR_i|$ samples in the bin i was divided by the sum of the samples of the whole series to obtain the bin proportion for each i segment, $P_i(L)$ in Equations (1) and (2). Then, the sum of the $[P_i(L)]^q$ was divided by the sum of the $[P_i(L)]^q$ of the whole series (Appendix A).

To obtain a direct estimation of the multifractal spectrum, the Chhabra and Jensen method considers three critical characteristics of bin proportion: (i) the singularity strength α quantifies the scaling between bin proportion P and bin length L by establishing the linear relationship logP~logL; (ii) using q moments to weight bin proportions allows quantifying heterogeneity in scaling exponents, thus providing a q-dependent range of α values; (iii) in addition, this heterogeneity is also sensitive to bin length. It follows that the (negative) Shannon entropy of bin proportions regressed against bin length quantifies this sensitivity (scaling exponent) in terms of the Hausdorff dimension, f. The method of Chhabra and Jensen is described as a direct determination of the multifractal spectrum, where DFA-based methods need to use a Legendre transformation [15,16].

As introduced in Section 1, the q-order generalization of the singularity strength α is the slope of the μ-weighted P against L on a log–log scale. In the same vein, the q-order generalization of the Hausdorff dimension f is the slope of the Shannon entropy (actually the negative Shannon entropy) against logL.

By considering the paired value of α and f at the given q value, the multifractal spectrum is obtained by plotting f(q)~α(q). For that first step, a range of q values [−20 +20] was used. However, it is advised to hold only q values that allow obtaining relative stability in the μ(q)logP~logP and μ(q)logμ(q)~logP linear relationships to establish the final multifractal spectrum. It is achieved by considering a benchmark, here a given value of R^2 as a threshold below which linearity cannot be an acceptable fitting model to quantify α and f. Although the most recommended threshold $R^2 > 0.9$ was used here, it is generally advised to choose the threshold value as a function of the analyzed series and the parametric data. Other standards exist to determine the boundaries of the multifractal spectrum that have been employed, for instance, with DFA-based methods of multifractal testing [17–20]. Again, all these considerations are nicely discussed in the recent tutorial, including the procedure for surrogate data testing [9].

2.4. Phase-Randomized Synthetic Data

The width of the multifractal spectrum (here MF) has its possible origin in non-linearly decomposable forms of tightly interwoven interactions involving processes across all scales on the one hand. On the other hand, multifractality can be observed in a case of interactions unfolding similarly but across several independent scales (see Figure 1 in [8]). It is possible to distinguish between across-scale origins, and independent-scale origins of multifractality by transforming the original series into phase-randomized surrogates that retain only the linear structure of the series. This is generally achieved using the iterative amplitude-adjusted Fourier-transform (IAAFT) routine proposed by Schreiber and Schmitz [21]. The MF of a finite set of IAAFT series (here, 50 surrogates obtained after 100 iterations to preserve the spectral amplitude despite phase randomization) is compared to the MF of the original series using a one-sample *t*-test [9].

2.5. DistEn

A recent publication shows an interest in computing DistEn to infer complexity in HRV series [13]. The workflow allowing the computation of this estimator is explained in the cited work, and the code used here was kindly provided by the authors. Computations of

DistEn were obtained here to analyze the complexity of the HRV series obtained at rest and during each cognitive task. Keeping in mind that any computation of entropy is affected by trends or drifts in the measured signal, a filtering pre-processing step is mandatory, and it should be added that the process is critical when it comes to the use of entropy markers [13,22]. Especially when analyzing nervous autonomic regulations of the heart rhythm, so-called very low frequencies (VLF < 0.04 Hz), despite obvious physiological backgrounds, are out of the spectrum of interest when it comes to capturing the entropy linked to short-term sympathovagal modulations. For this, a data-driven filtering method was used by employing empirical mode decomposition [23]. The method extracts the dominant oscillations in the measured signal based on local maxima to provide so-called intrinsic mode functions (imf) and a residual that does not have the property of an imf (meaning that it does not exhibit symmetric oscillations around zero, as did intrinsic modes, imfs). As it is generally unclear if subtracting only the residual is enough to detrend the signal adequately, the criterion here to provide a stable (not drifting) final signal to explore DistEn was a test of stationarity, and the RWS method as proposed by Porta et al. [24]. By progressively adding imf1 (highest frequency), with other imfs of lower frequencies (imf2, imf3, ...) and testing at each step how many experimental series reached reasonable stationary, a threshold was observed when summing three imfs (imf1-3) rather than four imfs (imf1-4) (Figure 1). So, DistEn was computed here on a series of pre-processed HRV retaining the first three imfs obtained with the emd procedure.

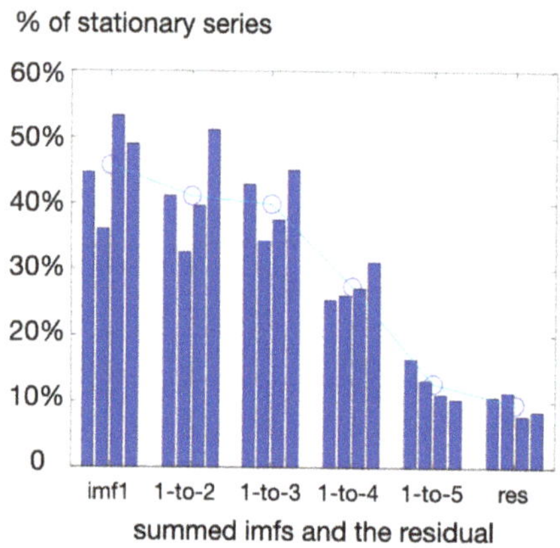

Figure 1. Percentage of stationary series identified via the RWS method when summing successive imfs (e.g. 1-to-2, then 1-to-3, then 1-to-4) obtained from prior HRV mode decomposition via the emd procedure. Imf1 contains the highest frequencies; imf4 adds low frequencies that could resemble trends identified as poor stationarity, given the short signal. The four bars indicate the four conditions, respectively: rest, Stroop, stop-signal, and go/no-go. Wide circles indicate the averaged % of stationary signals when grouping the four conditions.

The present computation of DistEn considered all existing two-sample segments and evaluated all the corresponding distances between two points of the series (embedding dimension m = 1). Then, to derive the Shannon entropy of all the distances that are computable in the HRV series, the empirical probability distribution function (ePDF) of the distances was calculated over 512 bins to derive the Shannon entropy of the distances, ShEn. DistEn was obtained by normalizing ShEn by log2(512), where 512 is the number of bins considered.

2.6. Statistics

Given the design of the experiment, the aim was to evaluate HRV estimators obtained in repeated cognitive tasks of different natures performed by all the participants, which means a four-repetition design in one sample of individuals. So, after testing for normality using the Shapiro–Wilk test to determine if the null hypothesis of composite normality is a reasonable assumption regarding the population distribution (based on *p*-value < 0.05), either a one-way ANOVA for repeated measures (null hypothesis true) or a Friedman test was used to compare the estimators among the four tasks.

3. Results

3.1. MF Estimates

Figure 2 shows a typical MF spectrum and compares the multifractal estimates for the measured RR increment series during each cognitive task. As main results, the figure highlights a different multifractal cardiovascular control behavior among cognitive tasks and resting conditions reflected in MF (χ^2 = 12.66, p = 0.005, post-hoc indicated on the figure) and in MF_{large} (χ^2 = 10.91, p = 0.012, post-hoc indicated on the figure). While cognitive inhibition required to perform the Stroop test was specifically reflected in the MF estimator, MF_{large} allowed distinguishing the go/no-go task from the other tasks. A higher MF_{large} during go/no-go indicated that large-sized fluctuations in HRV are specifically affected by the kind of cognitive–autonomic interplay in this particular condition. It has been suggested that the spectrum symmetry could be another estimator of interest [25]. The MF_{large}/MF_{small} ratio (*sym*) was computed to assess this symmetry. The value obtained for *sym* showed no difference (χ^2 = 2.63, p = 0.452) among the four conditions (rest 1.09 ± 0.32, Stroop 0.98 ± 0.28, stop-signal 1.03 ± 0.23, and go/no-go 0.94 ± 0.23).

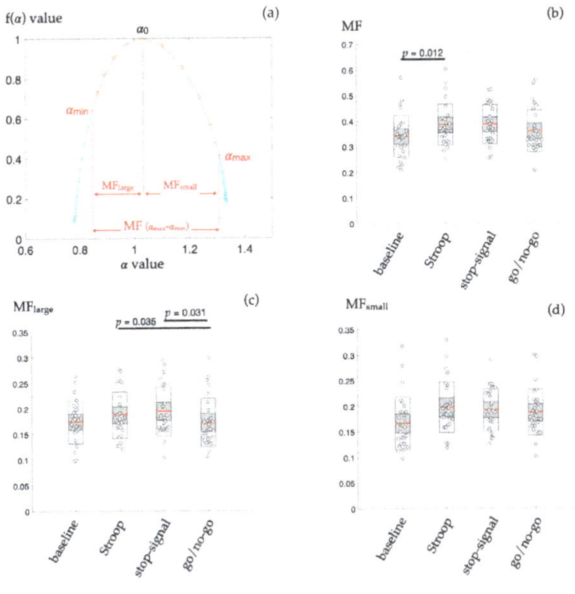

Figure 2. (**a**) A typical multifractal spectrum obtained by using a full range of q values [−20:20] (blue dots) where only q values providing R^2 > 0.9 (see methods) were retained (red dots) for computing the value of estimates MF, MF_{small}, and MF_{large}. (**b**–**d**) Values obtained from the three main estimates of the HRV behavior during each condition, baseline, and three cognitive tasks: MF, the multifractal spectrum width; MF_{large}, the width of the spectrum that is determined by magnifying large-sized fluctuations in the signal when using q > 0 moments.; MF_{small}, the width corresponding to q < 0 moments and magnification of small-sized fluctuations. Points are layed over a 1.96 SEM (95% confidence interval) in heavy grey and a 1 SD in light grey. The mean is indicated as a red line; *p*-values are indicated when the difference between the experimental conditions is significant in post-hoc tests.

3.2. Surrogate Data Testing

Phase-randomization of increment RR series, a method that aims to distinguish across-scales vs. scale-dependent origin of multifractality, indicated that across-scale behavior was not really dominant. In fact, the difference between the MF in measured series and MF in their linearized surrogates (IAAFT) did not reach significance, respectively, $p = 0.061$, $p = 0.091$, $p = 0.076$, and $p = 0.069$ for Stroop, stop-signal, and go/no-go. The corresponding t-statistic index was not higher either. In the same way, MF_{small} and MF_{large} values were not modified via the phase-randomization procedure (all $p > 0.05$).

3.3. DistEn

DistEn in the detrended HRV series showed almost identical values at rest (0.883 $\pm$ 0.015) and during each task: Stroop (0.872 $\pm$ 0.020), stop-signal (0.874 $\pm$ 0.020), and go/no-go (0.874 $\pm$ 0.019). Hence, despite a slightly higher value in resting condition, DistEn failed to distinguish task-dependent HRV complexity.

4. Discussion

Multifractal testing has been the subject of intense research in recent years to describe the complexity of control systems, including posture [12], cognition [10,26,27], movement [19,20,28], and cardiovascular control [17,18]. The present study, elaborating on a dataset obtained during cognitive tasking, aimed at exploring multifractal estimates of HRV complexity derived from a direct determination of the multifractal spectrum based on Shannon entropy metrics: the Chhabra and Jensen method [14]. Exploring new estimates is based on the intuition that different aspects of complexity can be reflected by specific markers, which depends on how entropy-based or fractal-based metrics are computed, as illustrated in recent studies [2,13]. Specifically, different components of the multifractal spectrum reflecting large-sized and small-sized fluctuations in HRV were evaluated here and confronted with cardiovascular responses to four different cognitive challenges that rely on distinctive cognitive–autonomic interplay (Figure 2). Cognitive models are obvious candidates to explore cardiovascular complexity since brain–heart interplay is generally more prompt to alter the complex dynamics of HRV rather than to produce a simple shift in the sympatho–vagal balance reflected in the power spectral density of HRV [1,29–31].

The main findings here, thanks to the application of the Chhabra and Jensen method to directly obtain the multifractal spectrum of HRV, were that: (i) during the Stroop task only, the width of the multifractal spectrum (MF) was greater than in baseline (resting) conditions (Figure 2b); (ii) the subpart of the MF spectrum highlighting the large-sized fluctuations in HRV (MF-large) mainly distinguished specific HRV dynamics during the go/no-go task (Figure 2c). It is worth noting that no such methodological sensitivity was reached in a previous analysis of this dataset [2] based on refined composite multiscale (sample) entropy, linear multiscale entropy (LMSE), conditional entropy computed with the binning model-free estimator (CEBi), conditional entropy computed with the kernel model-free estimator (CEKe), conditional entropy computed with the nearest neighbor model-free estimator (CENN), or the multifractal–multiscale detrended fluctuation analysis (MM-DFA). In fact, RCMSE and MM-DFA showed promising results, indicating that system complexity rather than signal randomness alone is modified, but estimates derived from said methods unveil singular HRV dynamics in one cognitive task each, separately. The ability of the present multifractal testing to provide estimates that distinguish several situations could be plaid for better performance, at least for the exploration of heart–brain interactions.

Applying the method of Chhabra and Jensen on increments series allowed a non-ambiguous determination of multifractal spectra, as illustrated in Figure 2. The predetermined choice of range in q moments (-20 to +20) allowed obtaining coherent values of q-dependent singularity strength α (x-axis) as well as coherent values of $f\alpha$ (y-axis) based on Shannon entropy conceptualization. The use of a correlation coefficient benchmark ensuring that stable values of α and $f\alpha$ are used to draw the multifractal spectrum allowed providing estimators MF, MF_{large}, and MF_{small} adequately. In this vein, combining better performance and non-ambiguous application, the present method is suggested to offer a reliable additional (but not exclusive) tool for exploring cardiovascular complexity.

Recently, distribution entropy was compared to sample entropy and fuzzy entropy in HRV dynamics, following the intuition that specific estimates of HRV might be less blind to complexity structures or more adequate to highlight a certain form of complexity [13]. Interestingly, the authors compared DistEn in white noise, pink noise, brown noise, and chaos to experimental HRV series from a human model to explore several forms of complexity. It was shown that DistEn highlights complexity in a chaotic series as well as in HRV series obtained in low-level spinal cord-injured (SCI) participants [32]. In our conditions, DistEn failed to provide a distinction among the cognitive tasks; the resting period before tasking also showed slightly but not significantly higher DistEn values (see Section 3.2). It is concluded that MF and DistEn may not provide the same type of information; although DistEn is an obvious marker of HRV complexity emerging from perturbated dynamics in vascular districts of SCI patients, this estimator might be less sensitive to changes in HRV complexity during cognitive tasks. In agreement with this intuition, sample entropy (RCMSE) distinguished the cognitive tasks in the previous analyses of the present data, and Castiglioni et al. [13] opposed sample entropy and DistEn. So, as hypothesized during the introduction of the present work, the notion that HRV markers derived from different nonlinear analyses can point to subtle neurophysiological dynamics in cardiovascular control coordination is strengthened by the present findings. Discrepancies among nonlinear analyses are not alarming but rather encourage us to consider that the dynamics of biological systems may take different forms of complexity.

The hypothesis that different kinds of emerging interactions in biological systems can be reflected in subtle multifractal properties is attractive. Nonlinear interactions are suspected to cohere in the large system linking cognitive, sensory, and motor processes, which has been documented in several studies, some of them being cited in the present manuscript. Less attention has been paid to subtilities in nonlinear dynamics of cardiovascular control. Mainly, it is unknown whether multifractality in HRV reflects intricate interactions across a wide range of scales or if scale-specific non-overlapped multifractal phenomena are finally reflected in a spectrum of similar width. Here, surrogate data testing using phase randomization to preserve only the linear attributes of the series leaves MF, MF_{small}, and MF_{large} unchanged. This would indicate that the strength of nonlinear multifractality in our conditions is low, so the multifractal organization mostly emerges from an assemblage of local interactivity spanning several scales [9]. In a sense, this echoes the need to combine a multifractal and a multiscale approach to HRV dynamics [17]. This observation offers another interesting point of view when considering why MF but not DistEn could distinguish cognitive tasks. The MF in our condition could be sensitive to changes in local interactivity during mental tasking, while DistEn might rather point to nonlinear behavior of the cardiovascular control when peripheral vascular beds have ineffective autonomic control. It could be added that MF in the present dataset analysis resembles RCMSE (sample entropy) in the previous exploration of the same dataset [2] since both estimates distinguish the HRV signature of cognitive interference during the Stroop test. On its side, MF-large, a specific aspect of MF where large-sized fluctuations are emphasized, is distinguished mainly from the go/no-go situation from other cognitive tasks. Large-sized fluctuations in a RR increments series rely dominantly on intense vagal outflow because only vagal arousals can have a rapid action on the sinus rhythm. Establishing the link between action–restraint (go/no-go testing) and multifractality in intense vagal outflow is out of the scope of the present study, but it is the perfect illustration that adding a multifractal approach based on Shannon entropy could shed light on particular aspects of cardiovascular control, not encountered in other approaches. Among alternative methods, multifractal DFA has been a widely used and reliable method to explore cardiovascular complexity. It is worth noting that DFA-based approaches evaluate the power-law form of the power spectrum, while the present approach evaluates the power-law form of an aggregate distribution of fluctuations. The former indicates the magnitude of oscillations at different observational scales, and the latter reflects the probability of different-sized fluctuations. Both approaches can make significant contributions to exploring cardiovascular complexity since the present reanalysis of the dataset illustrates that different kinds of mathematical views of the HRV series can help gain an improved understanding of complex neurophysiological mechanisms.

5. Conclusions

The observations in this work are consistent with the initial idea that there is no unique estimate(s) that can help researchers grasp the HRV complexity, but rather a number of mathematical views of the HRV series, each able to provide a reliable estimator of self-organized interactions in cooperating systems. This work is further evidence that entropy-based, multifractal-based, and other methods are not mutually exclusive but provide complementary information to understand the functioning of a complex networked system. In the absence of direct evidence to link a given marker to a given organization of cardiovascular control, our only weapon is default reasoning, which is strengthened by multiplying complexity explorations in specific contexts. MF, MF_{small}, and MF_{large} derived from applying the Chhabra and Jensen method to HRV could add value to this ongoing way of research.

Funding: This research received no external funding.

Institutional Review Board Statement: The study is a reanalysis of a previous dataset for which ethical statements have been provided.

Informed Consent Statement: Not applicable.

Data Availability Statement: Data can be supplied by the author on reasonable request.

Acknowledgments: I sincerely thank Paolo Castiglioni and Andrea Faini for having provided me with the code needed to compute DistEn. I also thank P. Bouny for sharing the code to test the presence of non-stationarities in the short HRV series, as in Porta et al. 2004 [24].

Conflicts of Interest: The author declares no conflict of interest.

Appendix A

A more detailed explanation about calculating bin (signal subparts) proportions (P) at a given bin length (L) is provided here. Indeed P~L and q-P~L relationships are the roots that allow the evaluation of α and f(α). A range of observational timescales is chosen, where each timescale corresponds to a subpart (here called bin) of the whole series containing a corresponding number of samples. For each timescale, the sum of the samples in the bin is divided by the sum of the samples of the whole series to obtain the bin proportion (P). An example is provided assuming that bins contain four samples (L = 4) each (RR increments in ms) and supposing that the whole series has eight samples (for simplicity). Actually, longer collected series allow us to reiterate the same calculation at several bin lengths (observational scales) chosen to be equally distant on a log scale (e.g., L = 8, L = 16, L = 32, . . . L = N/4).

The whole series = 48 4 13 14 6 28 2 12
bin1 = 48 4 13 14 bin2 = 6 28 2 12
sumbin1 = 79 sumbin2 = 48
sumseries = 79 + 48 = 127
P1 = 79/127 = 0.622 P2 = 48/127 = 0.378

The q moments are then introduced to modify each initial bin proportion and obtain new q-specific P; an example is provided assuming q = −4.

P1^−4 = 6.681 P2^−4 = 48.982
q-sum = 55.663
q-P1 = 6.681/55.663 = 0.120 q-P2 = 48.982/55.663 = 0.880

The above example (q = −4) illustrates that q < 0 values enhance small-sized fluctuations: here, P2 was only 0.378, but q-P2 reached 0.880. Inversely, P1 was 0.622, but q-P1 dropped to 0.12. The inverse is true; when using positive q moments (example here with q = 3) one obtains a magnification of large-sized fluctuations:

P1^3 = 0.241 P2^3 = 0.054

q-sum = 0.295

q-P1 = 0.241/0.295 = 0.817 q-P2 = 0.054/0.295 = 0.183

References

1. Blons, E.; Arsac, L.M.; Gilfriche, P.; Deschodt-Arsac, V. Multiscale Entropy of Cardiac and Postural Control Reflects a Flexible Adaptation to a Cognitive Task. *Entropy* **2019**, *21*, 1024. [CrossRef]
2. Bouny, P.; Arsac, L.M.; Touré Cuq, E.; Deschodt-Arsac, V. Entropy and Multifractal-Multiscale Indices of Heart Rate Time Series to Evaluate Intricate Cognitive-Autonomic Interactions. *Entropy* **2021**, *23*, 663. [CrossRef] [PubMed]
3. Costa, M.; Goldberger, A.L.; Peng, C.-K. Multiscale Entropy Analysis of Biological Signals. *Phys. Rev. E Stat. Nonlin. Soft Matter Phys.* **2005**, *71*, 021906. [CrossRef] [PubMed]
4. Peng, C.K.; Havlin, S.; Stanley, H.E.; Goldberger, A.L. Quantification of Scaling Exponents and Crossover Phenomena in Nonstationary Heartbeat Time Series. *Chaos* **1995**, *5*, 82–87. [CrossRef] [PubMed]
5. Goldberger, A.L.; Amaral, L.A.N.; Hausdorff, J.M.; Ivanov, P.C.; Peng, C.-K.; Stanley, H.E. Fractal Dynamics in Physiology: Alterations with Disease and Aging. *Proc. Natl. Acad. Sci. USA* **2002**, *99* (Suppl. S1), 2466–2472. [CrossRef] [PubMed]
6. Ihlen, E.A.F.; Vereijken, B. Multifractal Formalisms of Human Behavior. *Hum. Mov. Sci.* **2013**, *32*, 633–651. [CrossRef] [PubMed]
7. Ivanov, P.C.; Amaral, L.A.N.; Goldberger, A.L.; Havlin, S.; Rosenblum, M.G.; Struzik, Z.R.; Stanley, H.E. Multifractality in Human Heartbeat Dynamics. *Nature* **1999**, *399*, 461–465. [CrossRef] [PubMed]
8. Kelty-Stephen, D.; Palatinus, K.; Saltzman, E.; Dixon, J. A Tutorial on Multifractality, Cascades, and Interactivity for Empirical Time Series in Ecological Science. *Ecol. Psychol.* **2013**, *25*, 1–62. [CrossRef]
9. Dixon, J.A.; Holden, J.G.; Mirman, D.; Stephen, D.G. Multifractal Dynamics in the Emergence of Cognitive Structure. *Top Cogn. Sci.* **2012**, *4*, 51–62. [CrossRef]
10. Kelty-Stephen, D.G.; Furmanek, M.P.; Mangalam, M. Multifractality Distinguishes Reactive from Proactive Cascades in Postural Control. *Chaos Solitons Fractals* **2021**, *142*, 110471. [CrossRef]
11. Kelty-Stephen, D.G.; Lee, I.C.; Carver, N.S.; Newell, K.M.; Mangalam, M. Multifractal Roots of Suprapostural Dexterity. *Hum. Mov. Sci.* **2021**, *76*, 102771. [CrossRef] [PubMed]
12. Castiglioni, P.; Merati, G.; Parati, G.; Faini, A. Sample, Fuzzy and Distribution Entropies of Heart Rate Variability: What Do They Tell Us on Cardiovascular Complexity? *Entropy* **2023**, *25*, 281. [CrossRef] [PubMed]
13. Kelty-Stephen, D.G.; Lane, E.; Bloomfield, L.; Mangalam, M. Multifractal Test for Nonlinearity of Interactions across Scales in Time Series. *Behav. Res.* **2022**, *55*, 2249–2282. [CrossRef] [PubMed]
14. Chhabra, A.; Jensen, R. Direct Determination of the f(Alpha) Singularity Spectrum. *Phys. Rev. Lett.* **1989**, *62*, 1327–1330. [CrossRef]
15. Kantelhardt, J.W.; Zschiegner, S.A.; Koscielny-Bunde, E.; Bunde, A.; Havlin, S.; Stanley, H.E. Multifractal Detrended Fluctuation Analysis of Nonstationary Time Series. *Phys. A Stat. Mech. Its Appl.* **2002**, *316*, 87–114. [CrossRef]
16. Mukli, P.; Nagy, Z.; Eke, A. Multifractal Formalism by Enforcing the Universal Behavior of Scaling Functions. *Phys. A Stat. Mech. Its Appl.* **2015**, *417*, 150–167. [CrossRef]
17. Castiglioni, P.; Lazzeroni, D.; Coruzzi, P.; Faini, A. Multifractal-Multiscale Analysis of Cardiovascular Signals: A DFA-Based Characterization of Blood Pressure and Heart-Rate Complexity by Gender. *Complexity* **2018**, *2018*, e4801924. [CrossRef]
18. Castiglioni, P.; Faini, A. A Fast DFA Algorithm for Multifractal Multiscale Analysis of Physiological Time Series. *Front. Physiol.* **2019**, *10*, 115. [CrossRef]
19. Pratviel, Y.; Deschodt-Arsac, V.; Larrue, F.; Arsac, L.M. Fast Hand Movements Unveil Multifractal Roots of Adaptation in the Visuomotor Cognitive System. *Front. Physiol.* **2021**, *12*, 713076. [CrossRef]
20. Arsac, L.M. Multifractal Dynamics in Executive Control When Adapting to Concurrent Motor Tasks. *Front. Physiol.* **2021**, *12*, 662076. [CrossRef]
21. Schreiber, T.; Schmitz, A. Improved Surrogate Data for Nonlinearity Tests. *Phys. Rev. Lett.* **1996**, *77*, 635–638. [CrossRef] [PubMed]
22. Gow, B.; Peng, C.-K.; Wayne, P.; Ahn, A. Multiscale Entropy Analysis of Center-of-Pressure Dynamics in Human Postural Control: Methodological Considerations. *Entropy* **2015**, *17*, 7926–7947. [CrossRef]
23. Huang, N.E.; Shen, Z.; Long, S.R.; Wu, M.C.; Shih, H.H.; Zheng, Q.; Yen, N.-C.; Tung, C.C.; Liu, H.H. The Empirical Mode Decomposition and the Hilbert Spectrum for Nonlinear and Non-Stationary Time Series Analysis. *Proc. R. Soc. Lond. Ser. A Math. Phys. Eng. Sci.* **1998**, *454*, 903–995. [CrossRef]
24. Porta, A.; D'Addio, G.; Guzzetti, S.; Lucini, D.; Pagani, M. Testing the Presence of Non Stationarities in Short Heart Rate Variability. *Comput. Cardiol.* **2004**, *31*, 645–648. [CrossRef]
25. Aguilar-Molina, A.M.; Angulo-Brown, F.; Muñoz-Diosdado, A. Multifractal Spectrum Curvature of RR Tachograms of Healthy People and Patients with Congestive Heart Failure, a New Tool to Assess Health Conditions. *Entropy* **2019**, *21*, 581. [CrossRef]

26. Mirman, D.; Irwin, J.R.; Stephen, D.G. Eye Movement Dynamics and Cognitive Self-Organization in Typical and Atypical Development. *Cogn. Neurodyn.* **2012**, *6*, 61–73. [CrossRef]
27. Pratviel, Y.; Deschodt-Arsac, V.; Larrue, F.; Arsac, L.M. Tool Embodiment Is Reflected in Movement Multifractal Nonlinearity. *Fractal Fract.* **2022**, *6*, 240. [CrossRef]
28. Arsac, L.M.; Weissland, T. Multifractality in the Movement System When Adapting to Arm Cranking in Wheelchair Athletes, Able-Bodied Athletes, and Untrained People. *Fractal Fract.* **2022**, *6*, 176. [CrossRef]
29. Young, H.; Benton, D. We Should Be Using Nonlinear Indices When Relating Heart-Rate Dynamics to Cognition and Mood. *Sci. Rep.* **2015**, *5*, 16619. [CrossRef]
30. Dimitriev, D.A.; Saperova, E.V.; Dimitriev, A.D. State Anxiety and Nonlinear Dynamics of Heart Rate Variability in Students. *PLoS ONE* **2016**, *11*, e0146131. [CrossRef]
31. Blons, E.; Arsac, L.M.; Gilfriche, P.; McLeod, H.; Lespinet-Najib, V.; Grivel, E.; Deschodt-Arsac, V. Alterations in Heart-Brain Interactions under Mild Stress during a Cognitive Task Are Reflected in Entropy of Heart Rate Dynamics. *Sci. Rep.* **2019**, *9*, 18190. [CrossRef] [PubMed]
32. Castiglioni, P.; Merati, G.; Parati, G.; Faini, A. Decomposing the Complexity of Heart-Rate Variability by the Multifractal–Multiscale Approach to Detrended Fluctuation Analysis: An Application to Low-Level Spinal Cord Injury. *Physiol. Meas.* **2019**, *40*, 084003. [CrossRef] [PubMed]

entropy

Article

Spatio-Temporal Fractal Dimension Analysis from Resting State EEG Signals in Parkinson's Disease

Juan Ruiz de Miras [1,2,*], Chiara-Camilla Derchi [2], Tiziana Atzori [2], Alice Mazza [2], Pietro Arcuri [2], Anna Salvatore [2], Jorge Navarro [2], Francesca Lea Saibene [2,*], Mario Meloni [2,†] and Angela Comanducci [2,3,†]

1 Software Engineering Department, University of Granada, 18071 Granada, Spain
2 IRCCS Fondazione Don Carlo Gnocchi, 20148 Milan, Italy
3 Department of Engineering, Università Campus Bio-Medico di Roma, 00128 Rome, Italy
* Correspondence: demiras@ugr.es (J.R.d.M.); fsaibene@dongnocchi.it (F.L.S.); Tel.: +34-958-24-05-74 (J.R.d.M.)
† These authors contributed equally to this work.

Abstract: Complexity analysis of electroencephalogram (EEG) signals has emerged as a valuable tool for characterizing Parkinson's disease (PD). Fractal dimension (FD) is a widely employed method for measuring the complexity of shapes with many applications in neurodegenerative disorders. Nevertheless, very little is known on the fractal characteristics of EEG in PD measured by FD. In this study we performed a spatio-temporal analysis of EEG in PD using FD in four dimensions (4DFD). We analyzed 42 resting-state EEG recordings comprising two groups: 27 PD patients without dementia and 15 healthy control subjects (HC). From the original resting-state EEG we derived the cortical activations defined by a source reconstruction at each time sample, generating point clouds in three dimensions. Then, a sliding window of one second (the fourth dimension) was used to compute the value of 4DFD by means of the box-counting algorithm. Our results showed a significantly higher value of 4DFD in the PD group ($p < 0.001$). Moreover, as a diagnostic classifier of PD, 4DFD obtained an area under curve value of 0.97 for a receiver operating characteristic curve analysis. These results suggest that 4DFD could be a promising method for characterizing the specific changes in the brain dynamics associated with PD.

Keywords: fractal dimension; box-counting; EEG; Parkinson's disease; neurodegeneration

1. Introduction

The fractal dimension (FD) [1] is a quantitative measure of shape complexity with a large number of applications in characterizing neurodegenerative diseases from both medical imaging data [2] and electroencephalogram (EEG) signals [3]. Analyzing the complexity of EEG signals is crucial for understanding the underlying brain dynamics and detecting early signs of neurodegeneration. Correlation dimension [4] and, in particular, Higuchi's fractal dimension [5] are the most widely-used methods for computing the fractal dimension of EEG signals. Nevertheless, some other non-linear metrics, such as those related to the concept of entropy [6], are also commonly used to analyze the complexity of EEG signals. In [7], a novel methodology was presented for analyzing the complexity of the spatio-temporal dynamics in the brain based on the 3D fractal dimension (3DFD) and the 4D fractal dimension (4DFD) computed from EEG. The novelty of this approach relies on the use of source reconstruction from EEG to obtain a 3D representation of cortical activations over time, from which the values of 3DFD and 4DFD were then computed.

Parkinson's disease (PD) is a neurodegenerative disorder which affects approximately 1% of the world's elderly population [8]. Patients with PD suffer motor dysfunction symptoms such as gait disturbance, resting tremor, and muscle rigidity, and also non-motor symptoms such as cognitive impairment and neuropsychiatric symptoms such as anxiety, depression, and apathy, which can occur even many years before the motor symptoms [9].

Therefore, an early diagnosis is of great relevance in optimizing the clinical management of the disease [10].

Previous studies have shown that non-linear measures of EEG complexity can be of relevance in characterizing early neurodegeneration processes in PD. Müller et al. [11] showed that the correlation dimension of EEG is higher for PD patients compared to healthy subjects, especially during motor task performance. When comparing the performance of linear and non-linear methods in distinguishing resting-state EEG of PD patients from healthy subjects, non-linear measures, such as entropy, had a more powerful ability to differentiate between PD and healthy subjects, with higher entropy in PD [12]. Multiscale entropy was also used in [13] for exploring the characteristics of sleep EEG in patients with PD. The analysis of multiscale entropy revealed an increased complexity of the EEG signal during non-REM (rapid eye movement) sleep in PD compared to healthy controls. Furthermore, Han and colleagues found increased entropy (wavelet packet entropy) over the global frequency domain of resting-state EEG in patients with early PD compared to healthy subjects [14]. All these previous studies revealed that using complexity measures of EEG could be a useful tool for early diagnosis of PD. However, there is limited understanding of the fractal characteristics of EEG in PD, which our study wishes to address.

To this aim, using fractal dimension we analyzed the complexity of resting-state EEG signal in PD patients without dementia. As described above, most previous work in this topic has analyzed EEG using several measures of entropy, but very little is known about the fractal characteristics of EEG in PD. In our study, we followed a similar approach to the one presented in [7], adapted for the case of resting-state EEG, analyzing the spatio-temporal complexity in terms of the 4DFD of brain activations. The results demonstrated that patients with PD without dementia showed higher complexity values compared to the elderly, supporting the hypothesis that PD is characterized by a significant alteration in brain complexity and overall changes to the underlying organization of the brain. These findings suggest that 4DFD could be a valuable complexity measure for identifying early neurodegeneration in PD, even before it is captured by clinical scales for cognitive decline, with potential implications for enhancing early diagnosis and informing clinical practice in order to plan rehabilitation intervention.

2. Materials and Methods

2.1. Subjects

A total of 42 subjects participated in the study: 27 patients diagnosed with Parkinson's disease (PD group) (12 females and 15 males, with a mean age of 69.59 ± 6.74), and 15 healthy control subjects (HC group) with no symptoms or history of any neuropsychiatric disorder (7 females and 8 males, with a mean age of 67.53 ± 4.94). Demographics of both groups and clinical data of patients (see Table 1) were collected during face-to-face interviews by a neurologist and a neuropsychologist specialized in movement disorders.

Table 1. Study demographics and clinical data. Values expressed as mean ± standard deviation. Education given as years of school. UPDRS III: Unified Parkinson's Disease Rating Scale part III; MoCA: Montreal Cognitive Assessment; MMP: Mini-Mental Parkinson.

	PD	HC	Test, *p*-Value
N	27	15	
Sex (F:M)	12:15	7:8	$\chi^2 = 0.019, p = 0.89$ [a]
Age	69.59 ± 6.74	67.53 ± 4.94	$U = 249.5, p = 0.22$ [b]
Education	12.74 ± 3.79	12.80 ± 2.95	$U = 195.0, p = 0.84$ [b]
Disease duration	10.24 ± 6.78		
Hoehn and Yahr	2.41 ± 0.42		
UPDRS III	39.00 ± 9.79		
MoCA	25.01 ± 2.65		
MMP	28.62 ± 2.45		

[a] Chi-squared test. [b] Mann–Whitney test.

For each PD patient, the disease was diagnosed on the basis of their medical history and neurological examinations. Specifically, the diagnosis of PD was made following the Movement Disorder Society (MDS) clinical diagnostic criteria [15]. All patients also underwent a global cognitive assessment including Mini-Mental Parkinson (MMP) and Montreal Cognitive Assessment (MoCA) as recommended by the MDS PD-MCI study group [16]. Exclusion criteria were: a diagnosis of Parkinson's disease dementia (PD-D) based on MDS Task Force criteria [17], the presence of vascular Parkinsonism, drug-induced Parkinsonism, any suggestive features of a diagnosis of atypical Parkinsonism, concomitant neurologic and/or psychiatric diseases, and any other severe comorbidities which may significantly influence cognitive testing. As a global measure of the severity of motor symptoms, the MDS-Unified Parkinson's Disease Rating Scale part III (UPDRS III) [18] was also employed. A modified Hoehn and Yahr's scale [19] between 2 and 3 was also employed as a criterion to recruit patients with mild-to-moderate bilateral motor disability. Table 1 shows mean and standard deviation for all these scores in the PD group.

All subjects gave their informed consent for participation in the study (for details of the ethical protocol, please refer to the specific section below).

2.2. EEG Acquisition and Processing

During EEG recordings, both patients and healthy controls were seated in an upright or slightly reclined position in a quiet room. Spontaneous resting EEG recordings were acquired while at rest with their eyes open, not engaging in any task for a minimum of 5 min. The EEG session took place approximately 1–2 h after the administration of morning medications.

EEG was recorded with a Brainamp DC (Brain Products GmbH, Germany) equipped with 62 channels following the standard 10–20 montage. In all the recordings, reference and ground electrodes were located on the forehead and two additional channels were used to record the electrooculogram (EOG) in a diagonal montage to monitor for eye-blinking and saccades. Impedance at all electrodes was kept below 5 kOhm. Spontaneous EEG data were collected at 1000 Hz sampling rate and with a hardware filtering between 0.016 and 250 Hz. Due to the computational restrictions of further analysis, a resampling process to 500 Hz was performed.

EEG preprocessing was performed using MATLAB R2016b (Mathworks Inc., Natick, MA, USA) and custom-made scripts based on the EEGLab toolbox [20]. EEG was band-pass filtered (3rd order Butterworth, 0.5 Hz cutoffs, using the filtfilt MATLAB function) and notch-filtered (50 Hz harmonics up to 250 Hz). A trained researcher manually rejected artifactual epochs and bad electrodes. Artifactual electrodes were interpolated using spherical splines. Electrodes were then re-referenced to the average reference. Additional analysis of artifactual components through independent component analysis (ICA) decomposition allowed us to visually identify and reject components from ocular, muscular, and cardiac origin.

Source modelling was performed using the Brainstorm software [21], which is freely available for download online (http://neuroimage.usc.edu/brainstorm, accessed on 1 July 2023). First, a forward EEG model was computed using the boundary element method implemented in OpenMEEG [22]. We used the MNI/ICBM152 brain template of Brainstorm as the MRI anatomy [23]. Then, a source model of 15,002 current dipoles was obtained with the inverse sLORETA method implemented in Brainstorm [24]. Dipole orientations were constrained from normal to cortex, and the identity matrix was used as the noise covariance matrix. The EEG preprocessing process rejected a considerable amount of EEG signals, so finally a unique epoch consisting of the first 120 clean seconds of each recording was used. As a result of the source modelling process, a matrix of 15,002 (sources) $\times$ 60,000 (time samples) was obtained for each subject.

2.3. Fractal Dimension Computation

A 4D fractal dimension (4DFD) approach was used for measuring the spatio-temporal complexity of brain activations. We used the well-known box-counting algorithm [25] to estimate the 4DFD value of brain activations. By using the box-counting algorithm, the fractal dimension (FD) of a set $S \in \mathbb{R}^d$ can be computed as:

$$\mathrm{FD}(S) = \frac{\log(n(r))}{\log\left(\frac{1}{r}\right)} \tag{1}$$

where $n(r)$ is the number of boxes of scale r^d covering the set S. If S is not an ideal fractal, then the FD is estimated as the slope of the linear regression of $\log(n(r))$ vs. $\log(1/r)$.

In our study, the set S is composed of the 4D points defined by the 3D locations of cortical activations at each time sample (time is the fourth dimension). In order to obtain this 4D set, firstly, a binary 3D representation of the cortical activations at each time sample was obtained. As an example, Figure 1 shows this binarization process for three samples in the first second of one of the EEG recordings. A source was considered a cortical activation at a time sample if its absolute value was greater than the mean plus the standard deviation of the absolute values of that current for all time samples in the epoch (see Figure 1C,D). This threshold allowed us to select as cortical activations at a time sample those sources having a value significantly greater than the average in the epoch. Moreover, a sufficient number of brain activations were selected in this way, which guarantees a correct estimation of the 4D fractal dimension value obtained through the box-counting algorithm.

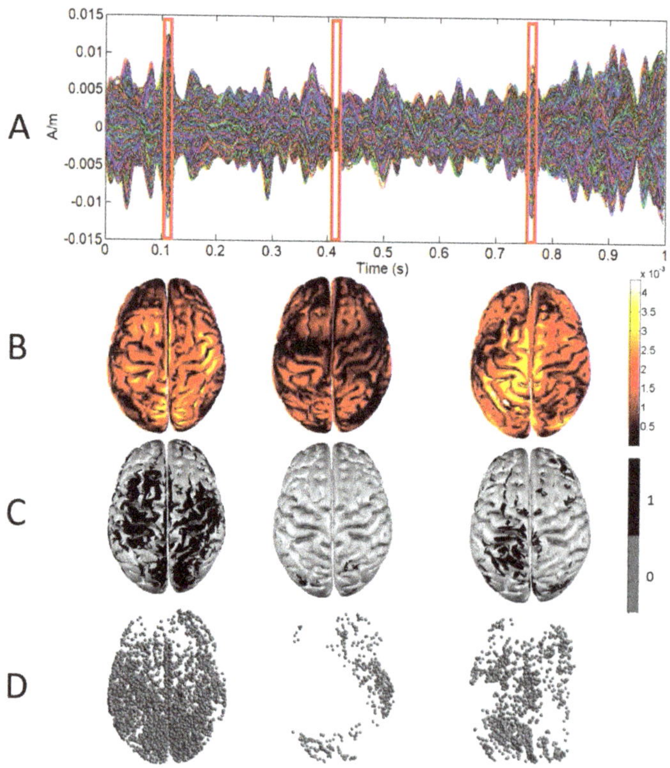

Figure 1. (**A**) Sources activity for one second. (**B**) 3D representation of sources activity at three different samples (red boxes). (**C**) 3D representation after binarization. (**D**) Point cloud defined by sources after binarization.

Once the 3D point clouds for all the time samples in the 120 s epoch had been computed, the epoch was divided using a sliding window of 1 s without overlapping. Then, the box-counting algorithm was used to obtain the 4DFD value for the 4D set defined in each window by putting together the 3D point clouds of the time samples for the corresponding second. In this way, a total of 120 4DFD values were obtained for each epoch. Figure 2 shows an example of the process of computing the FD of the brain activations using the box-counting algorithm. For the sake of clarity, the example shown in Figure 2 is presented in three dimensions. For the 4D case, the algorithm performed similarly with the same parameters, but having as input the 4D point clouds defined by all the 3D point clouds contained in one second.

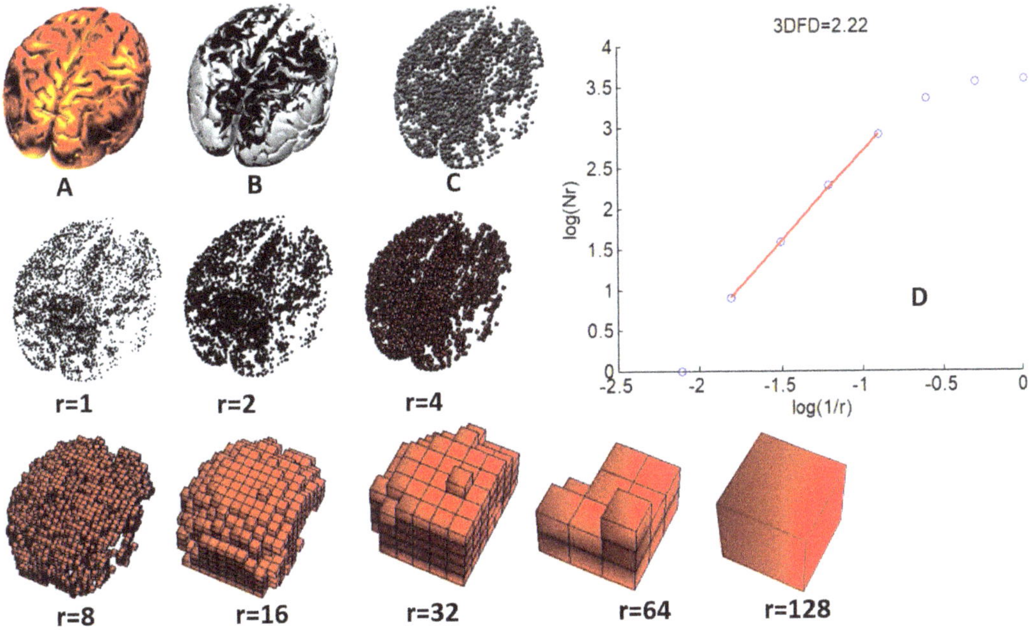

Figure 2. (**A**) Sources activity at a time sample. (**B**) Sources after binarization. (**C**) Point cloud defined by sources binarization. (**D**) 3DFD computation through the box-counting algorithm: log-log plot of number of boxes (N(r)) vs size (1/r) for voxelizations of sizes from r = 1 to r = 128. 3DFD computed as the slope of the regression line for box sizes from r = 8 to r = 64.

Computing the box-counting algorithm in 4D is a highly time-consuming task, so we used the CUDA/C++ source code of the GPU-optimized parallel version of the algorithm provided in [26]. Binarizing brain activations and obtaining the 3D point cloud representation from Brainstorm matrices containing source modelling were computed using home-made MATLAB scripts.

2.4. Statistical Analysis

Demographic variables were compared between groups using the nonparametric Mann–Whitney *U* test [27] (age and years of school) and Chi-squared test [28] (sex). In order to study differences in mean values of 4DFD and distributions of 4DFD over time between the PD and HC groups, nonparametric Mann–Whitney *U* tests were performed. Correlations between 4DFD and neuropsychological scores in the PD group were assessed using the nonparametric Spearman's rank correlation test (ρ) [29]. Since several variables were compared simultaneously, the Spearman's correlation test was configured using a Bonferroni post-hoc correction for multiple comparisons. In order to measure the performance of 4DFD as a classifier, we used a receiver operating characteristic (ROC) curve analysis [30]. In this analysis, the area under the ROC curve (AUC) was used to measure the classification accuracy of 4DFD.

Results of statistical tests were considered significant when the p-value obtained was below 0.05. All statistical analyses were performed in IBM SPSS 28.

3. Results

3.1. Demographic Results

The demographic and baseline clinical information are summarized in Table 1. Mann–Whitney and Chi-squared tests comparing demographic variables found no significant differences between the PD and HC groups in sex, age, and years of school (see Table 1). These results confirmed that both groups were matched regarding sex, age, and education.

3.2. 4DFD Comparison between HC and PD

Comparison between HC and PD groups based on 4DFD is shown in Figure 3. For each subject, the mean of the 120 4DFD values computed in the corresponding epoch was considered as a single 4DFD value. Results revealed that 4DFD was significantly higher in the PD group ($U = 393$, $p < 0.001$).

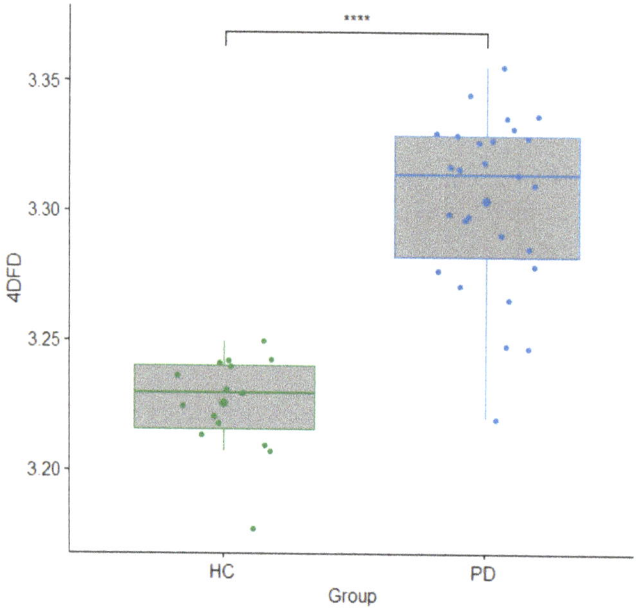

Figure 3. Boxplot with differences in 4DFD between the HC and PD groups. Each 4DFD value corresponds to the average of the 120 4DFD values computed for each subject, one 4DFD for each second in the epoch. p-value for Mann–Whitney U test (****: $p < 0.001$).

Six patients in the PD group exhibited tremor-related symptoms, so we conducted additional analyses in order to check whether the 4DFD increase in the PD group could be affected by this factor. Our analyses revealed that there was no significant difference in 4DFD values between PD patients with tremor and those without ($U = 87$, $p = 0.239$). Moreover, significant differences in 4DFD values were also found when comparing healthy subjects and PD patients without tremor ($U = 303$, $p < 0.001$).

We also compared the evolution of the 4DFD for the 120 s analyzed in each epoch (see Figure 4). Each point in Figure 4 represents the average of the 4DFD values for all the subjects of the group in the corresponding second. Each point is accompanied by the error bars showing the standard deviation of the mean. Results showed that the distribution of 4DFD values over time is clearly different between groups, with higher values for the PD group ($U = 14{,}400$, $p = 0.000$).

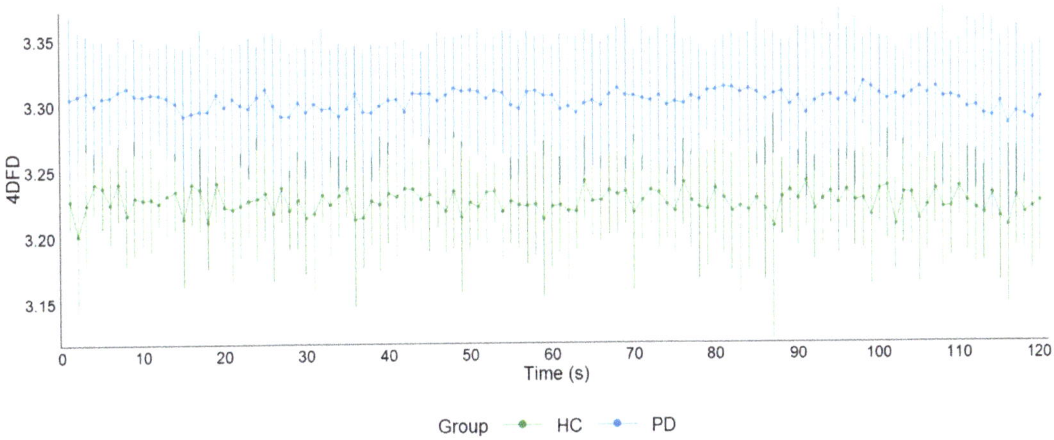

Figure 4. Average 4DFD evolution in 120 s for the PD and HC groups. Each point shows the average value of all the 4DFD computed for that second in the corresponding group. Error bars with standard deviation for each point are also shown.

3.3. 4DFD as Classifier for PD

In order to assess the performance of 4DFD as a diagnostic classifier for PD we used an ROC curve analysis (see Figure 5). In that analysis an AUC of 0.970 was achieved, meaning that 4DFD is an excellent classifier.

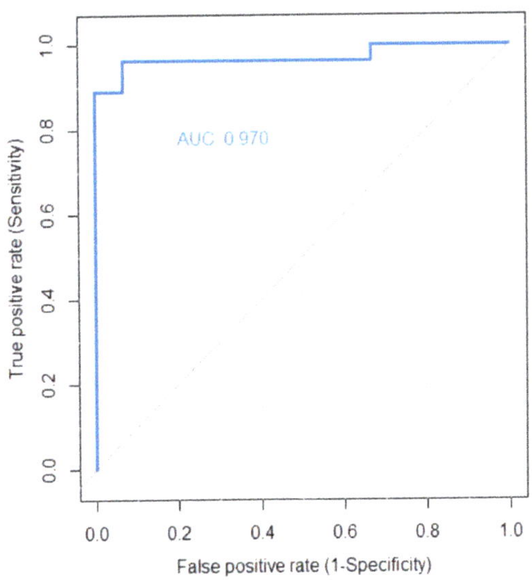

Figure 5. ROC curve analysis evaluating the performance of 4DFD as a classifier for PD. The area under the ROC curve (AUC) is 0.970.

3.4. Correlations between 4DFD and Motor and Neuropsychological Scores

Finally, in order to analyze the relationship between neuropsychological scores and 4DFD in the PD group, we calculated the Spearman correlation coefficient between all these variables with a post-hoc Bonferroni correction for multiple comparisons (see Figure 6). No significant correlations were found between 4DFD and scores of global cognitive impairments (as quantified by MMP and MoCA), but also with motor disability (UPDRS III and Hoehn and Yahr). Additionally, we performed a data reduction of the neuropsychological scores by using a principal component analysis to identify those factors which preserved the maximum variance in the data. The first principal component explained 90.1% of the variance, so we used the score of this first principal component for each PD subject as a new variable to compare. However, the Spearman correlation coefficient between the scores of the first principal component and the 4DFD measure was not significant either ($\rho = -0.118$, $p = 0.55$).

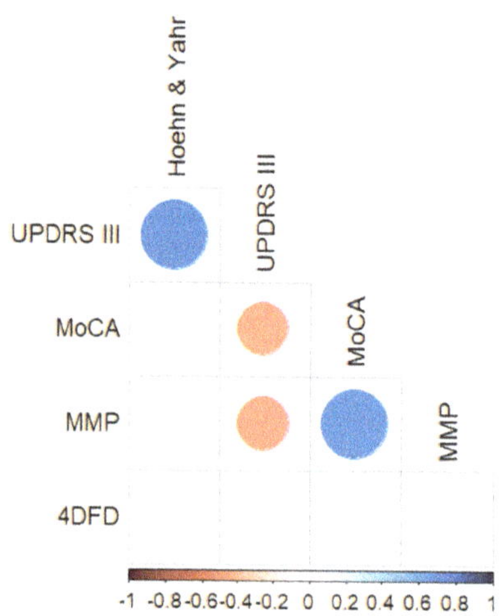

Figure 6. Significant Spearman correlations between neuropsychological scores and 4DFD for the PD group, corrected for multiple comparisons with the Bonferroni post-hoc method. Only statistically significant correlations ($p < 0.05$) are shown. UPDRS III: Unified Parkinson's Disease Rating Scale part III; MoCA: Montreal Cognitive Assessment; MMP: Mini-Mental Parkinson.

4. Discussion and Conclusions

In the last few years, several studies have proposed complexity measures of EEG as potential neurophysiological markers of early neurodegeneration in PD. These complexity measures were mainly related to the concept of entropy [11–14,31–33]. In the present paper we have also studied the complexity of EEG in PD, but our approach was based on analyzing the fractal structure of the brain activity evolution over time through the 4DFD, a measure which is sensitive to the dispersion of cortical activations.

We found that 4DFD was significantly higher in the PD group compared to the HC (see Figures 3 and 4), suggesting that the brain activity in patients suffering from PD presents a distribution over time more complex than in healthy subjects. Hence, with this study we verified that brain dynamics are modified in PD without dementia, and that the related brain activations over time occupy the 4D space in a more complex way than for the case of healthy subjects. A similar pattern of higher complexity in PD compared to healthy subjects has been observed in the majority of prior studies which analyzed complexity in terms of several different measures of entropy computed directly from the EEG signal [11–14,33]. This finding of increased complexity has been interpreted as reflecting a state of decreased organization which in turn leads to a decrease in information flow as a result of early alterations in cortical functioning and information processing.

Nevertheless, our approach based on the fractal dimension adds a spatial component which is not taken into account in those previous studies based on entropy. Roughly speaking, the higher the fractal dimension of a shape is, the more complex the object covers the space in which it is defined. In our study, the shape being analyzed was the 4D point cloud defined by the cortical brain activations over time. That means that the dynamics of brain activations in PD patients are distributed in a more complex (meaning more compact, complete, or widespread) way throughout the brain over time than for healthy controls. A possible explanation for this finding is that PD patients show an early impairment in their ability to regulate the flow of information and to allocate attention to filtering stimuli [34]. Therefore, an impairment of attentional processing may imply a more diffuse and changing distribution of brain activations in PD, as revealed by 4DFD. This increase in complexity may also reflect early compensatory mechanisms employed at the cortical level to maintain functionality despite the loss of dopamine-producing neurons. Additionally, the increase in fractal dimension could also be related to the presence of abnormal oscillatory patterns and altered network dynamics observed in Parkinson's patients. Therefore, further studies on complexity analysis of the specific brain regions associated with functional networks are needed. Two different strategies could be adopted: (1) following a similar 4DFD approach but applied only in those brain regions most affected by PD [35]; and (2) analyzing directly the functional networks using measures of fractal dimension of complex networks [36]. Our future efforts will focus on these directions and try to integrate neurophysiological with additional data types and neuroimaging findings. Moving forward, we recognize the importance of incorporating a broader spectrum of physiological and neuroimaging markers. Alongside our exploration of neurophysiological indicators, we see great potential in integrating cardiac data, such as EKG, into our analysis, and further enriching it with neuroimaging findings. This would provide a more holistic view of the physiological changes associated with PD, possibly unveiling meaningful correlations between the heart, brain, and structural changes as detected by neuroimaging.

There are also some other studies where PD patients presented reduced complexity of EEG compared to control subjects, such as the study of Yi et al. [31] and the one from Keller and colleagues [32]. In the first case, it was found that early-stage PD patients presented a lower value of permutation entropy of resting-state EEG than healthy controls. Similarly, Keller's study found a decreased complexity of resting-state EEG in the PD group, measured through Tsallis entropy, both compared to healthy subjects at baseline and after three years of cognitive decline. The most recent study that we found in analyzing the complexity of EEG in PD was the one presented by Pappalettera et al. [33]. Again, the results of this study showed that PD patients presented significantly higher values of entropy (approximate entropy) than elderly healthy controls when analyzing resting-state EEG signals. The variability in these findings could be attributed to the relatively few studies that concentrate on the significance of different complexity metrics in predicting early cognitive impairments, as well as the inclusion of PD patients who do not have dementia but already present pre-morbid minimal cognitive impairment (MCI) profiles.

4DFD also performed very well as a classifier (see Figure 5), separating PD and HC groups with very few exceptions. This means that 4DFD could be used as one of the features in advanced deep learning and machine learning methods for detecting PD [37,38].

No significant correlations were found between 4DFD and the severity of motor impairment in the PD group, nor were any correlations found with overall cognitive impairment scores. Thus, based on these findings, the spatio-temporal fractal dimension does not appear to be a suitable descriptor for the progression of motor and cognitive symptoms in PD. We hypothesize that, although 4DFD can detect general changes in brain dynamics between PD patients and healthy subjects, the subtle progressive neurodegeneration in PD may not be adequately captured by this whole-brain complexity measure. This could be attributed to the fact that the patients do not exhibit major cognitive impairments. Targeted follow-up studies could help clarify the role of fractal dimension as a predictor of progression into MCI and dementia.

Regarding previous studies using fractal analysis for assessing the complexity of EEG signal in PD, to the best of our knowledge there is only one study, the one presented by Muller and collaborators [11], based on correlation dimension. This previous study showed the ability of correlation dimension in differentiating PD and HC when subjects were performing motor tasks. However, no significant differences between groups were found for the resting-state EEG signal. Therefore, our approach based on 4DFD adds a new capacity of fractal analysis for differentiating PD from HC from resting-state EEG.

One requirement of the FD computation based on the box-counting algorithm is the necessity of binarizing the signal in order to construct the boxes at different scales [26]. This process of binarization naturally implies an information loss, which could affect more clearly in the case of using resting-state EEG signals instead of processing event or stimulus evoked potentials, where it is easier to select the brain activations based on significant changes in the signal due to the event or stimulus [7]. Therefore, further studies should be conducted using alternative methods for computing the fractal dimension which avoid the binarization of the signal. The differential box-counting algorithm (DBC) [39] is one of these methods, allowing the computation of the fractal dimension of gray-scale images. Our future efforts will focus on adapting the DBC algorithm to be used with the spatio-temporal 4D structures obtained from the resting-state EEG signal. In future research it may also be intriguing to investigate the potential of fractal dimension as a biomarker of minimal cognitive impairment, for example, by monitoring the cognitive progression of the disease and evaluating patients through longitudinal EEG recordings obtained at follow-up.

While our study has provided valuable insights into the complexity of resting-state EEG signals in PD patients, we acknowledge that our analysis of brain dynamics was confined to the resting-state. Future research would benefit from an extended analysis that includes active-state EEG, possibly during specifically designed cognitive tasks. Such an approach could provide insights into brain activity dynamics and complexity in PD and help unravel whether the observed changes in complexity are state-dependent and to what extent they are influenced by engagement in cognitive tasks.

In conclusion, in this study we have presented a novel methodology for performing the complexity analysis of resting-state EEG signals in PD. Our method is based on computing the 4DFD of the brain activations induced from the EEG signal, allowing us to perform in this way a spatio-temporal analysis. This complexity analysis based on the 4DFD of resting-state EEG obtained excellent performance in differentiating PD patients from healthy subjects. Our findings indicate that 4DFD is a promising tool for gaining deeper insights into the early cortical neurodegeneration characteristics of PD and to better plan rehabilitation treatment.

Author Contributions: Conceptualization, J.R.d.M. and A.C.; methodology, J.R.d.M., C.-C.D., A.M., A.C., M.M. and F.L.S.; software, J.R.d.M.; validation, J.R.d.M., C.-C.D. and A.C.; formal analysis, J.R.d.M.; investigation, J.R.d.M., C.-C.D., M.M., F.L.S., P.A., A.S., T.A. and A.C.; resources, J.N. and A.C.; writing—original draft preparation, J.R.d.M., C.-C.D. and A.C.; writing—review and editing, all authors; supervision, J.N. and A.C.; project administration, J.N. and A.C.; funding acquisition, J.R.d.M., J.N. and A.C. All authors have read and agreed to the published version of the manuscript.

Funding: This work is part of the research project PID2019-105145RB-I00 supported by the Spanish Government (MCIN/AEI/10.13039/501100011033). This work was also supported by the Italian Ministry of Health—(Ricerca Corrente 2022–2024). A.C. is also a PhD student enrolled in the National PhD in Artificial Intelligence, XXXVII cycle, course on Health and life sciences, organized by Università Campus Bio-Medico di Roma.

Institutional Review Board Statement: The study was conducted in accordance with the Declaration of Helsinki, and approved by Ethics Committee of IRCCS Fondazione Don Carlo Gnocchi (Italy) (protocol code 04_09/12/2020, approved in date 18 December 2020).

Data Availability Statement: The data presented in this study are available on request to fsaibene@dongnocchi.it. The data are not publicly available due to privacy restrictions.

Conflicts of Interest: The authors declare no conflict of interest.

References

1. Mandelbrot, B.B. *The Fractal Geometry of Nature*; WH Freeman Company: New York, NY, USA, 1983; Volume 51, ISBN 0716711869.
2. Ziukelis, E.T.; Mak, E.; Dounavi, M.E.; Su, L.; O'Brien, J.T. Fractal dimension of the brain in neurodegenerative disease and dementia: A systematic review. *Ageing Res. Rev.* **2022**, *79*, 101651. [CrossRef]
3. Lau, Z.J.; Pham, T.; Chen, S.H.A.; Makowski, D. Brain entropy, fractal dimensions and predictability: A review of complexity measures for EEG in healthy and neuropsychiatric populations. *Eur. J. Neurosci.* **2022**, *56*, 5047–5069. [CrossRef]
4. Grassberger, P.; Procaccia, I. Measuring the strangeness of strange attractors. *Phys. D Nonlinear Phenom.* **1983**, *9*, 189–208. [CrossRef]
5. Higuchi, T. Approach to an irregular time series on the basis of the fractal theory. *Phys. D Nonlinear Phenom.* **1988**, *31*, 277–283. [CrossRef]
6. Richman, J.S.; Moorman, J.R. Physiological time-series analysis using approximate entropy and sample entropy. *Am. J. Physiol. Circ. Physiol.* **2000**, *278*, H2039–H2049. [CrossRef] [PubMed]
7. Ruiz de Miras, J.; Soler, F.; Iglesias-Parro, S.; Ibáñez-Molina, A.J.; Casali, A.G.; Laureys, S.; Massimini, M.; Esteban, F.J.; Navas, J.; Langa, J.A. Fractal dimension analysis of states of consciousness and unconsciousness using transcranial magnetic stimulation. *Comput. Methods Programs Biomed.* **2019**, *175*, 129–137. [CrossRef] [PubMed]
8. Valls-Solé, J.; Valldeoriola, F. Neurophysiological correlate of clinical signs in Parkinson's disease. *Clin. Neurophysiol.* **2002**, *113*, 792–805. [CrossRef] [PubMed]
9. Jankovic, J. Parkinson's disease: Clinical features and diagnosis. *J. Neurol. Neurosurg. Psychiatry* **2008**, *79*, 368–376. [CrossRef]
10. Kalia, L.V.; Lang, A.E. Parkinson's disease. *Lancet* **2015**, *386*, 896–912. [CrossRef]
11. Müller, V.; Lutzenberger, W.; Pulvermüller, F.; Mohr, B.; Birbaumer, N. Investigation of brain dynamics in Parkinson's disease by methods derived from nonlinear dynamics. *Exp. Brain Res.* **2001**, *136*, 103–110. [CrossRef]
12. Pezard, L.; Jech, R.; Růžička, E. Investigation of non-linear properties of multichannel EEG in the early stages of Parkinson's disease. *Clin. Neurophysiol.* **2001**, *112*, 38–45. [CrossRef]
13. Chung, C.C.; Kang, J.H.; Yuan, R.Y.; Wu, D.; Chen, C.C.; Chi, N.F.; Chen, P.C.; Hu, C.J. Multiscale entropy analysis of electroencephalography during sleep in patients with parkinson disease. *Clin. EEG Neurosci.* **2013**, *44*, 221–226. [CrossRef]
14. Han, C.X.; Wang, J.; Yi, G.S.; Che, Y.Q. Investigation of EEG abnormalities in the early stage of Parkinson's disease. *Cogn. Neurodyn.* **2013**, *7*, 351–359. [CrossRef]
15. Postuma, R.B.; Berg, D.; Stern, M.; Poewe, W.; Olanow, C.W.; Oertel, W.; Obeso, J.; Marek, K.; Litvan, I.; Lang, A.E.; et al. MDS clinical diagnostic criteria for Parkinson's disease. *Mov. Disord.* **2015**, *30*, 1591–1601. [CrossRef] [PubMed]
16. Boel, J.A.; de Bie, R.M.A.; Schmand, B.A.; Dalrymple-Alford, J.C.; Marras, C.; Adler, C.H.; Goldman, J.G.; Tröster, A.I.; Burn, D.J.; Litvan, I.; et al. Level I PD-MCI Using Global Cognitive Tests and the Risk for Parkinson's Disease Dementia. *Mov. Disord. Clin. Pract.* **2022**, *9*, 479–483. [CrossRef]
17. Dubois, B.; Burn, D.; Goetz, C.; Aarsland, D.; Brown, R.G.; Broe, G.A.; Dickson, D.; Duyckaerts, C.; Cummings, J.; Gauthier, S.; et al. Diagnostic procedures for Parkinson's disease dementia: Recommendations from the movement disorder society task force. *Mov. Disord.* **2007**, *22*, 2314–2324. [CrossRef] [PubMed]
18. Goetz, C.G.; Tilley, B.C.; Shaftman, S.R.; Stebbins, G.T.; Fahn, S.; Martinez-Martin, P.; Poewe, W.; Sampaio, C.; Stern, M.B.; Dodel, R.; et al. Movement Disorder Society-sponsored revision of the Unified Parkinson's Disease Rating Scale (MDS-UPDRS): Scale presentation and clinimetric testing results. *Mov. Disord.* **2008**, *23*, 2129–2170. [CrossRef]
19. Hoehn, M.M.; Yahr, M.D. Parkinsonism. *Neurology* **1967**, *17*, 427. [CrossRef]
20. Delorme, A.; Makeig, S. EEGLAB: An open source toolbox for analysis of single-trial EEG dynamics including independent component analysis. *J. Neurosci. Methods* **2004**, *134*, 9–21. [CrossRef] [PubMed]
21. Tadel, F.; Baillet, S.; Mosher, J.C.; Pantazis, D.; Leahy, R.M. Brainstorm: A user-friendly application for MEG/EEG analysis. *Comput. Intell. Neurosci.* **2011**, *2011*, 1–13. [CrossRef]
22. Gramfort, A.; Papadopoulo, T.; Olivi, E.; Clerc, M. OpenMEEG: Opensource software for quasistatic bioelectromagnetics. *Biomed. Eng. Online* **2010**, *9*, 45. [CrossRef] [PubMed]
23. Fonov, V.; Evans, A.; McKinstry, R.; Almli, C.; Collins, D. Unbiased nonlinear average age-appropriate brain templates from birth to adulthood. *Neuroimage* **2009**, *47* (Suppl. S1). [CrossRef]
24. Pascual-Marqui, R.D. Standardized low-resolution brain electromagnetic tomography (sLORETA): Technical details. *Methods Find. Exp. Clin. Pharmacol.* **2002**, *24* (Suppl. D), 5–12. [PubMed]
25. Russell, D.A.; Hanson, J.D.; Ott, E. Dimension of strange attractors. *Phys. Rev. Lett.* **1980**, *45*, 1175–1178. [CrossRef]
26. Ruiz de Miras, J.; Posadas, M.A.; Ibáñez-Molina, A.J.; Soriano, M.F.; Iglesias-Parro, S. Fast computation of fractal dimension for 2D, 3D and 4D data. *J. Comput. Sci.* **2023**, *66*, 101908. [CrossRef]
27. Mann, H.B.; Whitney, D.R. On a Test of Whether one of Two Random Variables is Stochastically Larger than the Other. *Ann. Math. Stat.* **1947**, *18*, 50–60. [CrossRef]
28. Pearson, K.X. On the criterion that a given system of deviations from the probable in the case of a correlated system of variables is such that it can be reasonably supposed to have arisen from random sampling. *Lond. Edinb. Dublin Philos. Mag. J. Sci.* **1900**, *50*, 157–175. [CrossRef]
29. Spearman, C. The Proof and Measurement of Association between Two Things. *Am. J. Psychol.* **1904**, *15*, 72–101. [CrossRef]

30. Hanley, J.A.; McNeil, B.J. The meaning and use of the area under a receiver operating characteristic (ROC) curve. *Radiology* **1982**, *143*, 29–36. [CrossRef]
31. Yi, G.S.; Wang, J.; Deng, B.; Wei, X. Le Complexity of resting-state EEG activity in the patients with early-stage Parkinson's disease. *Cogn. Neurodyn.* **2017**, *11*, 147–160. [CrossRef]
32. Keller, S.M.; Gschwandtner, U.; Meyer, A.; Chaturvedi, M.; Roth, V.; Fuhr, P. Cognitive decline in Parkinson's disease is associated with reduced complexity of EEG at baseline. *Brain Commun.* **2020**, *2*, fcaa207. [CrossRef] [PubMed]
33. Pappalettera, C.; Miraglia, F.; Cotelli, M.; Rossini, P.M.; Vecchio, F. Analysis of complexity in the EEG activity of Parkinson's disease patients by means of approximate entropy. *GeroScience* **2022**, *44*, 1599–1607. [CrossRef] [PubMed]
34. Watson, G.S.; Leverenz, J.B. Profile of Cognitive Impairment in Parkinson's Disease. *Brain Pathol.* **2010**, *20*, 640–645. [CrossRef] [PubMed]
35. Leviashvili, S.; Ezra, Y.; Droby, A.; Ding, H.; Groppa, S.; Mirelman, A.; Muthuraman, M.; Maidan, I. EEG-Based Mapping of Resting-State Functional Brain Networks in Patients with Parkinson's Disease. *Biomimetics* **2022**, *7*, 231. [CrossRef]
36. Nahli, F.; Paramonov, A.; Soliman, N.F.; Aleisa, H.N.; Alkanhel, R.; Muthanna, A.; Ateya, A.A. Novel Path Counting-Based Method for Fractal Dimension Estimation of the Ultra-Dense Networks. *Intell. Autom. Soft Comput.* **2022**, *36*, 561–572. [CrossRef]
37. Aljalal, M.; Aldosari, S.A.; Molinas, M.; AlSharabi, K.; Alturki, F.A. Detection of Parkinson's disease from EEG signals using discrete wavelet transform, different entropy measures, and machine learning techniques. *Sci. Rep.* **2022**, *12*, 22547. [CrossRef]
38. Zhang, R.; Jia, J.; Zhang, R. EEG analysis of Parkinson's disease using time–frequency analysis and deep learning. *Biomed. Signal Process. Control* **2022**, *78*, 103883. [CrossRef]
39. Jiang, W.; Liu, Y.; Wang, J.; Li, R.; Liu, X.; Zhang, J. Problems of the Grid Size Selection in Differential Box-Counting (DBC) Methods and an Improvement Strategy. *Entropy* **2022**, *24*, 977. [CrossRef]

entropy

MDPI

Article

A Resource-Efficient Multi-Entropy Fusion Method and Its Application for EEG-Based Emotion Recognition

Jiawen Li [1,2,3], Guanyuan Feng [1], Chen Ling [1], Ximing Ren [1], Xin Liu [4], Shuang Zhang [5,6], Leijun Wang [1], Yanmei Chen [1,*], Xianxian Zeng [1,7,8] and Rongjun Chen [1,8,*]

1 School of Computer Science, Guangdong Polytechnic Normal University, Guangzhou 510665, China; lijiawen@gpnu.edu.cn (J.L.); fengguanyuan@gpnu.edu.cn (G.F.); lingchen@gpnu.edu.cn (C.L.); renximing@gpnu.edu.cn (X.R.); wangleijun@gpnu.edu.cn (L.W.); zengxianxian@gpnu.edu.cn (X.Z.)
2 Hainan Provincial Key Laboratory of Sports and Health Promotion, Hainan Medical University, Haikou 571199, China
3 Key Laboratory of Emergency and Trauma of Ministry of Education, The First Affiliated Hospital of Hainan Medical University, Haikou 571199, China
4 Department of Electrical and Computer Engineering, University of Macau, Macau 999078, China; yb87445@um.edu.mo
5 School of Life Science and Technology, University of Electronic Science and Technology of China, Chengdu 610056, China; zhang.s@njtc.edu.cn
6 School of Artificial Intelligence, Neijiang Normal University, Neijiang 641004, China
7 Guangdong Provincial Key Laboratory of Big Data Computing, The Chinese University of Hong Kong, Shenzhen (CUHK-Shenzhen), Shenzhen 518172, China
8 Guangdong Provincial Key Laboratory of Intellectual Property and Big Data, Guangdong Polytechnic Normal University, Guangzhou 510665, China
* Correspondence: chenyanmei@gpnu.edu.cn (Y.C.); chenrongjun@gpnu.edu.cn (R.C.)

Abstract: Emotion recognition is an advanced technology for understanding human behavior and psychological states, with extensive applications for mental health monitoring, human–computer interaction, and affective computing. Based on electroencephalography (EEG), the biomedical signals naturally generated by the brain, this work proposes a resource-efficient multi-entropy fusion method for classifying emotional states. First, Discrete Wavelet Transform (DWT) is applied to extract five brain rhythms, i.e., delta, theta, alpha, beta, and gamma, from EEG signals, followed by the acquisition of multi-entropy features, including Spectral Entropy (PSDE), Singular Spectrum Entropy (SSE), Sample Entropy (SE), Fuzzy Entropy (FE), Approximation Entropy (AE), and Permutation Entropy (PE). Then, such entropies are fused into a matrix to represent complex and dynamic characteristics of EEG, denoted as the Brain Rhythm Entropy Matrix (BREM). Next, Dynamic Time Warping (DTW), Mutual Information (MI), the Spearman Correlation Coefficient (SCC), and the Jaccard Similarity Coefficient (JSC) are applied to measure the similarity between the unknown testing BREM data and positive/negative emotional samples for classification. Experiments were conducted using the DEAP dataset, aiming to find a suitable scheme regarding similarity measures, time windows, and input numbers of channel data. The results reveal that DTW yields the best performance in similarity measures with a 5 s window. In addition, the single-channel input mode outperforms the single-region mode. The proposed method achieves 84.62% and 82.48% accuracy in arousal and valence classification tasks, respectively, indicating its effectiveness in reducing data dimensionality and computational complexity while maintaining an accuracy of over 80%. Such performances are remarkable when considering limited data resources as a concern, which opens possibilities for an innovative entropy fusion method that can help to design portable EEG-based emotion-aware devices for daily usage.

https://doi.org/10.3390/e27010096

Keywords: electroencephalography (EEG); multi-entropy fusion; brain rhythms; single-channel; emotion recognition

1. Introduction

Emotions are fundamental components of human psychological processes that shape thinking patterns, behavior, and decision making and play a critical role in social interaction, learning, and work. As humans increasingly rely on intelligent systems, the advancement of emotion recognition technologies is poised to revolutionize human–computer interaction. Thus, integrating emotional intelligence into such systems can significantly improve the efficiency of human–machine communication [1]. Furthermore, gaining a deeper understanding of the intrinsic mechanisms of emotions is beneficial for mental health, as it facilitates the development of accurate emotion monitoring methods while paving new pathways for the prevention and treatment of mental disorders such as depression and anxiety [2]. While emotions can be detected through external cues like facial expressions, speech, and other body language, social norms, personal contexts, and habits often obscure actual expressions, which are current challenges for precise emotion recognition [3].

Usually, a model capable of quantifying emotions is first needed to achieve emotion recognition. On the one hand, a discrete model divides emotions into a series of distinct emotional entities, such as happiness, contentment, sadness, and anger. Nonetheless, there still needs to be an academic consensus on the exact number of these basic emotions [4]. In addition, discrete emotion models struggle to describe continuous changes in emotions. On the other hand, a continuous model defines emotions through continuous dimensions, such as arousal and valence, better representing the variations in emotions. Specifically, valence refers to the pleasantness of emotion, ranging from negative (e.g., sadness or anger) to positive (e.g., happiness or contentment), and arousal denotes the level of emotional excitement, ranging from low (e.g., calmness or relaxation) to high (e.g., excitement or tension) [5]. Second, less masked inputs like physiological signals are preferred to improve emotion recognition accuracy. For instance, electroencephalography (EEG), electrooculography (EOG), and electrocardiography (ECG) signals are naturally generated by the body's systems. They directly reflect emotional reactions and are less easily influenced by human factors [6]. Typically, EEG signals, which record electrophysiological signals of neuronal activity in the brain, objectively contain response behaviors [7]. They provide a clear description of emotional expression, as their high-temporal-resolution record changes at the millisecond level, which makes them appropriate for detecting instantaneous variations in response to emotions triggered by specific stimuli or situations, demonstrating their portability and cost-effectiveness for the design of portable devices in emotion recognition [8].

As for EEG-based emotion recognition, data pre-processing, feature extraction, and classification are vital steps, and performance evaluation is essential for methodological validation. Initially, data pre-processing is necessary, as EEG signals are highly susceptible to noise and interference from other physiological signals. To this end, Independent Component Analysis (ICA) can denoise and remove artifacts [9]. In addition, raw EEG signals are segmented into several slices of varying lengths to include short-term fluctuations that sense emotional changes [10]. Therefore, pre-processed EEG signals provide a foundation for subsequent feature extraction, which aims to quantify and express the overt and latent information in the EEG signals. In this regard, these features are typically divided into three categories: time-domain (e.g., mean, standard deviation, and first-order difference [11]), frequency-domain (e.g., power spectral density [12]), and nonlinear (e.g., asymmetry, autocorrelation, zero-crossing rate, and entropy [13]) features. In light of the intricate and multifaceted nature of EEG signals, integrating different features to encapsulate their inherent characteristics is often imperative [14]. Moreover, the advance of deep learning has enabled the development of automated feature extractors based on Convolutional Neural Networks (CNNs) and Long Short-term Memory (LSTM) [15]. Classification aims to categorize emotional states based on extracted features. Generally, classifiers can be divided into two categories: conventional machine learning classifiers and deep learning models. Among conventional classifiers, Support Vector Machine (SVM), Random Forest (RF), k-nearest Neighbors (kNN) [16], and XGBoost [17–19] are widely utilized. For instance, SVM is particularly effective in addressing nonlinear problems, especially when dealing with limited datasets. RF enhances classification stability by combining the outputs of multiple decision trees. kNN is a nonparametric algorithm that makes no assumptions about the underlying data distribution that benefits biomedical signals like EEG, as it exhibits complex, nonlinear patterns across different emotions. XGBoost, a gradient-boosting framework, has demonstrated strong performance in EEG-based emotion recognition tasks by leveraging its efficiency and ability to handle imbalanced data. On the other side, deep learning models such as CNN, LSTM, Temporal Convolutional Network (TCN), and Graph Convolutional Neural Network (GCNN), are well-suited for handling high-dimensional data [20]. While deep learning models typically outperform conventional classifiers on large datasets, they usually require more computational costs and training time. Finally, the performance of the features and classifiers is assessed in a comparative study, where the most commonly used metrics include accuracy, precision, recall, F1 score, and the Receiver Operating Characteristic (ROC) curve [21]. Thus, a comprehensive evaluation ensures that the classifier or model demonstrates adequate generalization and stability for emotion recognition.

Recent studies have been reported based on the above steps. For example, Trujillo et al. [11] extracted 1086 time-domain and frequency-domain features from EEG signals and assessed them through feature selection combined with various classifiers. They claimed that the RF classifier paired with Kernel Principal Component Analysis (KPCA) offered an accuracy of 93.20%. Yu et al. [12] downsampled, filtered, and designed a short time window to analyze the temporal dependence in EEG signals by combining an attentional mechanism with LSTM, which provided accuracies of 85.40% and 74.26% on the SEED dataset and SEED-IV datasets, respectively. Zong et al. [19] constructed a hybrid FCAN-XGBoost model for EEG-based emotion recognition. It achieved 95.26% and 94.05% accuracy on DEAP and DREAMER, respectively. Fernandes et al. [22] investigated time-domain and frequency-domain features, employing deep learning models and conventional machine learning classifiers. They found that combining differential entropy with Dynamic Graph Convolutional Neural Networks (DGCNNs) yielded an accuracy of 89.97% across various frequencies. In comparison, kNN only achieved an accuracy of 73.23%, indicating the advantages of deep learning models in optimizing classification performance, although they necessitated more prolonged processing procedures. Song et al. [23] proposed an EEG-based Variational Instance-Adaptive Graph (V-IAG) method to address individual differences and uncertainty in the dynamic relationship between brain regions. By extracting the energy features of each EEG channel and constructing a fractional variational-instance adaptive graph combined with a multi-layer multi-graph neural network model, they obtained accuracies of 92.82% and 93.09% in the classification of valence and arousal, respectively. In addition, they stated that emotions exhibited a complex dynamic relationship with individualized modeling and brain regions and that integrating this relationship with graph neural networks can effectively improve the accuracy of emotion recognition. García-Hernández et al. [24] applied a genetic algorithm to select an optimal subset from the 2548 features, then compared the performances utilizing kNN, RF, and an Artificial Neural Network (ANN), which achieved accuracies of 90.06%, 93.62%, and 95.87%, respectively. Such results reveal that feature optimization and methodological comparison can beneficially identify the best approaches for EEG-based emotion recognition. Finally, Padhmashree and Bhattacharyya [25] adopted Multivariate Variational Mode Decomposition (MVMD) to extract modulated oscillations from multi-channel EEG signals, where the time-frequency images were generated by combining Joint Instantaneous Amplitude (JIA) and Joint Instantaneous Frequency (JIF). These time-frequency images were fed into a Residual Network (ResNet)-18 model to extract valuable features, followed by a softmax layer for emotion recognition. This framework offered classification accuracies of 99.03% for arousal and 97.75% for valence, demonstrating its ability to extract complex EEG features for emotion recognition.

The above studies reported various techniques for EEG-based emotion recognition, each with strengths, but they also presented drawbacks, especially in the case of a data resource-efficient classification. Many deep learning models, such as DGCNN and ResNet-18, demand substantial computational resources, limiting their deployment in resource-constrained environments. In addition, their complexity may lead to overfitting, particularly with smaller datasets, and the required extensive hyperparameter tuning can make them less practical compared to more straightforward solutions. Moreover, several studies could have improved the interpretability of advanced models, complicating the understanding of their decision-making processes. In this regard, multi-entropy fusion yields more interpretable results, as the entropy feature originates from information theory and mainly represents the uncertainty or complexity of an EEG system. It is also less sensitive to artifacts and noise, allowing it to present small fluctuations in brain activity, which are beneficial for analyzing emotional states and enhancing practical applications in real-world scenarios. However, prior works have yet to fully consider the effective achievement of multi-entropy fusion under conditions of limited data resources. To this end, this work proposes a resource-efficient multi-entropy fusion method for classifying emotional states, the novelty of which is the similarity measures of the Brain Rhythm Entropy Matrix (BREM), which involves multi-entropy feature extraction and similarity measure-based classification to improve the accuracy and interpretability of emotion recognition. For a better illustration, the overall framework is depicted in Figure 1.

The first step is data pre-processing, which comprises two operations: filtering and slicing. Filtering eliminates potential interference from ECG and EOG signals [26]. Then, to further optimize the data size and adapt it for portable devices, this work thoroughly investigates a series of segments with varying time lengths (5 s, 10 s, 20 s, and 30 s) for the identification of a suitable time window in the proposed method.

The second step is feature extraction, which contains three phases: obtaining brain rhythms by Discrete Wavelet Transform (DWT), extracting multi-entropy features, and producing the BREM through data fusion. As for emotion processing, negative emotions are primarily processed in the subcortical nuclei, temporal lobe, temporoparietal junction area, and inferior frontal gyrus [27]. Similarly, hyperactivity in the left prefrontal lobe is generally associated with positive emotions, while hyperactivity in the right prefrontal lobe is linked to negative emotions [28]. Such behaviors suggest that the forebrain plays a pivotal role in emotion recognition. In addition, there is a strong correlation between emotional states and brain rhythms: delta (0.5–4 Hz), theta (4–8 Hz), alpha (8–13 Hz), beta (13–30 Hz), and gamma (above 30 Hz) [29]. For instance, delta power increases when the subject is in a highly relaxed or emotionally depressed state [30]. Although time-domain and frequency-domain features provide deep insights into emotional changes by analyzing brain rhythms and brain regions, they are insufficient for indicating the subtle complexity of emotional shifts [31]. Therefore, the proposed method adopts entropy as a quantitative characteristic that extracts six types—Spectral Entropy (PSDE), Singular Spectrum Entropy (SSE), Sample Entropy (SE), Fuzzy Entropy (FE), Approximation Entropy (AE), and Permutation Entropy (PE)—from five brain rhythms and fuses them into a 5 × 6 matrix on a single channel, denoted as the BREM, which offers highly interpretable representations of emotional variations by considering the rhythmic characteristics in the entropy scale.

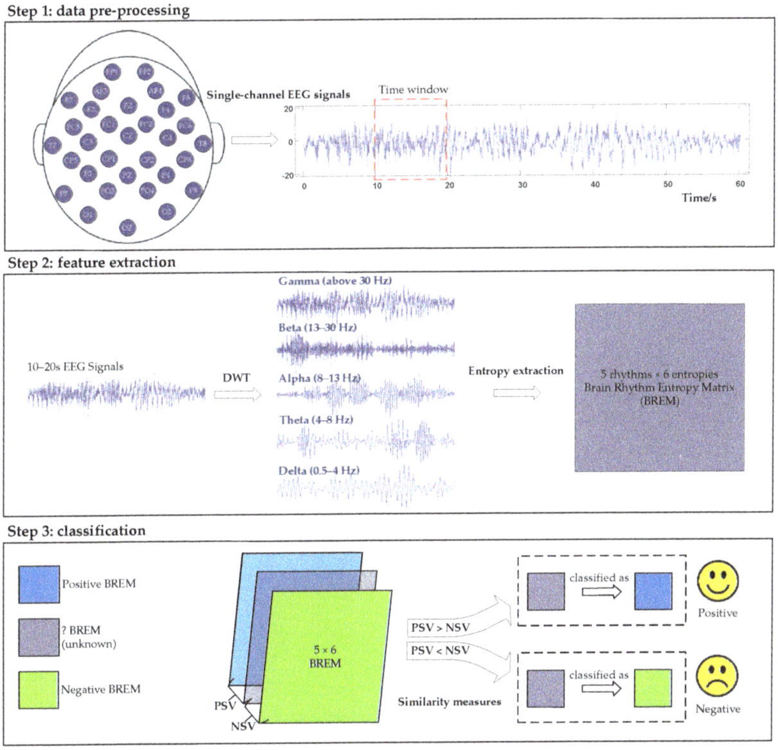

Figure 1. The overall framework of the proposed resource-efficient multi-entropy fusion method for EEG-based emotion recognition.

The third step is establishing a specialized classification method by applying the similarity measures of the BREM between the unknown testing data and positive/negative emotional samples. Specifically, a higher similarity indicates more significant information overlap, implying they are in the same category, while a lower similarity reveals they are less likely to be the same [32–34]. Since BREM includes the details regarding brain rhythms and entropies, similar BREMs show similar emotional states, providing an interpretable basis for EEG-based emotion recognition through the multi-entropy view. In addition, unlike other classification methods, such an approach does not rely on large amounts of data for training. Instead, it requires only positive and negative samples for similarity measures, making it more resource-efficient. Note that there are different methods for similarity measures, so selecting the appropriate one is vital for the classification task.

Finally, in the evaluation step, this work assesses the single-channel mode and the single-region mode for emotion recognition. The single-region mode is created by fusing the BREMs from the channels located in the same brain region, and the single-channel mode applies the BREM to one specific channel only. By evaluating the data inputs from a single channel or single region, the results provide an in-depth understanding of how the

BREM can be effectively improved for emotion recognition through the use of a few data resources to establish a resource-efficient multi-entropy fusion method. The contributions of this work are summarized as follows:

1. We propose the use of highly interpretable BREM data through a multi-entropy fusion approach to represent emotional EEG signals and the employment of a similarity-based classification approach to reduce the required sample size and the complexity of the training process.
2. We identify a suitable similarity measure method for classifying the BREMs from various emotional states, enhancing the performance of the proposed method.
3. We investigate the most suitable length (5 s, 10 s, 20 s, or 30 s) and the most resource-efficient data input mode (single-channel or single-region mode) to minimize data sources while maintaining an accuracy of over 80%.

The rest of this work is organized as follows: Section 2 describes the DEAP dataset and the valence–arousal emotional model. Subsequently, the details of the BREMs derived from multi-entropy fusion and the resource-efficient classification method through similarity measures are presented in Section 3. Section 4 shows the results and discussion, including statistical analysis, classification results, appropriate time-segment and channel results, a comparative study, and discussion. Finally, the conclusion and future work are contained in Section 5.

2. Experimental Dataset

The experimental data are from the DEAP dataset [35]. It contains EEG signals and corresponding emotion labels. The key information regarding the DEAP dataset is listed in Table 1. In total, 32 subjects (15 females and 17 males) participated in the experiment, with a mean age of 27.19 ± 4.45 years. Subjects were asked to watch 40 one-minute music videos with a Self-Assessment Manikin (SAM) to rate each music video according to their valence and arousal levels on a scale of 1–9. Moreover, different emotional states can be mapped into a coordinate system to form an emotional space consisting of valence and arousal, i.e., a valence–arousal model, where a valence score ≥ 5 indicates High Valence (HV) a score < 5 indicates Low Valence (LV). High Arousal (HA) and Low Arousal (LA) are similarly categorized based on score of 5 [36]. While watching music videos, EEG signals were recorded using a 32-channel system under a sampling frequency of 128 Hz. It is worth noting that EEG signals provided by the DEAP dataset were pre-processed with careful attention to minimizing noise, enabling their usage without artifact removal [37–39].

Table 1. Key information of the DEAP dataset.

Number of Subjects	32
Age	27.19 ± 4.45 years
Number of subjects	15 females + 17 males
Number of experimental trials per subject	40
Experimental stimuli	Music videos from YouTube
Time duration per trial	One minute
Number of EEG channels	32
Sampling frequency (Hz)	128

In addition, a balanced distributed sample is beneficial for methodological validation. The DEAP dataset exhibits a ratio of HA/HV (positive) to LA/LV (negative) approaching 1:1, reducing the risk of overfitting. Meanwhile, in order to investigate the resource-efficient data input mode based on a single channel or single region, the 32-channel system is categorized either by electrode positions, as shown in Figure 2a, or by brain regions in terms of anatomical distribution, as illustrated in Figure 2b. Here, each brain region is represented by a set of channels located within that region, with specific channels outlined in Table 2. Due to the low spatial resolution of electrodes, signals from a small surrounding area near the channels are also included in the regional representation.

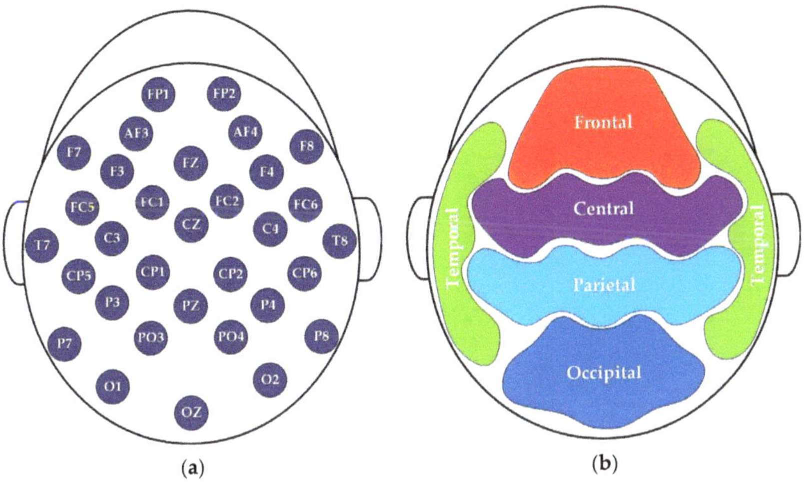

Figure 2. Channel and region locations in DEAP: (**a**) 32 EEG channels; (**b**) five brain regions.

Table 2. Specific EEG channels in each brain region.

Brain Region	EEG Channels
Frontal	FP1, FP2, AF3, AF4, F3, FZ, F4
Central	FC5, FC1, FC2, FC6, C3, CZ, C4
Parietal	CP5, CP1, CP2, CP6, P3, PZ, P4
Temporal	F7, F8, T7, T8, P7, P8
Occipital	PO3, PO4, O1, OZ, O2

3. Proposed Method

3.1. Feature Extraction

First, DWT is adopted to extract the brain rhythms from EEG. Unlike the commonly used transform approaches in signal processing, such as the Short-Time Fourier Transform (STFT), DWT offers multi-resolution analysis through time-frequency decomposition. This property allows DWT to adaptively focus on both high-frequency details and low-frequency trends in the signal, making it well-suited for the processing of EEG and sufficiently able to obtain local features at different time scales [40]. Furthermore, DWT's hierarchical decomposition enables it to preserve both time-domain and frequency-domain information, which is critical for analyzing the non-stationary nature of EEG. As a result, five brain rhythms, which correlate with diverse emotional states, are extracted. Specifically, DWT involves convolving the signal with a set of wavelet basis functions and extracting the rhythmic features from different frequency bands through multi-layer decomposition and reconstruction, as presented in (1):

$$X(j,k) = \sum_{n} x[n] 2^{j/2} \psi[2^j n - k] \quad (1)$$

where $X(j, k)$ denotes the wavelet coefficients, which consist of the low-frequency approximation coefficient (A_i) and the high-frequency detail coefficient (D_i); j represents the scale level; k is the displacement on scale j; and $x[n]$ represents the input EEG signals. Meanwhile, ψ represents the discretized wavelet basis function used in the analysis.

In this work, the Daubechies 4 (db4) wavelet is employed as the basis function for DWT due to its ability to balance time and frequency resolution well. Moreover, the compact support and symmetry of the db4 wavelet allow it to efficiently capture the non-stationary nature of EEG, making it well-suited for emotion recognition. An EEG signal from the DEAP dataset is applied to illustrate the DWT in the proposed method, with a sampling frequency of 128 Hz. Technically, emotional EEG signals can be decomposed and reconstructed through a four-level DWT to extract the five brain rhythms, as displayed in Figure 3.

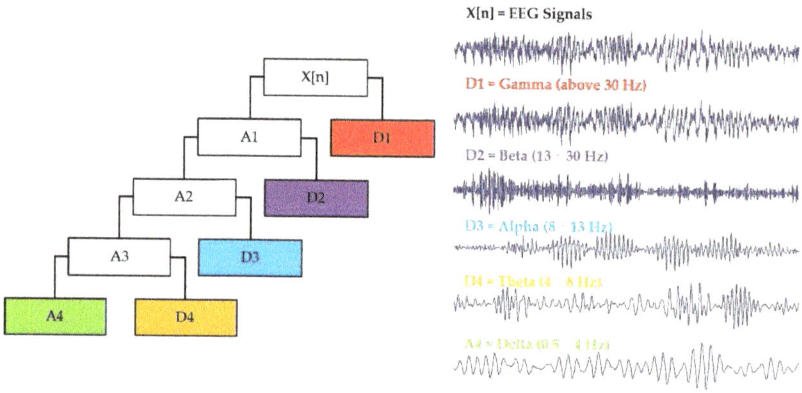

Figure 3. The 4-level DWT extracts five brain rhythms from emotional EEG signals.

Subsequently, entropy is viewed as a key, since it possesses the uncertainty, complexity, and regularity of EEG signals [41]. Mainly, with respect to multi-entropy fusion, PSDE, SSE, SE, FE, AE, and PE are extracted to indicate the dynamic characteristics of five brain rhythms comprehensively. Mathematically, their calculations are expressed as follows:

PSDE quantifies the uncertainty in the frequency distribution of a signal and helps analyze the complexity of rhythmic activities under various emotions [42]. It is derived from the Power Spectral Density (PSD), measuring how signal energy is distributed across different frequency components, as presented in (2):

$$PSDE = -\sum_{i=1}^{N} p(f_i) log(p + \varepsilon) \tag{2}$$

where $p(f_i)$ is the normalized PSD that uses Welch's method for energy metrics and ε is a value employed to avoid the $log(0)$ case.

SSE is based on a signal's Singular Value Decomposition (SVD). It mainly assesses the structural complexity of the time series rather than just its frequency content by examining the singular values obtained from the decomposition; in particular, it shows high sensitivity in processing brain activities during emotional states [43], as presented in (3):

$$SSE = \sum_{i=1}^{r} \widetilde{\sigma_i} log(\widetilde{\sigma_i} + \varepsilon) \tag{3}$$

where $\widetilde{\sigma_i}$ is the normalized singular value. If the singular value distribution is uniform, the signal is complex and has a larger value. On the contrary, if the signal is regular, it has a small value.

SE reflects the complexity and regularity of time-series signals by determining the likelihood that similar patterns will remain similar when the data are measured at different time points. It does not require data to be stationary, making it particularly suitable for physiological signals assessing complexity, like EEG [44]. In addition, it is sensitive to noise and outliers, as presented in (4):

$$SE(m,r) = -ln(\frac{\phi(m,r)}{\phi(m+1,r)}) \tag{4}$$

where m refers to the embedding dimension, r is the control threshold, and ϕ denotes the number of statistically similar vector pairs.

FE extends the concept of SE by incorporating fuzziness into the measure. It evaluates the uncertainty of the signal's state based on the degree of membership of patterns to sets. Thus, it not only captures subtle differences in patterns due to its fuzzy nature but also shows robustness to noise and variability in data [45], as presented in (5):

$$FE(m,n,r) = lim_{N\to\infty}[\phi^m(n,r) - \phi^{m+1}(n,r)] \tag{5}$$

where n represents the fuzzy weights.

AE assesses the randomness and variation patterns of time-series data by comparing sequences of points to detect patterns. It is sensitive to the choice of parameters such as the embedding dimension and tolerance [46], as shown in (6):

$$AE(m,r) = \lim_{N\to\infty}[\phi^m(r) - \phi^{m+1}(r)] \tag{6}$$

Lastly, *PE* evaluates the complexity of time series by analyzing the order of values rather than their actual magnitudes. It considers the patterns formed by the ranks of the data points and is robust to noise, so it effectively analyzes EEG signals that exhibit chaotic properties [47], as shown in (7):

$$PE(x) = -\sum_{j=1}^{k} P_j ln(P_j) \tag{7}$$

where P_j is the probability of occurrence of the j-th alignment pattern; meanwhile, the range of values of *PE* is [0, $\ln(N_p)$], with smaller values indicating a more structured signal and larger values referring to a more disordered signal.

3.2. Multi-Entropy Fusion

By integrating six extracted entropies, multi-entropy fusion offers a comprehensive understanding of changes in brain rhythms, particularly in the context of EEG-based emotion recognition. This approach leverages the strengths of each measure: SSE and PSDE complement each other by revealing the dynamic structure of EEG signals, capturing how brain activity evolves in the time scale; FE and AE delve deeper into the complexity and uncertainty of these signals, offering insights into the regularity and predictability of emotional responses; PE excels at identifying nonlinear features in EEG, which are often crucial for detecting subtle emotional states; and SE quantifies energy distribution across different frequency bands, showing a broad view of frequency-domain characteristics that are essential for assessing the stability and energy patterns of brain rhythms. This nuanced fusion not only enhances the understanding of EEG characteristics associated with emotional changes but also helps to identify specific brainwave patterns that correlate with distinct emotional states, making it a powerful solution in affective neuroscience. Thus, the proposed method incorporates feature-level fusion to generate the BREM.

Another aim is to investigate the single-channel mode and single-region mode for emotion recognition, resulting in variations in BREM generation for each mode. In the single-channel mode, BREM contains 30 features, representing a combination of five brain rhythms and six entropies for a single EEG channel. Therefore, the size in this mode is 5×6, as presented in (8). As for the single-region mode, the BREM is formed by fusing the channels within the same brain region. For example, regarding the occipital region, the BREMs from PO3, PO4, O1, O2, and OZ are combined, resulting in a larger matrix size of 5×30, as shown in (9). Following this approach, the BREMs of five brain regions are generated by fusing the corresponding EEG channels listed in Table 2, providing a comprehensive feature representation. This fusion strategy facilitates in-depth analysis of the channel number in emotion recognition by involving various BREMs, which helps to determine the appropriate data size for accomplishing a resource-efficient method.

$$BREM_{\text{single-channel}} = \begin{bmatrix} \gamma_{AE} & \gamma_{FE} & \gamma_{SE} & \gamma_{PE} & \gamma_{SSE} & \gamma_{PSDE} \\ \beta_{AE} & \beta_{FE} & \beta_{SE} & \beta_{PE} & \beta_{SSE} & \beta_{PSDE} \\ \alpha_{AE} & \alpha_{FE} & \alpha_{SE} & \alpha_{PE} & \alpha_{SSE} & \alpha_{PSDE} \\ \theta_{AE} & \theta_{FE} & \theta_{SE} & \theta_{PE} & \theta_{SSE} & \theta_{PSDE} \\ \delta_{AE} & \delta_{FE} & \delta_{SE} & \delta_{PE} & \delta_{SSE} & \delta_{PSDE} \end{bmatrix} \tag{8}$$

$$BREM_{\text{occipitalregion}} = \begin{bmatrix} BREM_{PO3} & BREM_{PO4} & BREM_{O1} & BREM_{O2} & BREM_{OZ} \end{bmatrix} \tag{9}$$

where γ, β, α, θ, and δ denote gamma, beta, alpha, theta, and delta, respectively.

3.3. Classification Method

Concerning biological sequences, the degree of similarity is a fundamental measure of the extent of resemblance between them. Inspired by this concept, this work puts forth a classification method based on the similarity measures of the BREM. For instance, unknown BREM data are compared with a positive BREM sample (HA/HV) and a negative sample (LA/LV) to determine similarity measures between them. If the unknown BREM exhibits a greater degree of similarity with the positive sample, this indicates that the testing data can be characterized as HA/HV. Then, four commonly used methods in this field, namely Dynamic Time Warping (DTW), Mutual Information (MI), the Spearman Correlation Coefficient (SCC), and the Jaccard Similarity Coefficient (JSC), are chosen to investigate in this work. The details are outlined as follows:

DTW is a nonlinear matching algorithm for comparing time-series signals that can perform dynamic alignment on the time axis to minimize the distance between two data points [48]. It mainly focuses on the shape of the time-series signals rather than their absolute values, which beneficially handle data of different lengths and temporal distortions, as shown in (10) and (11):

$$C(i,j) = d(A_i, B_j) + min\{C(i-1,j), C(i,j-1), C(i-1,j-1)\} \tag{10}$$

$$DTW(A,B) = C(n,m) \tag{11}$$

where A and B are two data point that are needed to calculate the similarity lengths of n and m; d is the local distance between A_i and B_j at points i and j, respectively; C is the cumulative distance matrix; and $C(n, m)$ is denoted as the value of the last element in the C matrix. The smaller the *DTW*, the higher the similarity between the BREMs.

MI measures the dependency between two data points, and it can demonstrate how much information they share [49]. Thus, by comparing the BREMs of different emotional states, the trend of entropy change can be indicated, as shown in (12):

$$MI(A;B) = H(A) + H(B) - H(A,B) \tag{12}$$

where H denotes the entropy of data points A and B and $H(A, B)$ represents the joint entropy between them.

SCC is a nonparametric measure of rank correlation that assesses how well a monotonic function can describe the relationship between two variables. It helpfully evaluates the degree to which the relationship between two variables can be expressed as a linear relationship after ranking the data [50], as shown in (13):

$$SCC = 1 - \frac{6\sum_{i=1}^{n}(A_i - B_i)^2}{n(n^2 - 1)} \tag{13}$$

Lastly, *JSC* is a measure of the similarity between two sets. It is commonly used to compare finite discrete sets by calculating the ratio of the intersection to the union of the sets, effectively identifying how much the two sets overlap [51]; its values range from 0 (no similarity) to 1 (entire similarity), as shown in (14):

$$JSC = 1 - \frac{|A \cap B|}{|A \cup B|} \tag{14}$$

To achieve a resource-efficient classification method using similarity, a template that serves as either a positive or negative sample is essential, functioning as a benchmark. Meanwhile, after thoroughly considering inter-individual variability and the differences in emotions elicited by watching music videos in the DEAP dataset, the Leave-One-Out Cross-Validation (LOOCV) approach is employed to choose the templates. For instance, for subject S1, out of 40 trials, 16 of them are LA, and 24 are HA. Thus, the BREMs from the first LA and the first HA trials are selected as positive and negative templates, respectively. In contrast, the remaining trials are applied as unknown testing data in the similarity measures to classify their emotional state as LA or HA. Next, the BREMs of the second LA and the first HA trials are selected as templates, followed by sequential cyclic analysis. Based on that, 16 × 24 evaluations are conducted, employing similarity measures per subject. In addition, as for evaluation metrics, the accuracy is defined as follows:

$$Accuracy = \frac{TN + TP}{TN + FN + TP + FP} \tag{15}$$

where *TP* refers to the correctly predicted positive examples, i.e., instances classified as HA or HV. *FP* represents the incorrectly predicted positive examples, i.e., HA or HV instances misclassified as LA or LV. *FN* denotes the incorrectly predicted negative examples, i.e., LA or LV instances misclassified as HA or HV. Finally, *TN* represents correctly predicted negative examples, i.e., instances correctly classified as LA or LV.

4. Results and Discussion

In this work, the experimental results of the proposed method are programmed by MATLAB. To facilitate reproducible research and positively affect the academic field, the codes are freely available at https://github.com/zyzc75/BREM-SIMILARITY.git (accessed on 18 January 2025).

4.1. Statistical Analysis

First, a qualitative analysis is performed to demonstrate the statistical significance of the selected entropy features in relation to emotional states. Specifically, Analysis of Variance (ANOVA) is a commonly used statistical verification method that evaluates differences in group means and variances. When the resulting p-value is equal to or smaller than the significance level (typically 0.05), the feature exhibits a significant difference concerning specific emotional variations. It is, therefore, suitable for classification tasks. Figure 4 presents two examples of ANOVA test box plots for the best entropy features that provide the highest accuracy from subject S1 in the DEAP dataset. In Figure 4a, the alpha-sample entropy (α_{SE}) in the P7 channel is selected for arousal classification. In contrast, in Figure 4b, the beta-singular spectrum entropy (β_{SSE}) in the O2 channel is chosen for valence classification. Their p-values are less than 0.05, so the statistical analysis indicates significant differences in the emotional categories.

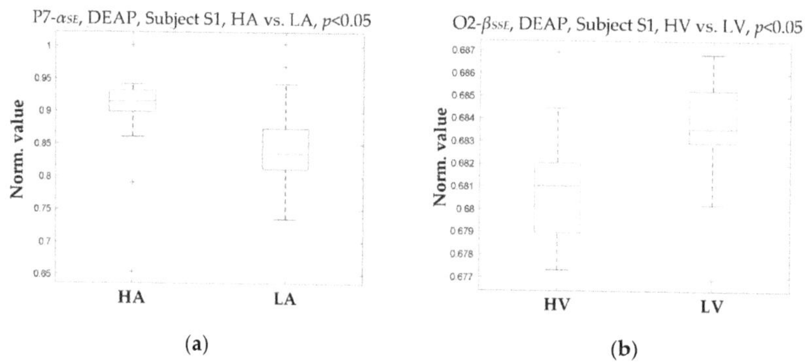

Figure 4. Two examples of ANOVA test box plots for the best entropy features that provide the highest accuracy from subject S1 in the DEAP dataset. (**a**) The alpha sample entropy (α_{SE}) in the P7 channel for arousal classification; (**b**) the beta singular-spectrum entropy (β_{SSE}) in the O2 channel for valence classification.

In Figure 4, higher values of P7-α_{SE} predominantly appear in the HA state for arousal classification. In comparison, lower values are found in the LA state, suggesting a positive correlation between arousal level and the presence of P7-α_{SE} in the BREM for subject S1. Similarly, the appearance of O2-β_{SSE} in the BREM for valence classification provides informative clues indicating valence levels. In this work, the BREM is not subjected to feature selection to retain the comprehensiveness of the entropy features while ensuring classification accuracy. Nonetheless, several features may not exhibit statistical significance for emotion recognition. Consequently, this statistical analysis only focuses on those entropy features selected based on specific channels with the highest accuracy rather than all extracted features.

Following this approach, ANOVA tests are conducted on the entropy features selected from each channel for all 32 subjects meeting the same conditions in the DEAP dataset. The overall results indicate that their p-values are all less than 0.05, showing trends similar to those of the two examples illustrated in Figure 4, demonstrating statistical significance. Such statistical results confirm the rationale for selecting valuable entropy features and provide optimal support for robust classification.

4.2. Classification Results

The mean ± standard deviation (%) of accuracy for arousal and valence classification employing different time windows, similarity measures, and data input modes are summarized in Table 3. The best results are underlined in bold.

Table 3. The classification results (mean ± standard deviation (%)) using different time windows, similarity measures, and data input modes.

Similarity Measure Method	Time Window (s)	Classification Accuracy (Mean ± Standard Deviation (%))			
		Single-Channel Mode		Single-Region Mode	
		Arousal	Valence	Arousal	Valence
DTW	30	80.92 ± 4.38	78.95 ± 3.27	78.29 ± 5.43	76.23 ± 3.81
	20	82.24 ± 4.28	79.85 ± 3.18	80.51 ± 5.08	78.54 ± 4.68
	10	83.06 ± 4.77	81.42 ± 3.40	80.92 ± 4.82	78.29 ± 3.41
	5	84.62 ± 4.39	**82.48 ± 2.88**	**83.06 ± 5.13**	**79.69 ± 2.93**
MI	30	81.25 ± 5.62	78.78 ± 3.89	79.36 ± 6.27	77.06 ± 3.68
	20	83.31 ± 4.80	79.36 ± 3.21	79.61 ± 5.71	76.97 ± 3.34
	10	83.72 ± 4.25	80.51 ± 2.40	81.42 ± 4.86	78.62 ± 3.39
	5	**84.79 ± 5.08**	81.99 ± 3.07	82.24 ± 4.87	79.61 ± 3.13
SCC	30	66.37 ± 12.45	62.34 ± 9.56	70.56 ± 13.17	68.42 ± 8.21
	20	65.87 ± 13.31	62.17 ± 8.71	70.64 ± 12.81	67.93 ± 8.06
	10	65.46 ± 14.38	62.58 ± 9.16	70.39 ± 13.33	68.09 ± 8.71
	5	68.75 ± 13.97	65.54 ± 7.33	71.96 ± 11.88	68.91 ± 8.41
JSC	30	59.38 ± 16.26	56.91 ± 9.70	59.38 ± 16.26	56.91 ± 9.70
	20	59.38 ± 16.26	56.91 ± 9.70	59.38 ± 16.26	56.91 ± 9.70
	10	59.38 ± 16.26	56.91 ± 9.70	59.38 ± 16.26	56.91 ± 9.70
	5	59.38 ± 16.26	56.91 ± 9.70	59.38 ± 16.26	56.91 ± 9.70

As shown in Table 3, SCC and JSC are less effective than the other methods, since the average accuracy of these two methods is below 72% and the standard deviation is around 10%. For example, JSC remains at 59.38 ± 16.26% and 56.91 ± 9.70% for 30 s time lengths for arousal and valence classification, respectively. The analysis demonstrates that, taking the arousal classification of subject S1 with 40 trials (16 LA and 24 HA) as an example, 15 LA and 23 HA samples need to be tested after selecting positive and negative samples. Employing JSC for similarity measures, all testing samples are misclassified as HA, resulting in a fixed classification accuracy of $23/(15 + 23) = 60.53\%$. This pattern is consistent across other subjects, so the mean and standard deviation remain fixed, regardless of changes in time windows, leading to performance failure. The failure of the JSC method is likely due to its limitations. JSC is suitable for measuring the degree of overlap between two sets, but the multi-entropy features in EEG signals are dynamic and complex. The changes under different emotional states are not simple set relationships or linear. As a result, it is not good enough to effectively capture such complex dynamic characteristics. In addition, the SCC mainly measures the rank correlation of two BREMs, i.e., their monotonic relationship. Nonetheless, emotional changes are often not monotonic, and EEG signals show nonlinear properties under different emotional states. Therefore, the SCC is improper for obtaining such changes in the BREM and only measures similarity according to the change in rank, making it ineffective at discerning the subtle differences between emotions, resulting in poorer performance.

On the other side, DTW adaptively addresses temporal data deformation. It effectively includes the nonlinear characteristics of change between disparate time segments in EEG signals, demonstrating high accuracy and stability in emotion recognition. MI reflects the relationship between EEG signals by quantifying the information shared between two variables, yielding promising results in emotion recognition. Consequently, nonlinear similarity measure methods, such as DTW and MI, are appropriate in this work. Overall, DTW shows stability in valence classification, as indicated by a smaller standard deviation. Meanwhile, it maintains balanced performance in arousal classification. Therefore, considering its accuracy and stability, DTW can be regarded as a suitable similarity measure method for the BREM to perform resource-efficient classification.

After settling the DTW, it is observed that as the time window progressively decreases, accuracy improves for both valence and arousal, accompanied by a reduction in the standard deviation. In particular, the performance of a 5 s time window is up to 80%, whether in a single channel or single region. To be precise, it improved by nearly 3% compared to a 30 s window and by almost 1% compared to both 10 s and 20 s windows. This improvement may be due to the instantaneous nature of emotional changes, which experience significant fluctuations over a relatively short period. On the other hand, a longer time window may contain more information but include additional irrelevant signals, which may reduce the prominence of these instantaneous fluctuations. As a result, a 5 s window is selected as the stationary segment for emotion recognition, consistent with previous work [52].

Next, the experimental results demonstrate that employing a single-channel input produces better outcomes than single-region input. Specifically, the arousal classification accuracy of the single-channel mode using DTW with a 5 s time window is 1.56% higher than that of the single-region mode. In comparison, the standard deviation is 0.74% lower. Similarly, in valence classification, the accuracy of the single-channel mode is 2.79% higher, with a standard deviation that is 0.05% lower. This enhancement is attributed to the single channel's ability to directly involve multi-entropy features associated with specific emotional dimensions, which helps to minimize potential interference and redundant information. Conversely, integrating multiple channels introduces complexity that may dilute the contribution of effective signals from individual channels. Although the difference is insignificant, single-channel data input reduces computational cost and model training time, accentuating the practical advantages of single-channel input in the proposed method, which is a primary concern when designing portable devices. Thus, the characteristics of the single-channel mode are the focus of the proposed method.

4.3. Appropriate Time-Segment and Channel Results

When selecting DTW as the similarity measure method for single-channel BREMs derived from a length of 5 s as input, the subsequent investigation concentrates on the appropriate time segment and channel for EEG-based emotion recognition. The results of 32 subjects from the DEAP dataset are listed in Table 4, based on which the best combination of each subject can be determined.

Table 4 reveals that the appropriate time segment and channel for arousal and valence classification vary among the subjects, implying individual differences in EEG responses to emotional stimuli. For arousal classification, the appropriate time segment is predominantly concentrated in the early stage of the experiment (0–20 s for 16 subjects) and the middle stage (20–40 s for 12 subjects), with the highest classification accuracy for S21, reaching 94.74% in 0–5 s. It also follows a similar trend for valence classification, where the highest classification accuracy of 89.47% appears in 20–25 s (S6) and 30–35 s (S23). In addition, Figure 5 depicts the statistical frequency of the optimal time segment for 32 subjects from the DEAP dataset. Such findings as those reported in Table 4 and Figure 5 align with the nature of the two emotional dimensions. Arousal reflects an individual's immediate response to emotional stimuli, which tends to be most prominent in the initial or middle periods [53]. In contrast, valence is associated with the individual's overall emotional evaluation and response, which may require more time for processing and assessment by the subject themselves [54].

Table 4. The appropriate time segments and channels of 32 subjects from the DEAP dataset for arousal and valence classifications.

Subject	Arousal			Valence		
	Channel	Time Segment (s)	Classification Accuracy (%)	Channel	Time Segment (s)	Classification Accuracy (%)
S1	T8	15–20	84.21	CP5	5–10	84.21
S2	FC5	30–35	81.58	O2	5–10	81.58
S3	PO4	10–15	89.47	FC2	20–25	84.21
S4	CP6	0–5	84.21	AF4	30–35	84.21
S5	PO3	5–10	81.58	P3	35–40	81.58
S6	CZ	20–25	81.58	F4	20–25	89.47
S7	PO3	15–20	84.21	FC6	5–10	86.84
S8	F4	15–20	81.58	P7	10–15	81.58
S9	CP1	5–10	78.95	F7	30–35	81.58
S10	F8	0–5	78.95	FC6	25–30	81.58
S11	P4	25–30	81.58	F3	5–10	81.58
S12	FC1	20–25	92.11	P3	10–15	78.95
S13	F4	0–5	92.11	PO3	0–5	81.58
S14	F7	5–10	86.84	AF3	0–5	81.58
S15	CP5	5–10	78.95	FC5	25–30	81.58
S16	P8	30–35	84.21	FC2	0–5	81.58
S17	C4	45–50	81.58	F8	50–55	81.58
S18	CP1	40–45	84.21	C4	0–5	81.58
S19	F3	20–25	84.21	T8	25–30	81.58
S20	P8	35–40	89.47	C3	10–15	81.58
S21	PO3	0–5	94.74	FZ	20–25	78.95
S22	FP1	15–20	84.21	CP1	40–45	81.58
S23	FZ	35–40	86.84	C3	30–35	89.47
S24	O2	25–30	92.11	PZ	35–40	81.58
S25	PO4	50–55	86.84	CP5	0–5	78.95
S26	FP1	50–55	84.21	F7	10–15	81.58
S27	CP1	0–5	81.58	F7	5–10	86.84
S28	F7	5–10	78.95	F4	20–25	81.58
S29	AF4	35–40	86.84	FC1	15–20	81.58
S30	PO3	35–40	78.95	FZ	50–55	86.84
S31	FC2	20–25	81.58	C4	20–25	84.21
S32	O2	0–5	89.47	F3	20–25	76.32

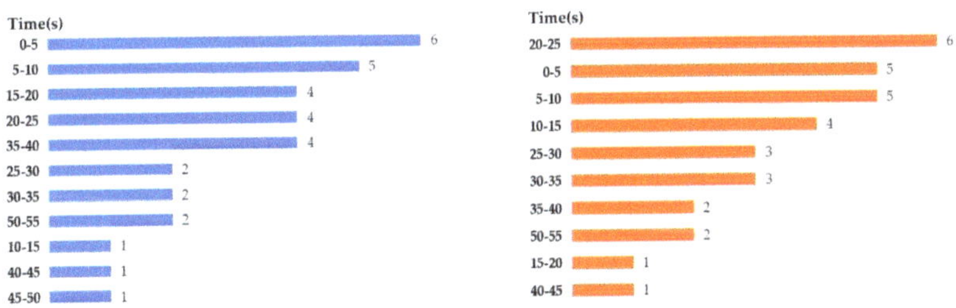

(**a**) (**b**)

Figure 5. Statistical frequency of the optimal time segment (s) for 32 subjects from the DEAP dataset: (**a**) arousal classification; (**b**) valence classification.

For better understanding, Figure 6 illustrates word clouds of the identified appropriate channels, indicating that most are located in the central, occipital, and frontal regions when conducting arousal and valence classifications. On the one hand, such regions are typically associated with sensory and attentional processing, given their role in arousal levels, which aligns with the function of the occipital region (responsible for visual processing), the central region (linked to motor control and sensory integration), and the frontal region (also involved in the cognitive evaluation of emotional stimuli, playing a central role in emotional regulation). Consequently, they facilitate the integration of the brain's immediate response to stimuli, contributing to effective arousal classification. On the other hand, regarding valence, the channel results exhibit slight differences, with higher occurrences of F7, C3, and C4, which aligns with the conclusion that the frontal region is associated with social behavior, decision making, and emotional control. Specifically, the central part of the brain is part of the motor cortex, which is involved in motor expression and is closely related to emotional expression, including facial expressions and body language. Moreover, the discovered channels are not identical among all subjects, indicating that cognitive processing and the cooperation of multiple brain regions are typically required to assess emotion. For example, the prefrontal cortex modulates activity in brain regions that produce emotions, with downregulation and upregulation enhancing activity in areas associated with emotional experiences, such as the amygdala and the prefrontal lobes [55]. The prefrontal lobe's location corresponds with that of F7, C3, and C4, situated on the left side of the frontal lobe and extending forward, implying a close relationship with the activity associated with downregulation and upregulation in emotions.

Figure 6. Word clouds of the representative channels for 32 subjects from the DEAP dataset: (**a**) arousal classification; (**b**) valence classification.

Based on the above observations, it is plausible to speculate that arousal classification relies more on brain regions associated with visual stimulus processing. In contrast, valence classification involves more excellent processing of social–emotional information and internal emotional regulation. Furthermore, activity in motor-related brain regions indicates a significant role of bodily expression in valence regulation. These findings provide vital insights for the further exploration of the application of single-channel EEG signals in emotion recognition, particularly regarding how specific emotional dimensions can be captured through various channels. Future research should explore the associations of such channels with other potential functional areas and their collaborative roles in the complex process of individual emotion recognition.

4.4. Comparative Study

To thoroughly investigate the benefits of the proposed method, a comparative study with recent works is performed, as outlined in Table 5. As seen, although the classification accuracies of the proposed method are somewhat lower than those of methods using multiple channels and complex deep learning networks, such as a fuzzy rank-based deep learning model [56], CNN [10,57], Multi-Layer Perceptron (MLP) [57], and LSTM [58,59], managing to deliver remarkable performance through single-channel data inputs, reducing dimensionality and computational cost. On the one hand, the CNN-based deep learning approach [57] achieves classification accuracies of 94.33% for arousal and 93.53% for valence, highlighting the potential of multi-channel configurations and deep learning networks in sophisticated emotion recognition. Other model-based networks like MLP [57] also display high accuracy, offering 94.25% for arousal and 93.39% for valence. On the other hand, conventional approaches, such as SVM [60], logistic regression [61], and Minimum Distance to Riemannian Mean (MDRM) [62], present comparatively lower performance. For instance, MDRM combined with Principal Component Analysis (PCA) [62] achieves only 57.42% for arousal and 64.06% for valence with 60 s data. Such disparities indicate the vital roles of feature selection and fusion in EEG-based emotion recognition. In this regard, the proposed method established by the BREM derived from multi-entropy fusion and similarity measures has been successfully implemented to recognize various emotions through a single-channel approach, emphasizing a balance between channel number and classification accuracy. Hence, such performances are more outstanding than those reported in previous studies, considering fewer data resources as a concern, facilitating the resource-efficient design of low-cost portable emotion-aware devices.

Table 5. Comparison with recent works.

	Time Window (s)	Number of Channel	Main Methodology	Classification Accuracy (%)	
				Arousal	Valence
Akhand et al. [10]	8	32	Connectivity feature map with CNN	91.66	91.29
Dhara et al. [56]	2	14	Fuzzy rank-based deep learning approach using Gompertz function	91.65	90.84
Kumar and Molinas [57]	1	32	Differential entropy with MLP	94.25	93.39
			Differential entropy with CNN	94.33	93.53
Gaddanakeri et al. [58]	60	14	Brain rhythms with LSTM (S1–S22)	82.40	78.28
			Brain rhythms with LSTM (S23–S32)	63.15	62.06
Singh et al. [59]	3	5	Grey Wolf Optimization (GWO) and LSTM with data augmentation	81.25	92.50
Jha et al. [60]	60	32	Brain rhythms with SVM	70.88	76.00
Pan et al. [61]	1	32	Logistic regression with Gaussian kernel and Laplacian prior	77.03	77.17
Al-Mashhadani et al. [62]	60	32	MDRM-PCA	57.42	64.06
This work	5	1	BREM from multi-entropy fusion and similarity measure by DTW	84.62	82.48

4.5. Discussion

First, the proposed method is validated and analyzed using the publicly available DEAP dataset in this work. The findings indicate that the classification accuracy gradually increases as the time length decreases. This trend is particularly evident when utilizing 5 s time segments, during which classification performance peaks. Such an observation suggests that shorter time segments help assess transient emotional responses. Longer time segments, despite containing more EEG signals, may include disturbances unrelated to the emotion, which could reduce the accuracy. The results from shorter time segments also exhibit lower standard deviations, revealing more stable outcomes for emotion recognition.

Second, when comparing various methods for measuring similarity levels between the BREMs derived from multi-entropy fusion, it becomes apparent that DTW is appropriate for this context, outperforming other similarity measures through its refined analytical capabilities. DTW excels in capturing nonlinear temporal variations, which are beneficial for identifying subtle emotional changes over time through the BREMs. Unlike SCC and JSC, which assume a more straightforward relationship between the signals, DTW can account for time shifts and distortions, making it highly effective for processing non-stationary biomedical signals such as EEG, where emotional states usually evolve dynamically.

Third, the investigation demonstrates that the inputs from individual channels achieve higher classification accuracy than those from the same region, reaffirming that individual channels can more directly capture EEG signals pertinent to emotions. Meanwhile, the appropriate time segments and channels for arousal and valence vary among subjects, reflecting the functional diversity of emotional dimensions across different brain parts. Thus, the proposed method provides valuable insights for choosing suitable time lengths and channels in EEG-based emotion recognition, laying the groundwork for future resource-efficient hardware designs.

Finally, this work exhibits several limitations despite demonstrating the advantage of multi-entropy fusion. The complexity and variability in emotional responses indicate that individual differences are significant. While several subjects exhibit better classification results with short time segments, others achieve better results with longer segments, such as 10 s. Future work could explore the association between individual differences and time lengths, particularly the potential links between emotional dimensions and individual neural responses regarding EEG-based entropy fusion views.

5. Conclusions

This work proposes an innovative resource-efficient multi-entropy fusion method for EEG-based emotion recognition in which the development of the BREM involves the extraction of five brain rhythms through DWT, followed by acquisition of multi-entropy features that encapsulate the inherent complexity and dynamics of EEG signals. Subsequently, DTW emerges as a suitable method for measuring similarities among BREM samples. The focused evaluation of time window analysis reveals that the 5 s EEG segment yields robust data for recognition. Although most methods select 32-channel or other multi-channel setups as input to capture abundant brain information to enhance emotion recognition performance, the proposed method using the single-channel mode provides a promising, resource-efficient way to simplify the EEG setup without compromising classification performance. The experimental results from the DEAP dataset demonstrate that single-channel data achieve 84.62% and 82.48% accuracy in arousal and valence classification tasks, respectively, underscoring the effectiveness of BREMs from multi-entropy fusion. Therefore, such performances are meaningful when considering fewer data resources as a concern, which opens the possibility of an entropy fusion method that can helps to design portable EEG-based emotion-aware devices for daily usage.

While the results are particularly interesting, further refinement and validation of the proposed method across diverse scenarios are desirable, especially with a single-channel EEG setup, which may not fully capture the complexity of emotional responses across different brain regions. Therefore, future work should explore ways to combine the resource-efficiency of single-channel EEG with the comprehensive spatial coverage provided by multi-channel EEG, improving performance for more nuanced emotional analysis. In addition, it is vital to focus on exploring real-time classification by integrating multimodal data and customizing the methodology to accommodate individual differences in emotional processing and validating its robustness using datasets beyond DEAP. Such efforts will assess the method's performance across diverse experimental conditions, enhancing its applicability in broader real-world scenarios.

Author Contributions: Conceptualization, J.L., G.F., C.L., X.R. and R.C.; Funding acquisition, J.L., S.Z., Y.C., X.Z. and R.C.; Methodology, J.L., G.F., C.L., X.R. and X.L.; Project administration, S.Z., L.W., Y.C. and X.Z.; Resources, J.L., X.L., Y.C., X.Z. and R.C.; Software, J.L., G.F., S.Z. and L.W.; Validation, J.L., G.F., X.L., Y.C. and R.C.; Writing—original draft, J.L., G.F., C.L. and X.R.; Writing—review and editing, X.L., S.Z., L.W., Y.C., X.Z. and R.C. All authors have read and agreed to the published version of the manuscript.

Funding: This work was supported in part by the National Natural Science Foundation of China under Grant 62072122; in part by the Special Projects in Key Fields of Ordinary Universities of Guangdong Province under Grant 2021ZDZX1087; in part by the Guangzhou Science and Technology Plan Project under Grants 2024B03J1361, 2023B03J1327, 2023A04J0361, and 202201011696; in part by the Research Fund of the Hainan Provincial Key Laboratory of Sports and Health Promotion and the Key Laboratory of Emergency and Trauma of the Ministry of Education under Grant HNYJZ2023010; in part by the Guangdong Province Ordinary Colleges and Universities Young Innovative Talents Project under Grants 2023KQNCX036 and 2022KQNCX038; in part by the Key Discipline Improvement Project of Guangdong Province under Grants 2022ZDJS015 and 2021ZDJS025; in part by the Open Research Fund of Guangdong the Provincial Key Laboratory of Big Data Computing under Grant B10120210117-OF08; in part by the Foundation of the 2023 Higher Education

Science Research Plan of the China Association of Higher Education under Grant 23XXK0402; in part by the Foundation of the Sichuan Research Center of Applied Psychology (Chengdu Medical College) under Grant CSXL-23305; in part by the Neijiang Philosophy and Social Science Planning Project under Grant NJ2024ZD014; in part by the Scientific Research Capacity Improvement Project of the Doctoral Program Construction Unit of Guangdong Polytechnic Normal University under Grants 22GPNUZDJS17 and 22GPNUZDJS31; in part by the Graduate Education Demonstration Base Project of Guangdong Polytechnic Normal University under Grant 2023YJSY04002; and in part by the Research Fund of Guangdong Polytechnic Normal University under Grants 2024XJJXCG002, 2024XJSFKC006, and 2022SDKYA015.

Institutional Review Board Statement: Not applicable.

Data Availability Statement: The datasets generated and/or analyzed during the current study are available at https://github.com/zyzc75/BREM-SIMILARITY.git (accessed on 18 January 2025).

Acknowledgments: The authors appreciate the special contributions of Digital Content Processing & Security Technology of the Guangzhou Key Laboratory.

Conflicts of Interest: The authors declare that they have no known competing financial interests or personal relationships that could have appeared to influence the work reported in this paper.

References

1. Padfield, N.; Ren, J.; Qing, C.; Murray, P.; Zhao, H.; Zheng, J. Multi-segment majority voting decision fusion for MI EEG brain-computer interfacing. *Cogn. Comput.* **2021**, *13*, 1484–1495. [CrossRef]
2. Fang, J.; Li, G.; Xu, W.; Liu, W.; Chen, G.; Zhu, Y.; Luo, Y.; Luo, X.; Zhou, B. Exploring abnormal brain functional connectivity in healthy adults, depressive disorder, and generalized anxiety disorder through EEG signals: A machine learning approach for triple classification. *Brain Sci.* **2024**, *14*, 245. [CrossRef] [PubMed]
3. Larradet, F.; Niewiadomski, R.; Barresi, G.; Caldwell, D.G.; Mattos, L.S. Toward emotion recognition from physiological signals in the wild: Approaching the methodological issues in real-life data collection. *Front. Psychol.* **2020**, *11*, 1111. [CrossRef]
4. Ortony, A. Are all "basic emotions" emotions? A problem for the (basic) emotions construct. *Perspect. Psychol. Sci.* **2022**, *17*, 41–61. [CrossRef] [PubMed]
5. Fu, Y.; Yuan, S.; Zhang, C.; Cao, J. Emotion recognition in conversations: A survey focusing on context, speaker dependencies, and fusion methods. *Electronics* **2023**, *12*, 4714. [CrossRef]
6. Jafari, M.; Shoeibi, A.; Khodatars, M.; Bagherzadeh, S.; Shalbaf, A.; García, D.L.; Gorriz, J.M.; Acharya, U.R. Emotion recognition in EEG signals using deep learning methods: A review. *Comput. Biol. Med.* **2023**, *165*, 107450. [CrossRef] [PubMed]
7. Li, J.; Feng, G.; Lv, J.; Chen, Y.; Chen, R.; Chen, F.; Zhang, S.; Vai, M.-I.; Pun, S.-H.; Mak, P.-U. A lightweight multi-mental disorders detection method using entropy-based matrix from single-channel EEG signals. *Brain Sci.* **2024**, *14*, 987. [CrossRef]
8. Chen, K.; Jing, H.; Liu, Q.; Ai, Q.; Ma, L. A novel caps-EEGNet combined with channel selection for EEG-based emotion recognition. *Biomed. Signal Process. Control* **2023**, *86*, 105312. [CrossRef]
9. Wu, M.; Hu, S.; Wei, B.; Lv, Z. A novel deep learning model based on the ICA and Riemannian manifold for EEG-based emotion recognition. *J. Neurosci. Methods* **2022**, *378*, 109642. [CrossRef]
10. Akhand, M.A.H.; Maria, M.A.; Kamal, M.A.S.; Murase, K. Improved EEG-based emotion recognition through information enhancement in connectivity feature map. *Sci. Rep.* **2023**, *13*, 13804. [CrossRef]
11. Trujillo, L.; Hernandez, D.E.; Rodriguez, A.; Monroy, O.; Villanueva, O. Effects of feature reduction on emotion recognition using EEG signals and machine learning. *Expert Syst.* **2024**, *41*, e13577. [CrossRef]
12. Yu, X.; Li, Z.; Zang, Z.; Liu, Y. Real-time EEG-based emotion recognition. *Sensors* **2023**, *23*, 7853. [CrossRef]
13. Padfield, N.; Ren, J.; Murray, P.; Zhao, H. Sparse learning of band power features with genetic channel selection for effective classification of EEG signals. *Neurocomputing* **2021**, *463*, 566–579. [CrossRef]
14. Chen, Y.; Wang, S.; Guo, J.F. DCTNet: Hybrid deep neural network-based EEG signal for detecting depression. *Multimed. Tools Appl.* **2023**, *82*, 41307–41321. [CrossRef]
15. Ramzan, M.; Dawn, S. Fused CNN-LSTM deep learning emotion recognition model using electroencephalography signals. *Int. J. Neurosci.* **2023**, *133*, 587–597. [CrossRef] [PubMed]
16. Álvarez-Jiménez, M.; Calle-Jimenez, T.; Hernández-Álvarez, M. A comprehensive evaluation of features and simple machine learning algorithms for electroencephalographic-based emotion recognition. *Appl. Sci.* **2024**, *14*, 2228. [CrossRef]

17. Parui, S.; Kumar, A.; Bajiya, R.; Samanta, D.; Chakravorty, N. Emotion recognition from EEG signal using XGBoost algorithm. In Proceedings of the IEEE 16th India Council International Conference (INDICON), IEEE, Rajkot, India, 13–15 December 2019; pp. 1–4.
18. Jimenez, I.A.C.; Olivetti, E.C.; Vezzetti, E.; Moos, S.; Celeghin, A.; Marcolin, F. Effective affective EEG-based indicators in emotion-evoking VR environments: An evidence from machine learning. *Neural Comput. Appl.* **2024**, *36*, 22245–22263. [CrossRef]
19. Zong, J.; Xiong, X.; Zhou, J.; Ji, Y.; Zhou, D.; Zhang, Q. FCAN–XGBoost: A novel hybrid model for EEG emotion recognition. *Sensors* **2023**, *23*, 5680. [CrossRef]
20. Islam, M.R.; Moni, M.A.; Islam, M.M.; Rashed-Al-Mahfuz, M.; Islam, M.S.; Hasan, M.K.; Hossain, M.S.; Ahmad, M.; Uddin, S.; Azad, A.; et al. Emotion recognition from EEG signal focusing on deep learning and shallow learning techniques. *IEEE Access* **2021**, *9*, 94601–94624. [CrossRef]
21. Pereira, E.T.; Gomes, H.M.; Veloso, L.R.; Mota, M.R.A. Empirical evidence relating EEG signal duration to emotion classification performance. *IEEE Trans. Affect. Comput.* **2021**, *12*, 154–164. [CrossRef]
22. Fernandes, J.V.M.R.; Alexandria, A.R.d.; Marques, J.A.L.; Assis, D.F.d.; Motta, P.C.; Silva, B.R.d.S. Emotion detection from EEG signals using machine deep learning models. *Bioengineering* **2024**, *11*, 782. [CrossRef]
23. Song, T.; Liu, S.; Zheng, W.; Zong, Y.; Cui, Z.; Li, Y.; Zhou, X. Variational instance—Adaptive graph for EEG emotion recognition. *IEEE Trans. Affect. Comput.* **2023**, *14*, 343–356. [CrossRef]
24. García-Hernández, R.A.; Celaya-Padilla, J.M.; Luna-García, H.; García-Hernández, A.; Galván-Tejada, C.E.; Galván-Tejada, J.I.; Gamboa-Rosales, H.; Rondon, D.; Villalba-Condori, K.O. Emotional state detection using electroencephalogram signals: A genetic algorithm approach. *Appl. Sci.* **2023**, *13*, 6394. [CrossRef]
25. Padhmashree, V.; Bhattacharyya, A. Human emotion recognition based on time-frequency analysis of multivariate EEG signal. *Knowl. Based Syst.* **2022**, *238*, 107867.
26. Li, J.W.; Barma, S.; Mak, P.U.; Chen, F.; Li, C.; Li, M.T.; Vai, M.I.; Pun, S.H. Single-channel selection for EEG-based emotion recognition using brain rhythm sequencing. *IEEE J. Biomed. Health Inform.* **2022**, *26*, 2493–2503. [CrossRef]
27. Li, X.; Zhang, Y.; Tiwari, P.; Song, D.; Hu, B.; Yang, M.; Zhao, Z.; Kumar, N.; Marttinen, P. EEG based emotion recognition: A tutorial and review. *ACM Comput. Surv.* **2022**, *55*, 1–57. [CrossRef]
28. Mir, M.; Nasirzadeh, F.; Bereznicki, H.; Enticott, P.; Lee, S. Investigating the effects of different levels and types of construction noise on emotions using EEG data. *Build. Environ.* **2022**, *225*, 109619. [CrossRef]
29. Vempati, R.; Sharma, L.D. EEG rhythm based emotion recognition using multivariate decomposition and ensemble machine learning classifier. *J. Neurosci. Methods* **2023**, *393*, 109879. [CrossRef] [PubMed]
30. Miljevic, A.; Bailey, N.W.; Murphy, O.W.; Perera, M.P.N.; Fitzgerald, P.B. Alterations in EEG functional connectivity in individuals with depression: A systematic review. *J. Affect. Disord.* **2023**, *328*, 287–302. [CrossRef] [PubMed]
31. Patel, P.; Balasubramanian, S.; Annavarapu, R.N. Cross subject emotion identification from multichannel EEG sub-bands using Tsallis entropy feature and KNN classifier. *Brain Inf.* **2024**, *11*, 7. [CrossRef]
32. Li, J.; Qiu, S.; Du, C.; Wang, Y.; He, H. Domain adaptation for EEG emotion recognition based on latent representation similarity. *IEEE Trans. Cogn. Dev. Syst.* **2020**, *12*, 344–353. [CrossRef]
33. Ma, Y.; Zhao, W.; Meng, M.; Zhang, Q.; She, Q.; Zhang, J. Cross-subject emotion recognition based on domain similarity of EEG signal transfer learning. *IEEE Trans. Neural Syst. Rehabil. Eng.* **2023**, *31*, 936–943. [CrossRef]
34. Chen, R.; Huang, H.; Yu, Y.; Ren, J.; Wang, P.; Zhao, H.; Lu, X. Rapid detection of multi-QR codes based on multistage stepwise discrimination and a compressed MobileNet. *IEEE Internet Things J.* **2023**, *10*, 15966–15979. [CrossRef]
35. Koelstra, S.; Muhl, C.; Soleymani, M.; Lee, J.S.; Yazdani, A.; Ebrahimi, T.; Pun, T.; Nijholt, A.; Patras, I. DEAP: A database for emotion analysis; using physiological signals. *IEEE Trans. Affect. Comput.* **2011**, *3*, 18–31. [CrossRef]
36. Agarwal, R.; Andujar, M.; Canavan, S. Classification of emotions using EEG activity associated with different areas of the brain. *Pattern Recognit. Lett.* **2022**, *162*, 71–80. [CrossRef]
37. Abdel-Hamid, L. An efficient machine learning-based emotional valence recognition approach towards wearable EEG. *Sensors* **2023**, *23*, 1255. [CrossRef]
38. Abdumalikov, S.; Kim, J.; Yoon, Y. Performance analysis and improvement of machine learning with various feature selection methods for EEG-based emotion classification. *Appl. Sci.* **2024**, *14*, 10511. [CrossRef]
39. Xu, F.; Pan, D.; Zheng, H.; Ouyang, Y.; Jia, Z.; Zeng, H. EESCN: A novel spiking neural network method for EEG-based emotion recognition. *Comput. Methods Programs Biomed.* **2024**, *243*, 107927. [CrossRef] [PubMed]
40. Kannadasan, K.; Veerasingam, S.; Shameedha Begum, B.; Ramasubramanian, N. An EEG-based subject-independent emotion recognition model using a differential-evolution-based feature selection algorithm. *Knowl. Inf. Syst.* **2023**, *65*, 341–377. [CrossRef]
41. Li, D.D.; Xie, L.; Chai, B.; Wang, Z.; Yang, H. Spatial-frequency convolutional self-attention network for EEG emotion recognition. *Appl. Soft Comput.* **2022**, *122*, 108740. [CrossRef]
42. Wei, M.; Liao, Y.; Liu, J.; Li, L.; Huang, G.; Huang, J.; Li, D.; Xiao, L.; Zhang, Z. EEG beta-band spectral entropy can predict the effect of drug treatment on pain in patients with herpes zoster. *J. Clin. Neurophysiol.* **2022**, *39*, 166–173. [CrossRef]

43. Sharma, L.D.; Bhattacharyya, A. A computerized approach for automatic human emotion recognition using sliding mode singular spectrum analysis. *IEEE Sens. J.* **2021**, *21*, 26931–26940. [CrossRef]
44. Patel, P.; Annavarapu, R.N. EEG-based human emotion recognition using entropy as a feature extraction measure. *Brain Inform.* **2021**, *8*, 20. [CrossRef] [PubMed]
45. Chen, F.; Zhao, L.; Li, B.; Yang, L. Depression evaluation based on prefrontal EEG signals in resting state using fuzzy measure entropy. *Physiol. Meas.* **2020**, *41*, 095007. [CrossRef] [PubMed]
46. Chen, T.; Ju, S.; Yuan, X.; Elhoseny, M.; Ren, F.; Fan, M.; Chen, Z. Emotion recognition using empirical mode decomposition and approximation entropy. *Comput. Electr. Eng.* **2018**, *72*, 383–392. [CrossRef]
47. Wang, Z.; Zhang, J.; He, Y.; Zhang, J. EEG emotion recognition using multichannel weighted multiscale permutation entropy. *Appl. Intell.* **2022**, *52*, 12064–12076. [CrossRef]
48. Chen, Y.; Xu, X.; Bian, X.; Qin, X. EEG emotion recognition based on ordinary differential equation graph convolutional networks and dynamic time wrapping. *Appl. Soft Comput.* **2024**, *152*, 111181. [CrossRef]
49. Wang, Z.M.; Hu, S.Y.; Song, H. Channel selection method for EEG emotion recognition using normalized mutual information. *IEEE Access* **2019**, *7*, 143303–143311. [CrossRef]
50. Gao, Z.; Cui, X.; Wan, W.; Zheng, W.; Gu, Z. Long-range correlation analysis of high frequency prefrontal electroencephalogram oscillations for dynamic emotion recognition. *Biomed. Signal Process. Control* **2022**, *72*, 103291. [CrossRef]
51. Mammone, N.; Ieracitano, C.; Adeli, H.; Bramanti, A.; Morabito, F.C. Permutation Jaccard distance-based hierarchical clustering to estimate EEG network density modifications in MCI subjects. *IEEE Trans. Neural Netw. Learn. Syst.* **2018**, *29*, 5122–5135. [CrossRef]
52. Chen, J.; Ro, T.; Zhu, Z. Emotion recognition with audio, video, EEG, and EMG: A dataset and baseline approaches. *IEEE Access* **2022**, *10*, 13229–13242. [CrossRef]
53. Herman, A.M.; Critchley, H.D.; Duka, T. The role of emotions and physiological arousal in modulating impulsive behaviour. *Biol. Psychol.* **2018**, *133*, 30–43. [CrossRef] [PubMed]
54. Ullah, R.; Amblee, N.; Kim, W.; Lee, H. From valence to emotions: Exploring the distribution of emotions in online product reviews. *Decis. Support Syst.* **2016**, *81*, 41–53. [CrossRef]
55. Min, J.; Nashiro, K.; Yoo, H.J.; Cho, C.; Nasseri, P.; Bachman, S.L.; Porat, S.; Thayer, J.F.; Chang, C.; Lee, T.-H.; et al. Emotion downregulation targets interoceptive brain regions while emotion upregulation targets other affective brain regions. *J. Neurosci. Off. J. Soc. Neurosci.* **2022**, *42*, 2973–2985. [CrossRef] [PubMed]
56. Dhara, T.; Singh, P.K.; Mahmud, M. A fuzzy ensemble-based deep learning model for EEG-based emotion recognition. *Cogn. Comput.* **2024**, *16*, 1364–1378. [CrossRef]
57. Kumar, M.; Molinas, M. Human emotion recognition from EEG signals: Model evaluation in DEAP and SEED datasets. In Proceedings of the First Workshop on Artificial Intelligence for Human-Machine Interaction (AIxHMI 2022), Udine, Italy, 2 December 2022; pp. 1–14.
58. Gaddanakeri, R.D.; Naik, M.M.; Kulkarni, S.; Patil, P. Analysis of EEG signals in the DEAP dataset for emotion recognition using deep learning algortihms. In Proceedings of the IEEE 9th International Conference for Convergence in Technology (I2CT), IEEE, Pune, India, 5–7 April 2024; pp. 1–7.
59. Singh, U.; Shaw, R.; Patra, B.K. A data augmentation and channel selection technique for grading human emotions on DEAP dataset. *Biomed. Signal Process. Control* **2023**, *79*, 104060. [CrossRef]
60. Jha, S.K.; Suvvari, S.; Kumar, M. EEG-based emotion recognition: An in-depth analysis using DEAP and SEED Datasets. In Proceedings of the 11th International Conference on Computing for Sustainable Global Development (INDIACom), IEEE, New Delhi, India, 28 February–1 March 2024; pp. 1816–1821.
61. Pan, C.; Shi, C.; Mu, H.; Li, J.; Gao, X. EEG-based emotion recognition using logistic regression with Gaussian kernel and Laplacian prior and investigation of critical frequency bands. *Appl. Sci.* **2020**, *10*, 1619. [CrossRef]
62. Al-Mashhadani, Z.; Bayat, N.; Kadhim, I.F.; Choudhury, R.; Park, J.-H. The efficacy and utility of lower-dimensional Riemannian geometry for EEG-based emotion classification. *Appl. Sci.* **2023**, *13*, 8274. [CrossRef]

entropy

MDPI

Article

Improved EEG-Based Emotion Classification via Stockwell Entropy and CSP Integration

Yuan Lu and Jingying Chen *

National Engineering Research Center for E-Learning, Central China Normal University, Wuhan 430079, China; ly21@jmu.edu.cn
* Correspondence: chenjy@mail.ccnu.edu.cn

Abstract: Traditional entropy-based learning methods primarily extract the relevant entropy measures directly from EEG signals using sliding time windows. This study applies differential entropy to a time-frequency domain that is decomposed by Stockwell transform, proposing a novel EEG emotion recognition method combining Stockwell entropy and a common spatial pattern (CSP). The results demonstrate that Stockwell entropy effectively captures the entropy features of high-frequency signals, and CSP-transformed Stockwell entropy features show superior discriminative capability for different emotional states. The experimental results indicate that the proposed method achieves excellent classification performance in the Gamma band (30–46 Hz) for emotion recognition. The combined approach yields high classification accuracy for binary tasks ("positive vs. neutral", "negative vs. neutral", and "positive vs. negative") and maintains satisfactory performance in the three-class task ("positive vs. negative vs. neutral").

Keywords: EEG; emotion recognition; Stockwell transform; Stockwell entropy; common spatial pattern

1. Introduction

Emotion recognition refers to the detection and analysis of human emotional states through technological means [1]. The electroencephalogram (EEG) technique measures brain activity by recording the electrical activity of cortical neurons [2]. Due to its non-invasive nature and real-time capabilities, EEG technology plays a crucial role in brain function research, particularly in studies that require the prolonged continuous monitoring of brain activity [3]. EEG signals are commonly represented in the form of waveforms, where distinct waveforms, such as alpha, beta, theta, and delta waves, correspond to different states of brain activity [4]. By analyzing key features of EEG signals, including frequency, power spectral density, and phase, researchers are able to investigate the functional state of the brain and discern various emotional states, such as happiness, sadness, and anxiety [5]. As a result, the potential of EEGs as a tool for emotion recognition is considered highly promising [6].

However, due to the asymmetric and non-stationary nature of electroencephalogram (EEG) signals [7], emotion recognition based on EEGs remains a complex scientific challenge [8]. Recently, learning approaches based on entropy measures have been demonstrated to be one of the most effective technical pathways for emotion-related EEG recognition [9]. Entropy, as a measure of information uncertainty, possesses a strong capability to extract clinically meaningful regularity information from EEG signals [10]. Entropy measures can be utilized to quantify the irregularity, randomness, and complexity of physiological signals [9]. Moreover, there is substantial and compelling evidence indicating that entropy-based metrics are highly effective for analyzing EEG signals [9,11]. For instance, Beatriz García-Martínez and Arturo Martínez-Rodrigo et al. were the first to introduce the application of three entropy-based metrics, namely, sample entropy, quadratic sample entropy, and distribution entropy, to differentiate between calm and negative-stress emotional states [12]. Ruo-Nan Duan and Jia-Yi Zhu et al. proposed differential entropy (DE) features and compared them with traditional energy spectrum (ES) features, demonstrating that DE features and their associated combinations offer superior performance for emotion recognition [11]. Yun Lu and Mingjiang Wang et al. introduced a dynamic entropy-based pattern-learning framework for recognizing inter-individual emotions from EEG signals, aiming to address the poor generalization capability of existing emotion recognition methods, which is caused by individual variability [9]. Wei-Long Zheng and Bao-Liang Lu utilized differential entropy features extracted from multi-channel EEG data to train a deep belief network for identifying positive, neutral, and negative emotions. They also investigated the weights of the trained deep belief network to explore key frequency bands and channels [13]. Research by Wu et al. demonstrates that Stockwell entropy, combined with the Hilbert transform, effectively detects events of interest (EoIs) compared to the standard Hilbert transform, achieving accurate identification of both EoIs and non-EoIs [14]. The experimental results indicated that neural features related to different emotions do, indeed, exist and exhibit commonalities across different experiments and individuals [13].

In this study, we propose combining Stockwell differential entropy with the CSP algorithm for the classification and recognition of emotional EEG signals. Specifically, differential entropy is first applied to the time-frequency domain decomposed by the Stockwell transform, followed by the use of the CSP algorithm to extract feature vectors corresponding to different emotional states. Additionally, this study further investigates how signal frequency and amplitude influence Stockwell entropy and its classification performance.

2. Related Work

The successful implementation of any EEG-based emotion recognition system relies on the accurate identification of features representing emotional states, which necessitates efficient feature extraction algorithms [15]. The extracted features must exhibit high discriminability to achieve superior recognition rates. The Stockwell transform is a time-frequency analysis tool that combines the advantages of short-time Fourier transform (STFT) and wavelet transform analyses, providing a detailed view of frequency variations over time [16,17]. CSP is another feature extraction technique that is widely used in biomedical applications. CSP employs a linear transformation method to project multi-channel EEG data into a low-dimensional space, thereby enhancing its discriminative capability for classifying EEG data across different categories [18,19]. K. Venu and P. Natesan proposed an approach called the HC+SMA-SSA scheme, which extracts features using improved Stockwell transform (ST) and CSP for motor imagery task classification, demonstrating superior performance in key metrics [20]. S. Sethi and R. Upadhyay et al. proposed a feature extraction method that integrates the Stockwell transform technique with CSP for designing motor imagery-based brain-computer interfaces, significantly improving the discriminability of motor imagery tasks [21]. Mausovi et al. combined wavelet transform (WT) with the CSP algorithm, achieving high classification accuracy in asynchronous offline brain–computer interface applications [22]. M.I. Chacon-Murguia and E. Rivas-Posada evaluated five feature extraction methods, including Stockwell, CSP + CWT, and CSP + ST, for classifying two types of motor imagery signals. Their proposed methods demonstrated superior performance compared to conventional approaches [23].

The main contributions of this paper are organized as follows: Section 3 introduces the EEG dataset used in this study, along with the Stockwell transform, Stockwell entropy, the

CSP algorithm, and the experimental procedure. Section 4 provides a detailed comparison of the emotion recognition performance of the combined Stockwell entropy and CSP approach across different frequency bands and emotional states, presenting both data analysis and experimental results. Section 5 offers an in-depth analysis of the experimental outcomes and the underlying reasons for this observed classification performance. The final section summarizes the key findings and conclusions of this study.

3. Dataset and Methods

3.1. Dataset

The SEED dataset is a publicly available dataset provided by Shanghai Jiao Tong University, which was primarily designed for research in affective computing [13]. This dataset comprises electroencephalogram (EEG) signals collected from 15 participants (7 males and 8 females), recorded using a 62-channel EEG system, arranged according to the international 10–20 standard, and covering major regions of the brain. During the experiments, the participants watched 15 carefully selected film clips (encompassing positive, neutral, and negative emotions), each lasting approximately 4 min, to induce coherent emotional responses. Consequently, each participant's data file contains 15 segments of preprocessed EEG data (channels × time-series data). The experiments were conducted three times for each participant, with intervals of approximately 1 week or longer between sessions [13].

3.2. Preprocessing

This study utilizes the "Preprocessed_EEG" brain electrical data files provided by the SEED dataset, as these EEG data files have already been downsampled and preprocessed. The data has been downsampled to 200 Hz and filtered using a 0–75 Hz bandpass filter [13]. On this basis, this study further applied a 0.1–46 Hz bandpass filter.

3.3. Methods

3.3.1. Stockwell Transform

Stockwell transform, also known as the S-transform, combines the advantages of the Fourier transform and the wavelet transform, providing a means for achieving the multi-resolution analysis of signals. It is particularly well-suited for the analysis of non-stationary signals [24,25].

Stockwell transform can be mathematically defined as follows:

$$S(\tau, f) = \int_{-\infty}^{+\infty} x(t)w(\tau - t)e^{-j2\pi ft}dt$$

In the given expression, $x(t)$ denotes the original signal, and the window function $w(t)$ is typically a Gaussian window. Here, τ represents the time position, and f represents the frequency. This integral expression provides the complex value of the signal at a given time and frequency, while the magnitude represents the energy level at that specific time and frequency [26,27].

The formula demonstrates that the Stockwell transform technique exhibits multi-resolution analysis capabilities [27]. Through the time-frequency decomposition of non-stationary signals, the technqiue generates a time-frequency representation of the signal [28]. The spectrum characteristics of the signal vary over time, thereby enabling the visualization of frequency component variations at different time points [29]. Unlike the short-time Fourier transform (STFT), the S-transform employs a frequency-dependent Gaussian window function, comprising a narrower window for high frequencies and a wider window for low frequencies [27,30]. Consequently, at high frequencies, the S-transform achieves higher time resolution through a narrower time window, while at low frequencies, it maintains better frequency resolution using a wider time window [31]. This adaptive window function enables the S-transform to more accurately capture transient features in non-stationary signals, making it a valuable tool for analyzing such signals in numerous fields including EEG and ECG [32–34].

3.3.2. Stockwell Transform Entropy

Shannon entropy serves as a measure of information uncertainty. Time-frequency signals processed using the Stockwell transform technique yield a new feature vector by computing the differential entropy in the time domain for each frequency band through a sliding time window approach. This feature vector is termed "Stockwell entropy" [14].

By calculating Stockwell entropy, we can quantify the complexity and randomness of EEG signals across different frequency bands over certain time periods. The computational process of Stockwell entropy using a sliding window approach is depicted in Figure 1.

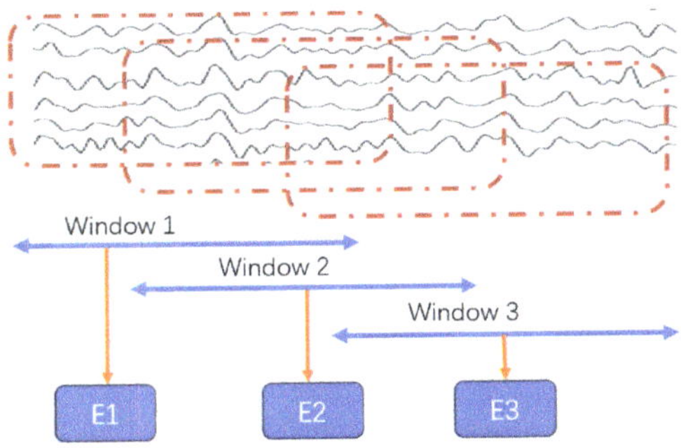

Figure 1. Calculation process of Stockwell entropy with sliding windows.

Under the assumption that the time series **X** follows a Gaussian distribution $\mathcal{N}(\mu,\sigma^2)$, the formula for differential entropy shows that, for a fixed-length EEG sequence, the differential entropy in a specific frequency band equals the logarithm of the energy spectrum [35]:

$$f(X) = -\int_{-\infty}^{\infty} p(x)\cdot\log(p(x))dx = \frac{1}{2}\log(2\pi e\sigma^2) \tag{1}$$

where: $p(x) = \frac{1}{\sqrt{2\pi\sigma^2}}e^{-\frac{(x-\mu)^2}{2\sigma^2}}$

Suppose that in a certain epoch, the EEG signal of one channel, after employing Stockwell transform, has a time-series of length N in the time-frequency signal, denoted as $S = [s_1, s_2, \ldots, s_N]$, and a sliding window with a width of k, where $k \leq N$. Then, the data processing process of the sliding window on the time-series in the time-frequency signal is as follows.

For each position i, where $1 \leq i \leq N - k + 1$, the sub-sequence covered by the sliding window is $S_i = [s_i, s_{i+1}, \ldots, s_{i+k-1}]$.

Step 1. Calculate the average value of all elements within the window S_i.

Step 2. Calculate the standard deviation of the sub-sequence covered by the sliding window S_i.

Step 3. Apply Formula (1) to each sub-sequence covered by the sliding window, to obtain a new sequence $T = [t_1, t_2, \ldots, t_{N-k+1}]$. The new formula can be expressed as follows:

$$T = f([s_i, s_{i+1}, \ldots, s_{i+k-1}]) = \frac{1}{2}\log(2\pi e\sigma_i^2)$$

The obtained sequence $T = [t_1, t_2, \ldots, t_{N-k+1}]$ is Stockwell entropy.

3.3.3. Common Spatial Pattern

The CSP is a feature extraction algorithm that highlights spatial distribution patterns in the EEG signals associated with specific tasks [36,37]. The core principle of the CSP algorithm involves finding a set of spatial filters that maximize the variance between two types of trials [38]. When these spatial filters are applied to the original EEG signals, the signals are projected into a new feature space, generating features that are optimized for classification. In this transformed space, the samples from different categories become more distinguishable [39,40]. Additionally, CSP reduces the dimensionality of the original EEG data by extracting discriminative features while preserving classification-relevant information. Finally, the extracted feature vectors are fed into machine learning classifiers, such as support vector machines (SVM) or random forests, for training and testing to classify the different task states. In this study, Stockwell entropy is processed using the CSP algorithm to generate eight distinct feature vectors for subsequent analysis [40].

3.3.4. Feature Extraction Process Combining Stockwell Entropy and CSP

To obtain certain features of EEG activity based on Stockwell entropy and CSP, in this study, the EEG time periods for one experiment for each subject (including 15 EEG segments) were first decomposed into several epochs. In this study, the length of each epoch was set to 3 s, and each epoch served as a sample, corresponding to an emotional state code (1—positive, 0—neutral, and −1—negative). Subsequently, time-frequency decomposition based on Stockwell transform was performed on each channel of every epoch [41]. For this paper, the 'epochs.compute_tfr' function from the MNE library was used with the following parameters: 'method = "stockwell"', 'freqs = (0.1, 46.0)', and 'width = 1' [42]. Then, the coefficients of the Stockwell transform were grouped, and the time-frequency domain was divided into six different frequency sub-bands, namely, the full frequency band (0.1, 46) Hz, Delta (0.1, 4) Hz, Theta (4, 8) Hz, Alpha (8, 12) Hz, Beta (12, 30) Hz, and Gamma (30, 46) Hz frequency bands. Next, the differential entropy value was calculated for each "frequency sub-band-time" unit on each channel of every epoch. The result of this step was to obtain a new "channel × frequency-sub-band × Stockwell entropy" matrix. Subsequently, the CSP algorithm was executed separately on each frequency sub-band to obtain the corresponding emotional feature vectors of each frequency sub-band on each channel [41]. At the same time, the effect of dimensionality reduction was also achieved. Finally, the generated feature vectors were input into classifiers such as SVM or random forest for the classification of emotional states. Figure 2 shows the flowchart of the Stockwell entropy-CSP feature extraction method.

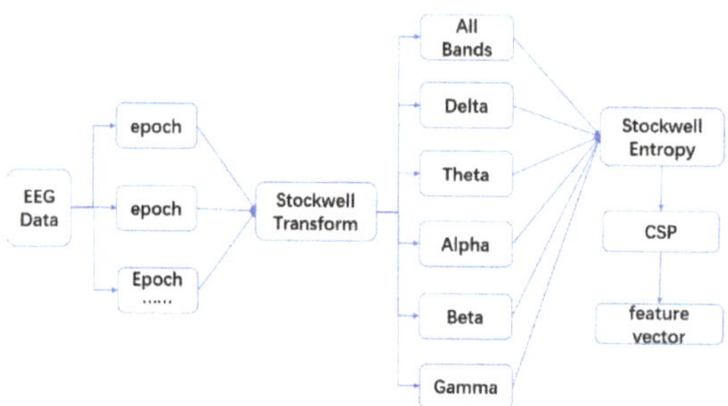

Figure 2. Flowchart of the Stockwell entropy-CSP feature extraction method.

3.4. Classification

In the first step of this study, stratified k-fold cross-validation was adopted to evaluate the effectiveness of the "Stockwell entropy–CSP combination model"; at the same time, this could avoid data leakage in the second-step classification comparison. This study used the most common stratified 5-fold cross-validation technique, dividing the training dataset into 5 subsets. In each iteration, a different subset was used as the test set, and the remaining part was used as the training set. Firstly, the "positive, neutral, and negative" emotions of each subject were combined. Three groups of experiments, namely, "Positive vs. Neutral", "Positive vs. Negative", and "Neutral vs. Negative", respectively, were designed to verify the model's performance. Meanwhile, the influence of different frequency bands on the classification of emotional states was examined. Then, the frequency band with the best evaluation effect was tested under different window states and for different emotional classifications. The final result of the experiment was an average recognition rate of 15 subjects in the three groups of experiments.

In the second step, two classifiers—SVM and random forest (RF)—were employed to compare the model's classification performance on the test set data. SVM performs exceptionally well in high-dimensional spaces and can effectively address nonlinear classification problems using kernel functions. In contrast, random forest, as an ensemble learning method, demonstrates strong robustness against noise and overfitting [42]. By comparing the recognition rates of these two classifiers, the stability and reliability of the classification results were comprehensively validated. Four experimental setups were designed: Positive vs. Neutral vs. Negative, Positive vs. Neutral, Positive vs. Negative, and Neutral vs. Negative. In each setup, the data were randomly divided into training and test sets, with 80% being allocated to the training set and the remaining 20% to the test set. The final results were reported as the average recognition rates across the four experimental groups for all 15 subjects.

4. Results

4.1. Impact of Different Frequency Bands on Emotional State Binary Classification

As described in Section 3.4, this experiment employed stratified five-fold cross-validation. Three experimental setups—Positive vs. Neutral, Positive vs. Negative, and Neutral vs. Negative—were examined to assess the impact of different frequency bands on emotional state classification. The impact of different frequency bands is evaluated for each experimental setup. The classifier employed here is an RBF-kernel SVM, with its parameters set to kernel = 'rbf', C = 20, and Gamma = 'scale'.

Table 1 summarizes the recognition accuracy and standard deviation obtained using stratified five-fold cross-validation for three binary classification tasks: Positive vs. Neutral, Negative vs. Neutral, and Positive vs. Negative. The results are reported across different window widths (W_5, W_10, W_20) and frequency bands (full frequency band, Delta, Theta, Alpha, Beta, and Gamma). Abbreviations are defined as follows: Pos = positive, Neu = neutral, and Neg = negative.

Table 1. Classification of three emotional combinations according to different frequency bands and different window widths.

Bands	W_5%			W_10%			W_20%		
	Neu vs. Neg	Pos vs. Neg	Pos vs. Neu	Neu vs. Neg	Pos vs. Neg	Pos vs. Neu	Neu vs. Neg	Pos vs. Neg	Pos vs. Neu
All	90.1 ± 1.1	95.2 ± 0.9	95.8 ± 0.8	90.2 ± 1.2	95.3 ± 0.9	95.9 ± 0.8	90.3 ± 1.2	95.5 ± 0.8	95.9 ± 0.7
Delta	70.9 ± 1.6	71.9 ± 1.8	71.1 ± 2	70.9 ± 1.8	72.1 ± 1.7	71 ± 1.9	70.7 ± 1.7	72.2 ± 1.6	70.9 ± 2.1
Theta	67.7 ± 1.9	71.5 ± 1.7	71.6 ± 1.8	67.9 ± 1.7	71.3 ± 1.6	71.9 ± 1.7	67.8 ± 2	71.1 ± 1.9	71.6 ± 1.7
Alpha	72.2 ± 2	78.9 ± 1.5	79.4 ± 1.6	72.4 ± 2.2	79 ± 1.6	79.1 ± 1.5	72.3 ± 2.3	79 ± 1.8	79.2 ± 1.5
Beta	87.1 ± 1.3	94.3 ± 1.1	94.1 ± 1	87 ± 1.3	94.3 ± 1.1	94.1 ± 1	87.1 ± 1.3	94.4 ± 0.9	94.1 ± 1
Gamma	92.8 ± 1.2	96.2 ± 0.9	96.7 ± 0.8	92.9 ± 1.3	96.2 ± 0.8	96.7 ± 0.7	92.8 ± 1.3	96.2 ± 0.8	96.8 ± 0.8

Figure 3 illustrates the recognition accuracy of pairwise classifications among three emotion categories (Positive vs. Neutral, Negative vs. Neutral, Positive vs. Negative) across different frequency bands: the full frequency band, Delta, Theta, Alpha, Beta, and Gamma. The x-axis denotes the frequency bands, while the y-axis indicates the recognition accuracy. Specifically, Figure 3a corresponds to a sliding window width of 5 during Stockwell entropy calculation, Figure 3b corresponds to a sliding window width of 10, and Figure 3c corresponds to a sliding window width of 20.

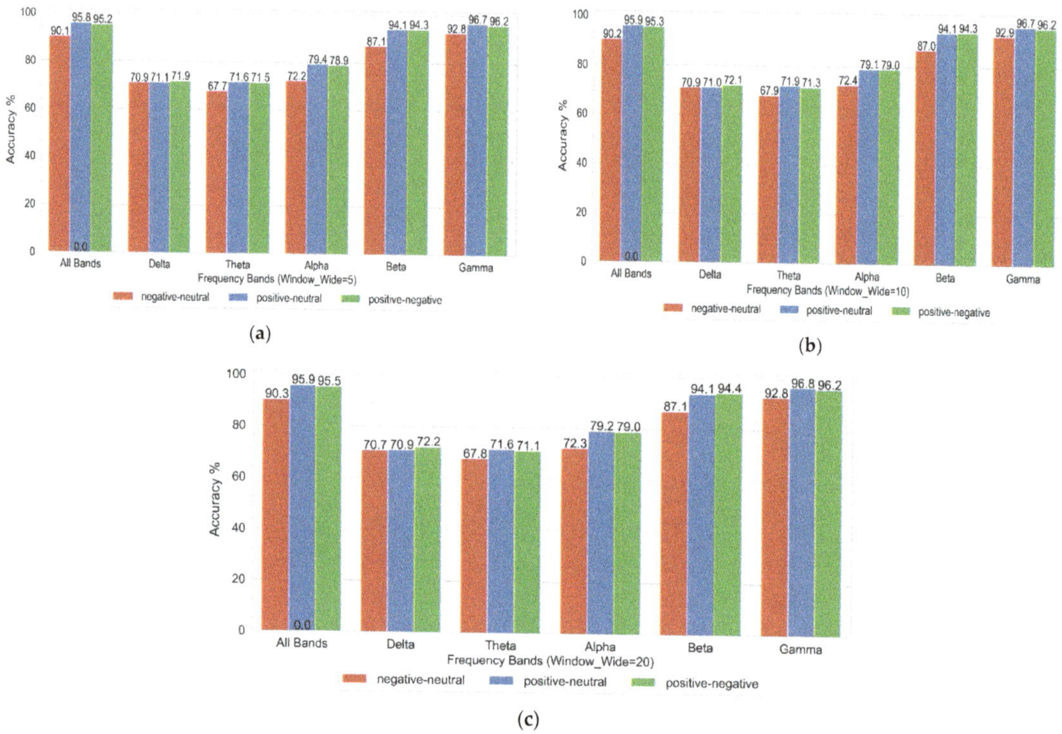

Figure 3. Classification of the combinations of three emotional combinations under different frequency bands: (**a**) when the width of the sliding window is 5; (**b**) when the width of the sliding window is 10; (**c**) when the width of the sliding window is 20.

Table 1 and the three images in Figure 3 indicate that the Gamma frequency band achieved the highest accuracy across all three emotional combinations, with the full frequency band following closely. The classification accuracies of the Delta and Theta frequency bands were generally low, indicating significant differences in classification performance across the different frequency bands and tasks. Selecting an appropriate frequency band is crucial for enhancing the accuracy of EEG-based emotion classification.

4.2. Influence of Gamma Frequency Band on the Classification of Four Emotional Combinations

Since the Gamma frequency band exhibited the highest accuracy, the classification performance of this band was further evaluated for four emotional combinations: three-class classification (Positive, Negative, and Neutral), Positive vs. Neutral, Negative vs. Neutral, and Positive vs. Negative. This evaluation was conducted under different sliding window widths (W_5, W_10, and W_20). The classifier employed was an RBF-kernel SVM with the parameters set as follows: kernel = 'rbf', C = 20, and Gamma = 'scale', as shown in

Figure 4. The x-axis denotes the window width, while the y-axis indicates the recognition accuracy. W_5, W_10, and W_20 correspond to sliding window widths of 5, 10, and 20, respectively, during Stockwell entropy calculation.

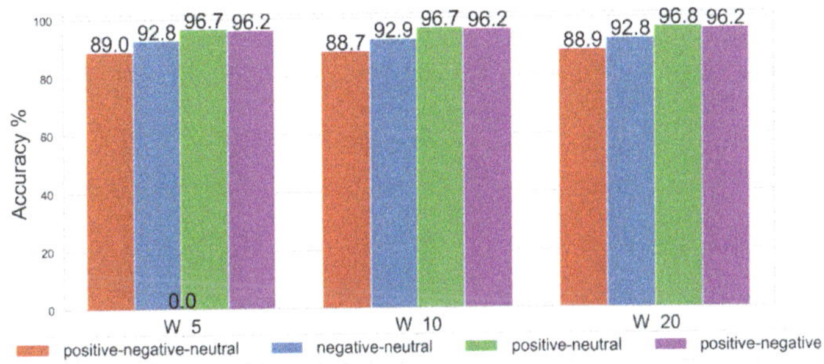

Figure 4. Classification of the combinations of four emotional states.

Figure 4 shows that, for the Gamma frequency band, increasing the window width from W_5 to W_20 had minimal impact on classification accuracy, thereby demonstrating the relatively stable performance of the Stockwell entropy-CSP algorithm in EEG-based emotion recognition. The accuracy of the Gamma frequency band in two-class classification tasks (Positive vs. Neutral, Negative vs. Neutral, and Positive vs. Negative) was generally high, with the lowest accuracy reaching 92.8%. Moreover, the accuracies of Positive vs. Neutral and Positive vs. Negative both remained above 96%. When the window width was 20, the classification accuracy of Positive vs. Neutral reached its highest value of 96.8%. Generally, the CSP algorithm is more suitable for two-class classification scenarios. However, as shown in Figure 4, the CSP algorithm was also applied to evaluate the three-class classification case (Positive vs. Negative vs. Neutral). Although the three-class classification accuracy was lower than that for the two-class case, the lowest level of accuracy still reached 88.7%.

As shown above, as the window width increased from 5 to 20, the model's classification accuracy tended to improve. However, increasing the window width had a limited impact on classification accuracy. Most classification tasks demonstrated consistent performance across all window widths. For two-class tasks (Positive vs. Neutral, Negative vs. Neutral, and Positive vs. Negative), high accuracy was achieved at all window widths. The three-class classification task (Positive vs. Negative vs. Neutral) achieved lower accuracy than the two-class tasks but remained above 88% across all window widths.

4.3. *Influence of Different Classification Methods on Emotional State Recognition*

In the first two sections, the model's performance was evaluated using stratified five-fold cross-validation. In this section, the classification performance of two classifiers—RBF-SVM (kernel = 'rbf', C = 20, Gamma = 'scale') and random forest (n_estimators = 256)—on the test set data was compared. In the experiment, the emotional states of each subject (positive, neutral, and negative) were used to design four experimental setups: Positive vs. Neutral vs. Negative, Positive vs. Neutral, Positive vs. Negative, and Neutral vs. Negative. In each setup, the data were randomly divided into training and test sets, with 80% of the data allocated to the training set and the remaining 20% to the test set. The final results were reported as the average recognition rates across the four experimental groups for all 15 subjects.

The experimental results demonstrate that the varying window widths have minimal impact on classification accuracy. Figure 5 compares the classification accuracy across different frequency bands for binary tasks (Positive vs. Neutral, Negative vs. Neutral, and

Positive vs. Negative) and the ternary task (Positive vs. Negative vs. Neutral). The analysis was performed using a sliding window width of 20. The x-axis denotes the frequency bands (full frequency band, Delta, Theta, Alpha, Beta, and Gamma), while the y-axis indicates the classification accuracy. Figure 5 reveals that the overall performance of SVM and RF was similar across all frequency bands. RF outperformed SVM in the Delta, Theta, and Alpha bands, whereas SVM achieved slightly higher accuracy in the Gamma band. Notably, the Gamma band yielded the highest accuracy under both classifiers. Binary classification tasks generally achieved greater accuracy than ternary tasks. Nevertheless, both classifiers exceeded 88% accuracy in the Gamma band for ternary classification tasks.

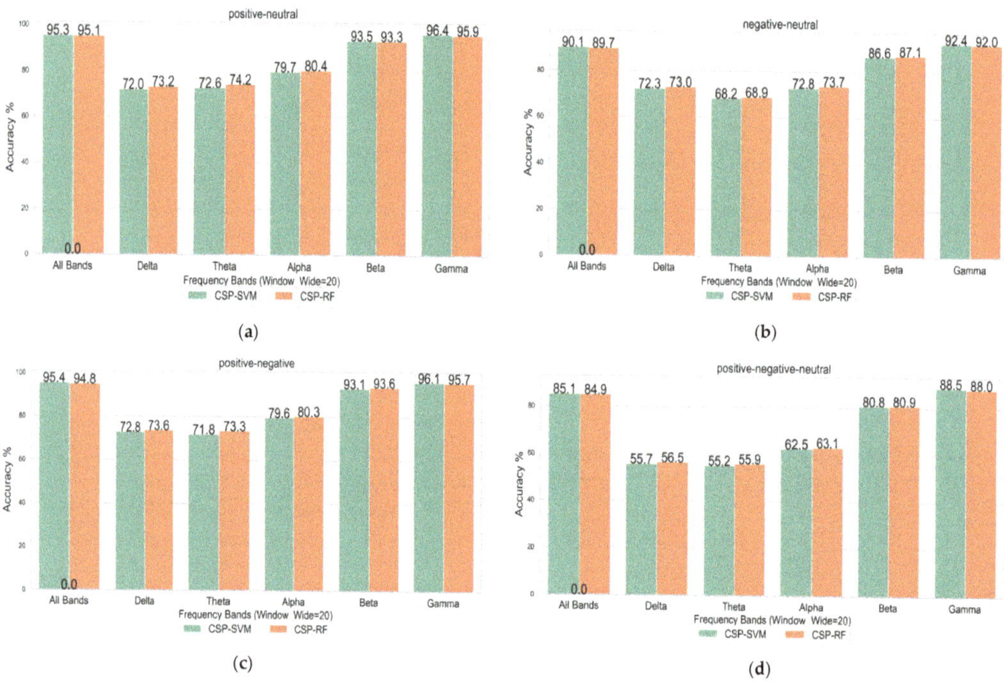

Figure 5. Classification of emotional states by SVM and RF in different frequency bands: (**a**) Positive vs. Neutral; (**b**) Negative vs. Neutral; (**c**) Positive vs. Negative; (**d**) Positive vs. Negative vs. Neutral.

5. Discussion

5.1. Experimental Conclusions

This study's innovative nature is demonstrated by applying differential entropy to the time-frequency domain decomposed by the Stockwell transform and by proposing an EEG-based emotion extraction method that integrates Stockwell entropy with the CSP algorithm.

The experiments demonstrate the following:

1. The classification accuracy was highest in the Gamma frequency band.
2. Increasing the sliding window width from W_5 to W_20 had a minimal impact on classification accuracy, demonstrating that the Stockwell entropy–CSP algorithm exhibits relatively stable performance in EEG-based emotion recognition.
3. The accuracy of binary classification tasks—namely, Positive vs. Neutral, Negative vs. Neutral, and Positive vs. Negative—was generally high. Among these tasks, Positive vs. Neutral and Positive vs. Negative achieved the highest recognition rates.
4. Although the CSP algorithm is more suitable for binary classification tasks, it also demonstrated strong performance in the three-class task (Positive vs. Negative vs. Neutral).

In prior neurophysiological research, the overall body of evidence also supported an association between Gamma oscillations and emotional states [43]. Yang et al. (2020) observed increased Gamma connection density between the prefrontal, temporal, parietal, and occipital regions during emotional processing [44]. Luther et al. (2023) reported increased Gamma power over posterior areas for unpleasant compared to pleasant and neutral pictures [45]. These findings suggest that Gamma oscillations play a role in various aspects of emotional processing, including the perception of emotional stimuli, the cognitive regulation of emotions, and the interaction between emotions and other cognitive processes [46].

5.2. Analysis of the Reasons Behind the Experimental Results

5.2.1. Influence of Signal Frequency and Amplitude on Stockwell Entropy

This study primarily focuses on the analysis of swept-frequency signals. To better simulate the behavior of Stockwell entropy for real EEG signals, the frequency range of the swept-frequency signal was set to 0–46 Hz, with a sampling rate of 200 Hz. Figure 6 illustrates the influence of different signal states on Stockwell entropy. Figure 6a demonstrates the effect of increasing frequency with constant amplitude on Stockwell entropy. As the frequency of the swept-frequency signal increases, the entropy values of high-frequency signals gradually stabilize, whereas those of low-frequency signals exhibit significant fluctuations. Figure 6b depicts the scenario where both signal frequency and amplitude increase. The results indicated that the entropy values of high-frequency signals increased approximately linearly and remained relatively concentrated, while those of the low-frequency signals fluctuated significantly. Figure 6c represents the scenario where signal frequency increases while amplitude decreases. It can be seen that the entropy values of high-frequency signals decreased approximately linearly and remained relatively concentrated, whereas those of low-frequency signals continued to fluctuate significantly. Collectively, Figure 6a–c demonstrates that as the frequency increases, the entropy values across different window widths converge toward stability.

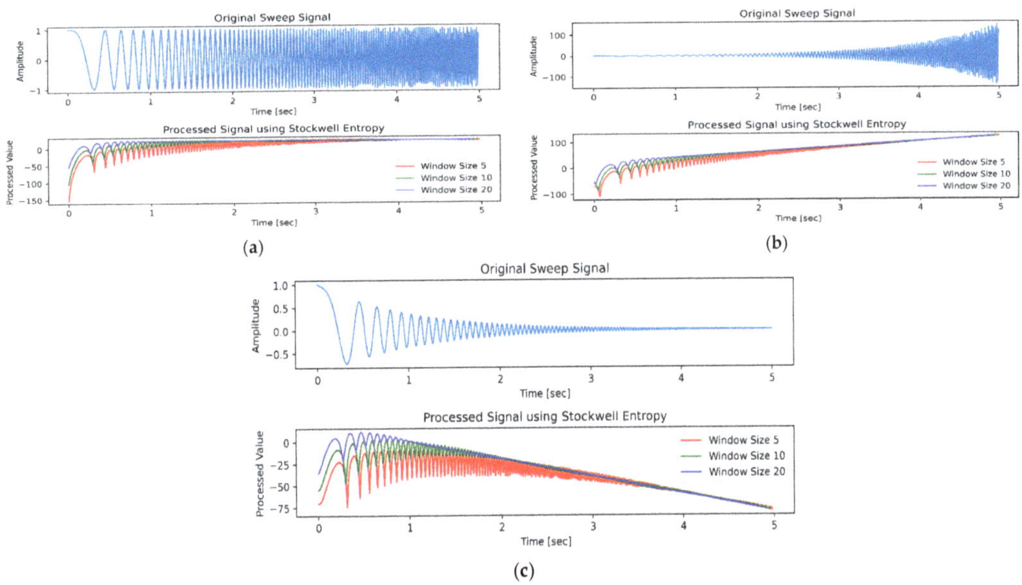

Figure 6. *Cont.*

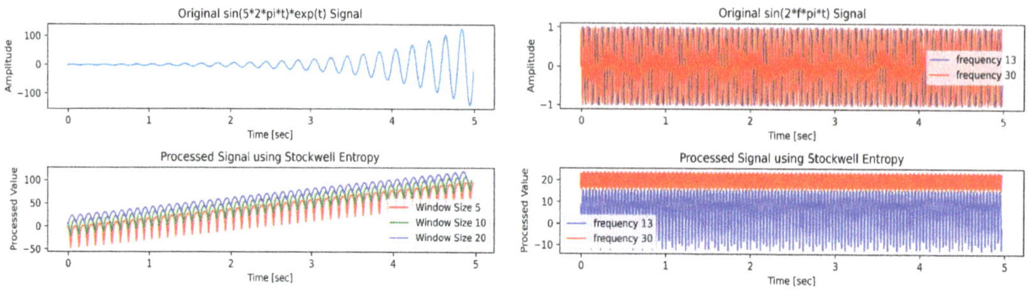

(**d**) (**e**)

Figure 6. Influence of signal frequency and amplitude on Stockwell entropy: (**a**) the frequency of the swept-frequency signal increases; (**b**) both the frequency and amplitude of the swept-frequency signal increase; (**c**) the frequency of the swept-frequency signal increases, while its amplitude decreases; (**d**) the frequency of the signal is fixed, while the amplitude increases; (**e**) the amplitude of the signal is fixed, but the frequencies are different.

As shown in Figure 6d,e, we used the controlled variable method to analyze the effects of amplitude and frequency changes on Stockwell entropy. Figure 6d illustrates the scenario where the signal amplitude increased while the frequency remained constant, with the signal frequency fixed at 5 Hz. The results indicated that when the signal frequency was held constant and the amplitude increased, the entropy value exhibited oscillatory growth, accompanied by significant fluctuations. Figure 6e depicts the scenario where the signal amplitude remained unchanged while the frequency varied. It is evident that low-frequency signals exhibit substantial fluctuations in entropy values, whereas high-frequency signals demonstrate relatively stable behavior.

Table 2 compares the standard deviations of Stockwell entropy for the signal $\sin(2\pi ft)$ across different frequencies and window widths (W = 5, 10, 20). The table lists only the standard deviations at the boundary frequencies of different frequency bands. The results indicate that as the frequency increased, the standard deviations generally decreased. When the frequency reached the Gamma band, the standard deviations reported in Table 2 typically dropped below 1, thereby showing that Stockwell entropy values are relatively stable and suitable for effective emotional classification. Figure 6e compares the results for frequencies of 13 Hz and 30 Hz.

Table 2. Standard deviations of Stockwell entropy at different frequencies with different window widths.

Bands	Frequency	Window = 5 Std	Window = 10 Std	Window = 20 Std
Delta	1	16.62	15.43	13.66
	3	14.66	12.29	9
Theta	4	13.82	11.09	7.26
	7	11.93	8.17	3.04
Alpha	8	11.37	7.33	1.88
	12	9.39	4.4	0.76
Beta	13	8.94	3.75	0.59
	30	2.82	0.36	0
Gamma	31	2.51	0.61	0.24
	36	1.03	0.84	0.29
	41	0.23	0.22	0.21
	42	0.43	0.41	0.33
	43	0.62	0.55	0.32
	44	0.79	0.63	0.17
	45	0.93	0.64	0.04

From Figure 6 and Table 2, the following conclusions can be drawn:

1. The entropy values of high-frequency signals are relatively stable, while those of low-frequency signals fluctuate significantly.
2. Both an increase and a decrease in amplitude can cause changes in the entropy values of high-frequency signals, and these changes are approximately linear. Therefore, the entropy values of high-frequency signals respond well to amplitude changes and can be used to detect such changes.
3. As the frequency increases, the values under different window conditions tend to stabilize, indicating that the selection of window width has little impact on the Stockwell entropy values of high-frequency signals. This demonstrates that Stockwell entropy values are highly stable for classification and recognition.

These conclusions indicate that the entropy values in the high-frequency Gamma band remain relatively stable, thereby contributing to the identification of emotional states. This is why the classification accuracy in this band is relatively high.

5.2.2. Influence of the CSP Algorithm on Emotional State Classification

In this study, the CSP algorithm was applied to decompose Stockwell entropy into eight components. This study uses the CSP component features of the "Positive vs. Negative" binary classification state in the Gamma frequency band of the 15th subject as an example. For clarity, Figure 7 displays the curves of three representative components. It can be observed from Figure 7 that negative emotions correspond to low values on the orange curve and high values on the blue curve, whereas positive emotions correspond to high values on the orange curve and low values on the blue curve. Thus, the CSP-transformed Stockwell entropy features exhibit high discriminability for emotional EEG signals, effectively distinguishing between different emotional states and achieving high recognition rates.

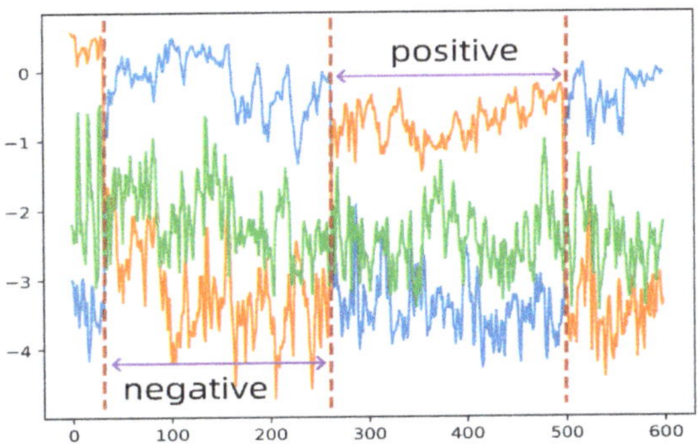

Figure 7. CSP components in the "Positive vs. Negative" state.

5.3. *Deficiencies and Prospects*

In this study, an emotion recognition method combining Stockwell entropy and CSP was compared with the findings of previous studies by Ruo-Nan Duan, Wei-Long Zheng, and Bao-Liang Lu. Ruo-Nan Duan and Jia-Yi Zhu investigated the classification accuracy of four features—the energy spectrum (ES), differential entropy (DE), differential asymmetry (DASM), and rational asymmetry (RASM)—for the Positive vs. Negative binary classification task. The average classification accuracies of these features were 76.56%, 84.22%, 80.96%, and 83.28%, respectively [11]. In contrast, the classification accuracies of the SVM and RF classifiers in the Gamma frequency band for the same task reached 96.1% and 95.7%, respectively. Wei-Long Zheng, Bao-Liang Lu, and their colleagues developed an EEG-based emotion recognition model using deep belief networks (DBNs). The DBNs were trained on differential entropy features extracted from multi-channel EEG data. The SEED dataset was utilized to classify three emotional states: positive, neutral, and negative, achieving a maximum accuracy of 86.65%. The average classification accuracies of DBN, SVM, logistic regression (LR), and K-nearest neighbors (KNN) were 86.08%, 83.99%, 82.70%, and 72.60%, respectively [13]. In this study, the classification accuracies of the SVM and RF classifiers for the three-class task in the Gamma frequency band reached 88.5% and 88.0%, respectively.

Although the combination of Stockwell entropy and CSP has demonstrated strong performance in classifying emotional EEG signals, this study still has certain limitations. In existing emotion recognition research, individual differences remain a significant challenge. Emotion-related signal patterns exhibit substantial inter-individual variability, potentially achieving high performance in within-subject tests but performing poorly in cross-subject scenarios. The final experimental results reported in this study represent the average recognition rates across 15 subjects; however, the emotional data used for each classification test were derived from the same individual. Therefore, future work will focus on enabling individual-independent emotion recognition and refining the proposed method accordingly. Additionally, investigating the spatial distribution of entropy features in the brain may help identify the key cortical regions associated with emotional changes, which could serve as a promising direction for future research.

6. Conclusions

This study proposes a method for extracting emotional EEG features by integrating Stockwell entropy with the CSP technique. Stockwell transform, a time-frequency analysis tool, provides a localized time-frequency representation of non-stationary EEG signals, while Stockwell entropy captures entropy features in the time-frequency domain. The CSP algorithm, a widely used feature extraction method, achieves dimensionality reduction and is frequently applied to spatial filtering in binary motor imagery tasks within the brain–computer interface (BCI). This study integrates these two techniques to enhance the accuracy and stability of emotional EEG recognition.

The experimental results indicate that the proposed combined method exhibits both excellent classification performance and strong stability for emotional EEG signals in the Gamma (30–46 Hz) frequency band. The method achieves high classification accuracy in binary classification tasks, including Positive vs. Neutral, Negative vs. Neutral, and Positive vs. Negative. Furthermore, it also achieves satisfactory classification results in the three-class task (Positive vs. Negative vs. Neutral). Additionally, this study investigates and analyzes the impact of varying window sizes and frequency bands on classification accuracy.

Author Contributions: Conceptualization, Y.L.; methodology, Y.L.; validation, Y.L.; data curation, Y.L.; writing—original draft preparation, Y.L.; writing—review and editing, Y.L. and J.C. All authors have read and agreed to the published version of the manuscript.

Funding: This research was supported by the National Natural Science Foundation of China under Grant 62377018, the National Key R\&D Program of China under Grant 2024YFC3308300, the self-determined Research Funds of CCNU from the Colleges' Basic Research and Operation of MOE under Grants CCNU24ZZ154.

Institutional Review Board Statement: Not applicable.

Data Availability Statement: Further inquiries can be directed to the corresponding authors.

Conflicts of Interest: The authors declare no conflicts of interest.

References

1. Alarcão, S.M.; Fonseca, M.J. Emotions Recognition Using EEG Signals: A Survey. *IEEE Trans. Affect. Comput.* **2019**, *10*, 374–393. [CrossRef]
2. Im, C.; Seo, J.M. A review of electrodes for the electrical brain signal recording. *Biomed. Eng. Lett.* **2016**, *6*, 104–112. [CrossRef]
3. Abdulridha, F.; Albaker, B.M. Non-invasive real-time multimodal deception detection using machine learning and parallel computing techniques. *Soc. Netw. Anal. Min.* **2024**, *14*, 97. [CrossRef]
4. Wankar, R.V.; Shah, P.; Sutar, R. Feature extraction and selection methods for motor imagery EEG signals: A review. In Proceedings of the 2017 International Conference on Intelligent Computing and Control (I2C2), Coimbatore, India, 15–16 June 2017; pp. 1–9.
5. Chanel, G.; Rebetez, C.; Bétrancourt, M.; Pun, T. Emotion Assessment From Physiological Signals for Adaptation of Game Difficulty. *IEEE Trans. Syst. Man, Cybern. Part A Syst. Hum.* **2011**, *41*, 1052–1063. [CrossRef]
6. Shi, S.; Liu, W. Interactive Multi-Agent Convolutional Broad Learning System for EEG Emotion Recognition. *Expert Syst. Appl.* **2025**, *260*, 125420. [CrossRef]
7. Wang, C.; Li, Y.; Wang, L.; Liu, S.; Yang, S. A study of EEG non-stationarity on inducing false memory in different emotional states. *Neurosci. Lett.* **2023**, *809*, 137306. [CrossRef]
8. Kim, J.Y.; Kim, H.G. Emotion Recognition Using EEG Signals and Audiovisual Features with Contrastive Learning. *Bioengineering* **2024**, *11*, 997. [CrossRef]
9. Lu, Y.; Wang, M.; Wu, W.; Han, Y.; Zhang, Q.; Chen, S. Dynamic entropy-based pattern learning to identify emotions from EEG signals across individuals. *Measurement* **2020**, *150*, 107003. [CrossRef]
10. Yao, L.; Lu, Y.; Qian, Y.; He, C.; Wang, M. High-Accuracy Classification of Multiple Distinct Human Emotions Using EEG Differential Entropy Features and ResNet18. *Appl. Sci.* **2024**, *14*, 6175. [CrossRef]
11. Duan, R.N.; Zhu, J.Y.; Lu, B.L. Differential Entropy Feature for EEG-based Emotion Classification. In Proceedings of the 6th International IEEE EMBS Conference on Neural Engineering (NER), San Diego, CA, USA, 6–8 November 2013; pp. 81–84.
12. García-Martínez, B.; Martínez-Rodrigo, A.; Zangróniz Cantabrana, R.; Pastor García, J.M.; Alcaraz, R. Application of Entropy-Based Metrics to Identify Emotional Distress from Electroencephalographic Recordings. *Entropy* **2016**, *18*, 221. [CrossRef]
13. Zheng, W.-L.; Lu, B.-L. Investigating Critical Frequency Bands and Channels for EEG-based Emotion Recognition with Deep Neural Networks. *IEEE Trans. Auton. Ment. Dev.* **2015**, *7*, 162–175. [CrossRef]
14. Wu, M.; Wan, T.; Wan, X.; Fang, Z.; Du, Y. A New Localization Method for Epileptic Seizure Onset Zones Based on Time-Frequency and Clustering Analysis. *Pattern Recognit.* **2020**, *111*, 107687. [CrossRef]
15. Moukadem, A.; Bouguila, Z.; Abdeslam, D.O.; Dieterlen, A. A new optimized Stockwell transform applied on synthetic and real non-stationary signals. *Digit. Signal Process.* **2015**, *46*, 226–238. [CrossRef]
16. Zhong, X.; Liu, G.; Dong, X.; Li, C.; Li, H.; Cui, H.; Zhou, W. Automatic Seizure Detection Based on Stockwell Transform and Transformer. *Sensors* **2024**, *24*, 77. [CrossRef]
17. Dash, S.; Dash, D.K.; Tripathy, R.K.; Pachori, R.B. Time–frequency domain machine learning for detection of epilepsy using wearable EEG sensor signals recorded during physical activities. *Biomed. Signal Process. Control* **2025**, *100*, 107041. [CrossRef]
18. Blanco-Díaz, C.F.; Guerrero-Mendez, C.D.; Delisle-Rodriguez, D.; Jaramillo-Isaza, S.; Ruiz-Olaya, A.F.; Frizera-Neto, A.; de Souza, A.F.; Bastos-Filho, T. Evaluation of temporal, spatial and spectral filtering in CSP-based methods for decoding pedaling-based motor tasks using EEG signals. *Biomed. Phys. Eng. Express* **2024**, *10*, 035003. [CrossRef]
19. Jiang, Q.; Zhang, Y.; Ge, G.; Xie, Z. An Adaptive CSP and Clustering Classification for Online Motor Imagery EEG. *IEEE Access* **2020**, *8*, 156117–156128. [CrossRef]
20. Venu, K.; Natesan, P. Hybrid optimization assisted channel selection of EEG for deep learning model-based classification of motor imagery task. Biomedizinische Technik. *Biomed. Eng./Biomed. Tech.* **2023**, *69*, 125–140. [CrossRef]
21. Sethi, S.; Upadhyay, R.; Singh, H.S. Stockwell-common spatial pattern technique for motor imagery-based Brain Computer Interface design. *Comput. Electr. Eng.* **2018**, *71*, 492–504. [CrossRef]
22. Mousavi, E.A.; Maller, J.J.; Fitzgerald, P.B.; Lithgow, B.J. Wavelet Common Spatial Pattern in asynchronous offline brain computer interfaces. *Biomed. Signal Process. Control* **2011**, *6*, 121–128. [CrossRef]
23. Chacon-Murguia, M.I.; Rivas-Posada, E. Feature Extraction Evaluation for Two Motor Imagery Recognition Based on Common Spatial Patterns, Time-Frequency Transformations and SVM. In Proceedings of the 2020 International Joint Conference on Neural Networks (IJCNN), Glasgow, UK, 19–24 July 2020.
24. Poh, K.K.; Marziliano, P. Analysis of Neonatal EEG Signals using Stockwell Transform. In Proceedings of the International Conference of the IEEE Engineering in Medicine & Biology Society, Lyon, France, 22–26 August 2007; pp. 594–597.
25. Kulkarni, D.; Dixit, V.V. EEG-based emotion classification Model: Combined model with improved score level fusion. *Biomed. Signal Process. Control* **2024**, *95*, 106352. [CrossRef]
26. Gupta, B.; Verma, A.K. Linear canonical Stockwell transform and the associated multiresolution analysis. *Math. Methods Appl. Sci.* **2024**, *47*, 9287–9312. [CrossRef]
27. Stockwell, R.G.; Mansinha, L.; Lowe, R.P. Localisation of the complex spectrum: The S Transform. *IEEE Trans. Signal Process.* **1996**, *44*, 998–1001. [CrossRef]

28. Bajaj, A.; Kumar, S. Design of Novel Time–Frequency Tool for Non-stationary α-Stable Environment and its Application in EEG Epileptic Classification. *Arab. J. Sci. Eng.* **2024**, *49*, 15863–15881. [CrossRef]
29. Shin, Y.; Hwang, S.; Lee, S.-B.; Son, H.; Chu, K.; Jung, K.-Y.; Lee, S.K.; Park, K.-I.; Kim, Y.-G. Using spectral and temporal filters with EEG signal to predict the temporal lobe epilepsy outcome after antiseizure medication via machine learning. *Sci. Rep.* **2023**, *13*, 22532. [CrossRef]
30. Jibon, F.A.; Miraz, M.H.; Khandaker, M.U.; Rashdan, M.; Salman, M.; Tasbir, A.; Nishar, N.H.; Siddiqui, F.H. Epileptic seizure detection from electroencephalogram (EEG) signals using linear graph convolutional network and DenseNet based hybrid framework. *J. Radiat. Res. Appl. Sci.* **2023**, *16*, 3. [CrossRef]
31. Guerrero, C.M.; Trigueros, A.M.; Franco, J.I. Time-frequency EEG analysis in epilepsy: What is more suitable? In Proceedings of the Fifth IEEE International Symposium on Signal Processing and Information Technology, Athens, Greece, 18–21 December 2005; pp. 202–207. [CrossRef]
32. Rao, B.M.; Kumar, A.; Bachwani, N.; Marwaha, P. Detection of atrial fibrillation based on Stockwell transformation using convolutional neural networks. *Int. J. Inf. Technol.* **2023**, *15*, 1937–1947. [CrossRef]
33. Kumar, N.; Raj, S. An Adaptive Scheme for Real-Time Detection of Patient-Specific Arrhythmias Using Single-Channel Wearable ECG Sensor. *IEEE Sens. Lett.* **2024**, *8*, 7001504. [CrossRef]
34. Burnos, S.; Hilfiker, P.; Sürücü, O.; Scholkmann, F.; Krayenbühl, N.; Grunwald, T.; Sarnthein, J. Human intracranial high frequency oscillations (HFOs) detected by automatic time-frequency analysis. *PLoS ONE* **2014**, *9*, e94381. [CrossRef]
35. Shi, L.-C.; Jiao, Y.-Y.; Lu, B.-L. Differential entropy feature for EEG-based vigilance estimation. In Proceedings of the 2013 35th Annual International Conference of the IEEE Engineering in Medicine and Biology Society (EMBC), Osaka, Japan, 3–7 July 2013; pp. 6627–6630. [CrossRef]
36. Jiang, Q.; Zhang, Y.; Hu, X.; Wang, W.; Ge, G.Y. Geometry-aware Common Spatial Patterns for Motor Imagery-based Brian-Computer Interfaces. *IAENG Int. J. Appl. Math.* **2024**, *54*, 1476–1489.
37. Jiang, X.; Meng, L.; Chen, X.; Wu, D. A CSP-based retraining framework for motor imagery based brain-computer interfaces. *Sci. China Inf. Sci.* **2024**, *67*, 189403. [CrossRef]
38. Grosse-Wentrup, M.; Buss, M. Multiclass Common Spatial Patterns and Information Theoretic Feature Extraction. *IEEE Trans. Biomed. Eng.* **2008**, *55*, 1991–2000. [CrossRef] [PubMed]
39. Mishuhina, V.; Jiang, X. Complex Common Spatial Patterns on Time-Frequency Decomposed EEG for Brain-Computer Interface. *Pattern Recognit.* **2021**, *115*, 107918. [CrossRef]
40. Lu, Y.; Wang, H.; Lu, Z.; Niu, J.; Liu, C. Gait pattern recognition based on electroencephalogram signals with common spatial pattern and graph attention networks. *Eng. Appl. Artif. Intell.* **2025**, *141*, 109680. [CrossRef]
41. Gramfort, A.; Luessi, M.; Larson, E.; Engemann, D.A.; Strohmeier, D.; Brodbeck, C.; Goj, R.; Jas, M.; Brooks, T.; Parkkonen, L.; et al. MEG and EEG data analysis with MNE-Python. *Front. Neurosci.* **2013**, *7*, 267. [CrossRef]
42. Huang, Y.; Zha, W.X. Comparison on Classification Performance Between Random Forests and Support Vector Machine. *Software* **2012**, *33*, 107–110.
43. Kang, J.-H.; Jeong, J.W.; Kim, H.T.; Kim, S.H.; Kim, S.-P. Representation of cognitive reappraisal goals in frontal gamma oscillations. *PLoS ONE* **2014**, *9*, e113375. [CrossRef]
44. Yang, K.; Tong, L.; Shu, J.; Zhuang, N.; Yan, B.; Zeng, Y. High Gamma Band EEG Closely Related to Emotion: Evidence From Functional Network. *Front. Hum. Neurosci.* **2020**, *14*, 89. [CrossRef]
45. Luther, L.; Horschig, J.M.; van Peer, J.M.; Roelofs, K.; Jensen, O.; Hagenaars, M.A. Oscillatory brain responses to emotional stimuli are effects related to events rather than states. *Front. Hum. Neurosci.* **2023**, *16*, 868549. [CrossRef]
46. Kang, J.H.; Ahn, H.M.; Jeong, J.W.; Hwang, I.; Kim, H.T.; Kim, S.H.; Kim, S.P. The modulation of parietal gamma oscillations in the human electroencephalogram with cognitive reappraisal. *Neuroreport* **2012**, *23*, 995–999. [CrossRef]

entropy

MDPI

Article

Dynamic Evolution of EEG Complexity in Schizophrenia Across Cognitive Tasks

Rosa Molina [1], Yasmina Crespo-Cobo [1,*], Francisco J. Esteban [2], Ana Victoria Arias [3], Javier Rodríguez-Árbol [1], Maria Felipa Soriano [4], Antonio J. Ibáñez-Molina [1] and Sergio Iglesias-Parro [1]

1 Department of Psychology, University of Jaén, 23071 Jaén, Spain; rmlopez@ujaen.es (R.M.); aibanez@ujaen.es (A.J.I.-M.); siglesia@ujaen.es (S.I.-P.)
2 Department of Experimental Biology, University of Jaén, 23071 Jaén, Spain; festeban@ujaen.es
3 Department of Social Psychology, National Distance Education University (UNED), 28040 Madrid, Spain; avarias@psi.uned.es
4 Mental Health Department, San Agustín University Hospital, 33401 Linares, Spain; mariaf.soriano.sspa@juntadeandalucia.es
* Correspondence: ycrespo@ujaen.es

Abstract: Schizophrenia is characterized by widespread disruptions in neural connectivity and dynamic modulation. Traditional EEG analyses often rely on static or averaged measures, which may overlook the temporal evolution of neural complexity across cognitive demands. This study employed Higuchi Fractal Dimension, a non-linear measure of signal complexity, to examine the temporal dynamics of EEG activity across five cortical regions (central, frontal, occipital, parietal, and temporal lobes) during an attentional and a memory-based task in individuals diagnosed with schizophrenia and healthy controls. A permutation-based topographic analysis of variance revealed significant differences in neural complexity between tasks and groups. In the control group, results showed a consistent pattern of higher neural complexity during the attentional task across the different brain regions (except during a few moments in the temporal and occipital regions). This pattern of differentiation in complexity between the attentional and memory tasks reflects healthy individuals' ability to dynamically modulate neural activity based on task-specific requirements. In contrast, the group of patients with schizophrenia exhibited inconsistent patterns of differences in complexity between tasks over time across all neural regions. That is, differences in complexity between tasks varies across time intervals, being sometimes higher in the attentional task and other times higher in the memory task (especially in the central, frontal, and temporal regions). This inconsistent pattern in patients can explain reduced task-specific modulation of EEG complexity in schizophrenia, and suggests a disruption in the modulation of neural activity on function of task demands. These findings underscore the importance of analyzing the temporal dynamics of EEG complexity to capture task-specific neural modulation.

Keywords: schizophrenia; electroencephalography; Higuchi fractal dimension; neural complexity; temporal dynamics

1. Introduction

The brain's complex dynamics are inherently non-linear, with non-linearity arising even at the cellular level due to threshold and saturation phenomena in neuronal behavior [1]. These non-linear properties extend to large-scale neural networks, where interactions among neurons give rise to emergent behaviors that cannot be fully captured by linear models. Electroencephalogram (EEG) signals are particularly valuable in neuroscience research due to their high temporal resolution, which enables the capture of rapid neural dynamics on the millisecond scale. This precision is crucial for studying the brain's non-linear and transient processes, as cognitive functions such as attention, memory, and sensory integration often unfold over very short time intervals [2]. Historically, EEG analyses have heavily relied on linear methods focused on spectral power or averaged measures across time and frequency domains. While these approaches have provided critical insights into brain function, they often fail to capture the full complexity and temporal variability of neural processes, particularly those underlying cognitive tasks [3,4].

To address these limitations, non-linear methods have emerged as powerful tools for quantifying the irregularity and complexity of EEG signals. These measures are particularly adept at capturing dynamic changes in brain activity that occur during both physiological states and pathological conditions. For instance, fractal dimension analysis has been applied to study the spatiotemporal complexity of brain activity in Alzheimer's disease [5], epilepsy [6], and schizophrenia [4]. Such methods provide unique insights into how neural networks adapt—or fail to adapt—to cognitive demands. Higuchi's fractal dimension is a nonlinear measure used to quantify the complexity of time-series data, such as EEG signals. This method evaluates the self-similarity and irregularity of a signal over multiple scales, providing a robust representation of its dynamic properties. Unlike linear methods, Higuchi's fractal dimension is particularly well-suited for nonstationary data, as it captures the intricate temporal patterns inherent in brain activity without requiring transformations into the frequency domain [7,8].

In the context of schizophrenia, EEG studies have consistently highlighted widespread disruptions in neural connectivity and dynamic modulation, particularly during tasks requiring attention or memory retrieval [9,10]. These disruptions are thought to stem from impairments in the brain's ability to dynamically reconfigure its functional networks in response to changing cognitive demands. However, many traditional EEG studies rely on static or averaged measures, which may obscure subtle but meaningful temporal fluctuations in neural complexity that are critical for understanding these impairments. For example, averaged power spectral analyses or event-related potentials (ERPs), while informative, often fail to capture the non-linear and transient nature of brain activity [2,11].

Recent advances in non-linear time-series analysis have provided powerful tools for exploring the temporal dynamics of brain activity. Unlike linear methods, which assume proportionality and constant relationships between variables, non-linear methods capture irregularities and emergent behaviors characteristic of brain activity [1]. HFD quantifies the fractal complexity of EEG signals directly in the time domain, offering a sensitive measure of how neural complexity evolves dynamically over time [7]. This method is particularly well-suited for studying schizophrenia, where abnormalities in neural complexity have been linked to core cognitive deficits, such as impaired working memory and attentional control [4,12]. A study of Iglesias-Parro et al. [13] showed that schizophrenic patients, when compared with control participants, exhibited less differentiated EEG complexity at different conscious states (external attention vs. mind wandering). This finding suggests that patients tend to show the same overall complexity pattern regardless of the specific mental state that they present. In the present study, we aim to explore whether this reduction of differentiation holds for attention and memory processes and whether a temporal evolution of complexity analysis adds new light on understanding EEG complexity patterns in schizophrenic patients.

The attention task consisted of presenting images with local landscapes to participants, and they needed to detect whether a small red dot appeared on the screen. In the memory task, the same type of images were presented, but in this case, participants were instructed to recall an event from their memory. Hence, our study builds on this growing body of research by applying HFD to analyze the temporal evolution of EEG complexity across different cortical regions during attentional (Dots) and memory-based (Remember) tasks in individuals diagnosed with schizophrenia (SCZ) and healthy controls (CTRL). By focusing on temporal evolution rather than averaged measures, we aim to capture the dynamic modulation of neural networks underlying task-specific processing. We hypothesize that individuals with schizophrenia will exhibit reduced task-specific modulation of EEG complexity and less consistent patterns of neural complexity over time compared to controls. This approach highlights the importance of moving beyond traditional static analyses to better understand the disrupted dynamics of brain activity in schizophrenia.

In this study, we conceptualize complexity as the brain's capacity to generate diverse, adaptive, and organized patterns of activity. This concept aligns with theoretical frameworks that describe the brain as a dynamic system operating at the edge of order and chaos. Such a balance is essential for functional efficiency and cognitive flexibility, allowing neural networks to adapt dynamically to changing cognitive demands. By employing Higuchi's fractal dimension, we aim to quantify this balance and explore how it varies across tasks (Dots vs. Remember) and groups (controls vs. schizophrenia). The primary objective of this study was to identify significant differences in brain activity between the experimental conditions Dots and Remember using Higuchi's complexity metric. These differences were analyzed across temporal windows (Steps) and brain regions (topologies) using a permutation-based topographic analysis of variance (TANOVA). This approach was inspired in [14], allowing for robust statistical testing of differences while accounting for the spatial and temporal structures of EEG data. To complement the TANOVA findings, the correlation between the complexity of different brain regions across tasks (Dots and Remember) was analyzed in both the control and schizophrenia groups.

2. Method

2.1. Participants

The sample included 21 individuals diagnosed with schizophrenia (SCZ) attending the Mental Health Day Unit at the University St. Agustin Hospital (Spain) and 17 healthy controls (HC).

In the SCZ group, 19% were women and 81% were men. The mean age was 36.476 (*SD* = 8.256, *median*: 34 [*min*: 25, *max*: 51]). For the SCZ participants, the inclusion criteria were an ICD-10 diagnosis of schizophrenia (F20) or a psychotic disorder (F23). The educational level of the participants was categorized into three categories: basic (primary and secondary education), medium (general certificate of education, certificate of higher education), and high (university degree). According to these categories, the sample of participants with SCZ included 57.1% participants with a basic educational level, 28.6% with a medium level, and 14.3% with a high level.

In the HC group, 41% of participants were women and 59% were men. The mean age was 41.7 (*SD* = 13.1, *median*: 38 [*min*: 25, *max*: 66]). This group included 35.3% with a basic educational level, 41.2% with a medium level, and 23.5% with a high level.

There was neither a significant association between group and education level (χ^2 (4, N = 38) = 6.209, p = 0.184), nor between sex and group (χ^2 (1, N = 38) = 2.237, p = 0.135), nor did the groups differ significantly with respect to age (t = 1.500; p = 0.142).

Exclusion criteria for all groups were as follows: concurrent diagnosis of neurological disorder, concurrent diagnosis of substance abuse, history of developmental disability, inability to sign informed consent, or vision disorders (vision disorders which, although corrected by glasses or contact lenses, suppose a loss of visual acuity, e.g., cataracts). In addition, an exclusion criterion for the HC group was the diagnosis of a mental disorder (according to verbal reports from participants). All participants gave their written informed consent according to the Declaration of Helsinki and the Ethics Committee on Human Research of the Hospital approved the study.

2.2. Procedure

The experiment was carried out in one session in the laboratory of the Mental Health Department at San Agustín Hospital in Linares. All participants provided written informed consent before participating in the study, in accordance with the principles outlined in the Declaration of Helsinki. The study protocol was approved by the Human Research Ethics Committee of the hospital. Participants were seated at a distance of approximately 70 cm from a computer screen. A set of 64 active electrodes was then placed on the participants' scalps, with electrode impedance maintained below 5 kΩ.

Participants in this study performed two tasks while their EEG activity was recorded with an AD rate of 500 Hz. The experimental tasks were carried out in the same order for all participants. First, they performed a sustained attention task, and subsequently, within the same experimental session and after a 5 min break, they completed a recall task.

In the sustained attention task (Dots) participants were shown a series of neutral images (landscapes) on a computer screen. Each image was displayed for 10 s. In 50% of the trials, a red dot appeared randomly at any location on the screen. Participants were instructed to press a designated key as soon as they detected the red dot. The task comprised 40 trials in total: 20 trials in which the red dot was present and 20 trials without it. The trials with a dot and without a dot were presented in random order. Following the dot detection task, participants completed a memory task (Remember). During this phase, the same set of images was presented again on the screen for 10 s each. Participants were instructed to observe each image and allow their thoughts to wander freely to any memory, regardless of whether it was related to the content of the image. This task consisted of 20 trials.

Each trial was divided into 40 temporal windows of 2 s each, referred to henceforth as steps. Each step overlapped 90% of the time with the previous one. These steps represent consecutive time intervals during which Higuchi's fractal dimension (HFD) was calculated for EEG signals recorded from the 63 electrode channels referenced to the Cz electrode site. The trials were grouped by experimental conditions (Dots and Remember), and the electrodes were further categorized into five brain regions or topologies: frontal, central, occipital, parietal, and temporal. Below is a detailed description of how the channels were assigned to each topology (see Appendix A).

The frontal topology included 16 electrodes positioned over the anterior region of the scalp. These regions have been associated with higher-order cognitive functions such as decision-making, attention, and working memory [15]. The central topology comprised 13 electrodes located near the central midline of the scalp. This region is primarily involved in motor control and somatosensory processing [16]. This grouping ensured that activity from both hemispheres and the central sulcus was adequately represented. The parietal topology included 16 electrodes positioned over the parietal cortex, which is associated with sensory integration, spatial reasoning, and attention [17]. The occipital topology consisted of 8 electrodes placed over the posterior region of the scalp, corresponding to areas responsible for visual processing [18]. The temporal topology included 10 electrodes located over the lateral temporal regions of the scalp. These regions are involved in auditory processing, language comprehension, and memory functions [19,20].

The assignment of the channels to topologies was based on standard EEG electrode placement systems [10–20] and their correspondence to underlying cortical regions. This grouping facilitated region-specific analyses while maintaining consistency with established neuroanatomical landmarks [21].

2.3. Data Processing, Analysis, and Results

Data processing was conducted using Brain Vision Analyzer (Version 2.2.2., Brain Products GmbH, Gilching, Germany), EEGLAB (Version 2023.0, Swartz Center for Computational Neuroscience, San Diego, CA, USA), and MATLAB (R2024b, MathWorks, Natick, Massachusetts, MA, USA). Blinks and other artifacts were removed using Infomax ICA (version 2.2.0.0, https://www.mathworks.com/matlabcentral/fileexchange/38300-pca-and-ica-package (accessed on 5 May 2018)). The Infomax ICA algorithm is based on the principle of maximizing entropy. Specifically, it seeks to optimize the representation of the data by finding an unmixing matrix that maximizes the mutual independence of the sources. Unlike other dimensionality reduction techniques, such as principal component analysis (PCA), which decorrelates data but assumes orthogonality, Infomax ICA focuses on achieving statistical independence without imposing constraints on orthogonality. ICA components containing artifacts were identified by visual inspection of the scalp topography, power spectra, and raw activity from all components. Once the noisy components were selected, they were removed from the original signals.

The cleaned EEG signals were then used as inputs for a custom MATLAB script developed to calculate Higuchi's fractal dimension (HFD) as a measure of signal complexity over time. This non-linear approach provides insight into brain activity dynamics by capturing variations in signal complexity [7]. We selected HFD due to its computational efficiency and relevance for analyzing nonstationary signals. This metric was chosen over other nonlinear measures, such as entropy-based methods or recurrence quantification analysis (RQA) because it provides a direct and interpretable measure of signal complexity while being computationally less demanding. HFD is a measure of irregularity for discrete time series. The algorithm obtains new series by sampling the original signal at different intervals (k). For each k, the lengths, L(k), of the signals are calculated, normalizing the sums of the differences of the values, with a distance of k and a starting point m (m = 1, 2,. . . , k). Finally, a double logarithmic plot, ln L(k) vs. ln k, is used to estimate the actual dimension value of the signal. The range of values for HFD lies between 1 and 2. Higher HFD values indicate greater EEG complexity, often associated with increased neural activity, greater neural network engagement, and higher cognitive demands. Conversely, lower HFD values reflect more regular or predictable activity, typically observed during resting states or under reduced cognitive load. For this experiment, we calculated a single HFD estimation for each trial using a sliding window procedure ($n = 40$ windows or steps).

To statistically evaluate the differences in neural complexity between experimental tasks and groups, we implemented linear mixed-effects models (LMMs) using the R package lme4 (version, 1.1-36 R Core Team, Global). Separate models were fitted for healthy controls and individuals with schizophrenia to account for group-specific neural dynamics. Each model incorporated Task (Dots vs. Remember) and Topology (brain region: central, frontal, occipital, parietal, temporal) as fixed effects, along with their interaction term (Task × Topology), to assess how task-related differences in complexity varied across cortical regions. To account for repeated measures and individual variability, random intercepts were included for participants (ID), allowing baseline complexity levels to vary between individuals. Additionally, random slopes for Task nested within ID were incorporated to model intra-individual variability in task-related complexity differences.

Next, to examine how task-related neural complexity evolved across both cortical regions and temporal windows, we implemented a permutation-based topographic analysis of variance (TANOVA) inspired by [14]. This method extends traditional ANOVA by explicitly modeling the interaction between Task (Dots vs. Remember) and Topology (brain regions) across successive temporal steps (1–40), thereby capturing spatiotemporal dynamics often obscured in static analyses based on averages across time.

HFD values were calculated for 40 steps separately for each of the five brain regions. To assess statistical significance, condition labels (Dots/Remember) were permuted 1000 times within each Step–Topology pair, generating a null distribution of differences. Observed differences were considered significant if they exceeded the 95th percentile of this distribution ($p < 0.05$, two-tailed). The TANOVA's design allowed us to test whether task effects depended on specific spatiotemporal contexts (i.e., Task × Topology × Step interactions), a critical feature for detecting dynamic reconfigurations in neural networks.

To control for Type I error inflation from multiple comparisons (40 steps × 5 topologies = 200 tests per group), we applied Benjamini–Hochberg false discovery rate (FDR) correction ($q < 0.05$) [22]. This approach ensured the robust identification of regions and time points where task demands significantly modulated neural complexity while preserving the hierarchical structure of EEG data across space and time.

To further explore the relationships between brain regions and to complement the findings from the TANOVA, we conducted correlation analyses. These analyses were performed to assess the strength and direction of functional connectivity between regions (central, frontal, occipital, parietal, and temporal) during the Dots and Remember tasks for both the control and schizophrenia groups. By calculating pairwise correlations of complexity values across regions, we aimed to capture patterns of synchronization and interaction that could provide additional insights into task-specific network dynamics and group differences.

We have structured our analyses and results into two main sections. First, we present dynamic analyses that explicitly consider the temporal dimension of complexity differences between experimental tasks for controls. Second, we introduce the equivalent analyses for the participants with schizophrenia.

2.3.1. Control Participants

A linear mixed-effects model (LMM) was conducted to evaluate differences in EEG complexity (Higuchi fractal dimension) between the attentional and memory-based tasks in the healthy control participants. The model included fixed effects for Task, Topology, and their interaction, with random intercepts and slopes for Task nested within participants to account for individual variability.

The analysis revealed a non-significant main effect of Task, $F(1, 16) = 2.27$, $p = 0.151$, indicating no overall differences in neural complexity between the tasks across participants. However, a significant main effect of Topology was observed, $F(4, 2{,}608{,}158) = 1858.04$, $p < 0.001$, showing pronounced variability in complexity across brain regions. Importantly, the interaction between Task and Topology was statistically significant, $F(4, 2{,}608{,}158) = 44.86$, $p < 0.001$, indicating that task-related differences in complexity were region-specific.

For each topology, we computed the mean Higuchi fractal dimension across all subjects and trials for each task (Dots and Remember). These mean values illustrate the temporal evolution of complexity within a typical trial. The results are shown in Figure 1.

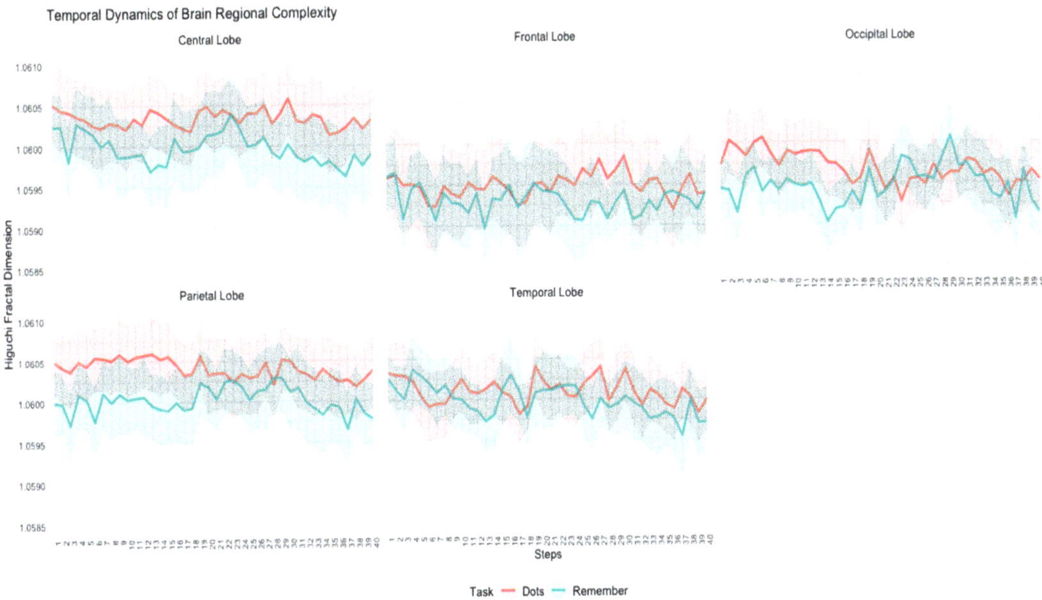

Figure 1. Temporal dynamics of Higuchi's fractal dimension across brain regions (central, frontal, occipital, parietal, and temporal lobes) for the control group during attentional (Dots) and memory (Remember) tasks. The solid lines represent the mean HFD values, while the shaded areas indicate the standard error of the mean (SEM).

Next, HFD values were calculated for 40 steps separately for each of the five brain regions. These differences provided a measure of the observed effect size for each temporal window and brain region.

A permutation-based approach was employed to test the statistical significance of the observed differences. Condition labels (Dots and Remember) were randomly shuffled within each experimental condition (Step and Topology). For each permutation, the mean difference was recalculated. This process was repeated 1000 times to generate a null distribution of differences. The null distribution allowed us to estimate the likelihood that the observed differences occurred by chance.

HFD values were calculated for 40 steps separately for each of the five brain regions, and p-values were calculated as the proportion of permuted differences that were greater than or equal to the absolute value of the observed difference. To control for Type I error due to multiple comparisons, the false discovery rate (FDR) correction was applied using the Benjamini–Hochberg procedure [22].

The results can be visualized as bar plots, with each bar representing the observed difference for a specific combination of Step and Topology (see Figure 2). Significant differences were highlighted with red asterisks above or below the corresponding bars, depending on the sign. The visualization included separate panels for each brain region (Topology).

As can be seen in Figures 1 and 2, the control participants exhibited a consistent temporal pattern of higher complexity during the attentional task compared to the memory task. This is especially evident in the central, frontal, and parietal regions. In the occipital and temporal regions, there were a few specific moments when complexity was higher during the memory task, but for most periods, complexity was higher in the attentional task.

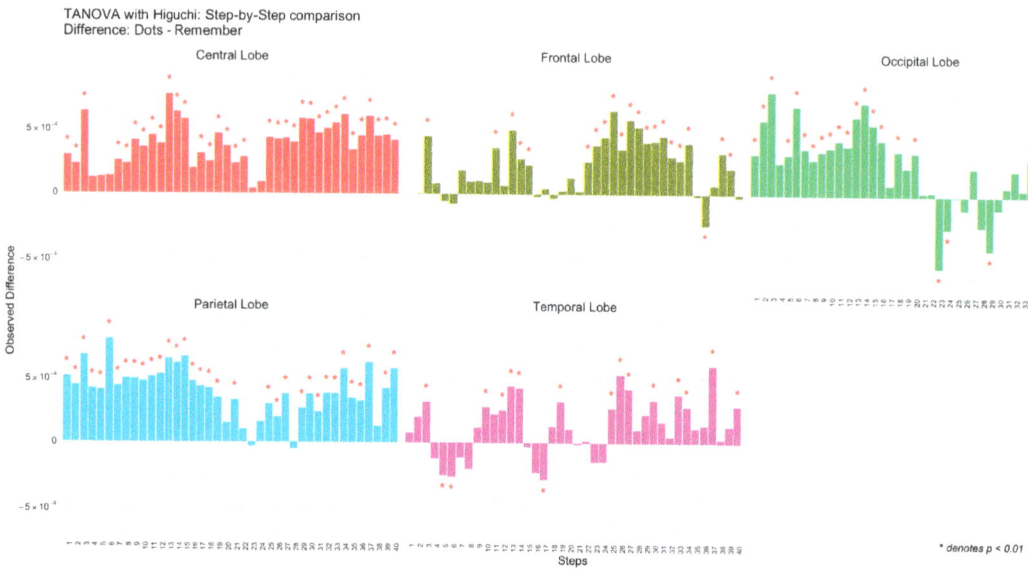

Figure 2. Differences in observed Higuchi complexity between tasks (Dots vs. Remember) for the control group across brain regions (central, frontal, occipital, parietal, and temporal lobes) and steps.

2.3.2. Participants with Schizophrenia

Replicating the mixed model framework used for the controls, patients exhibited a non-significant main effect of Task, $F(1, 20) = 0.007$, $p = 0.931$, indicating no overall differences in neural complexity between the attentional and memory tasks across participants. However, a significant main effect of Topology was observed, $F(4, 3{,}283{,}510) = 3648.86$, $p < 0.001$, showing substantial variability in complexity across brain regions. Importantly, the interaction between Task and Topology was statistically significant, $F(4, 3{,}283{,}510) = 57.598$, $p < 0.001$, highlighting that task-related differences in complexity depended on specific brain regions.

Next, we replicated the same permutation-based topographic analysis of variance (TANOVA) methodology previously applied to the control group to examine the temporal evolution of Higuchi's fractal dimension (HFD) during task performance across brain regions. Specifically, for each topology, the mean Higuchi fractal dimension across all subjects and trials was computed for each task (Dots and Remember). The results for this sample are presented in Figure 3.

Next, HFD values were calculated for 40 steps separately for each of the five brain regions.

The results are presented as bar plots (Figure 4), where each bar represents the observed difference for a specific combination of Step and Topology. Significant differences are marked with red asterisks. As with the control group, separate panels are included for each brain region (Topology) to facilitate comparison. Positive differences indicate greater complexity during the attentional task (Dots), while negative differences reflect greater complexity during the memory task (Remember).

In Figures 3 and 4, the patterns of complexity differences between tasks in the patient group demonstrate considerable variability. Specifically, in all brain regions, the complexity values during the attentional task (Dots) are higher than those observed during the memory task (Remember) at certain temporal steps, while at other steps, the opposite occurs (Figure 5). This temporal variability does not follow a consistent or statistically significant pattern across time. The inconsistency is particularly pronounced in the central, frontal, and temporal regions, where fluctuations between tasks appear more irregular and lack a clear trend. These findings suggest that, in patients, the dynamic modulation of brain complexity between tasks is less structured compared to what might be expected in a more organized neural system.

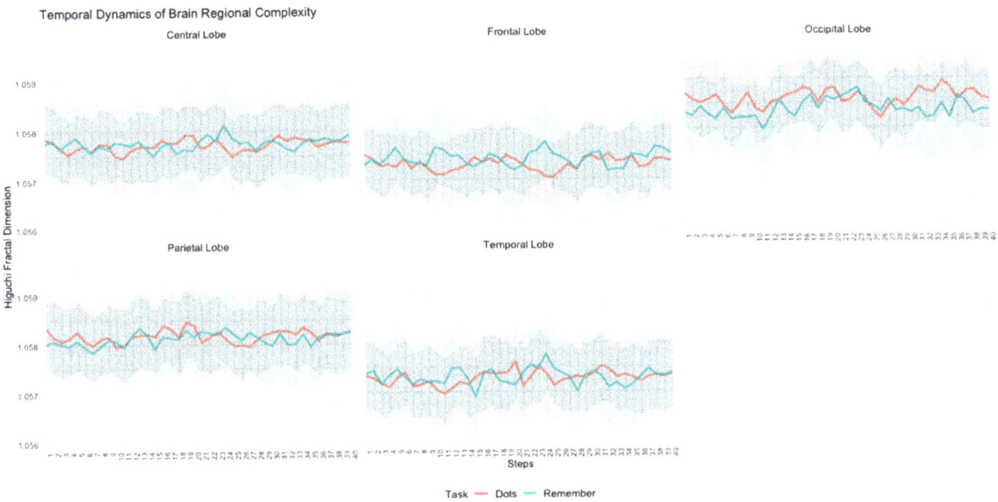

Figure 3. Temporal dynamics of Higuchi's fractal dimension across brain regions (central, frontal, occipital, parietal, and temporal lobes) for the schizophrenia group during attentional (Dots) and memory (Remember) tasks. The solid lines represent the mean HFD values, while the shaded areas indicate the standard error of the mean (SEM).

Figure 4. Differences in observed Higuchi complexity between tasks (Dots vs. Remember) for the schizophrenia group across brain regions (central, frontal, occipital, parietal, and temporal lobes) and steps.

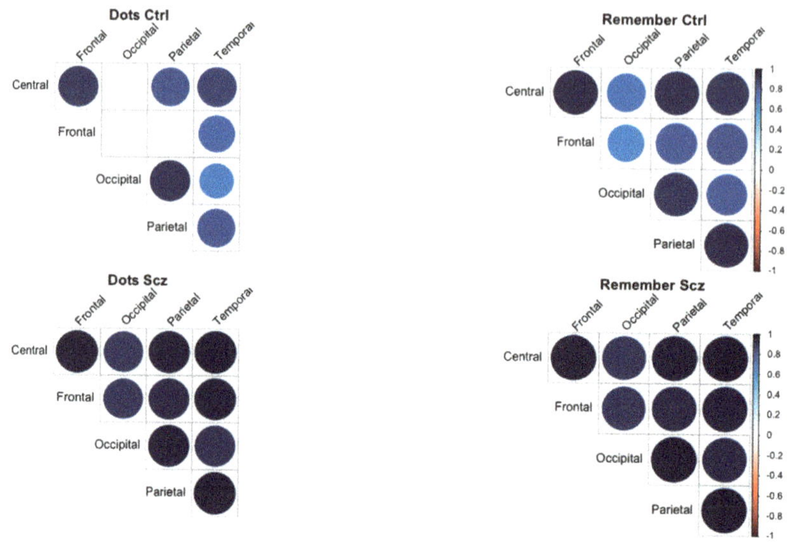

Figure 5. Correlation matrices between brain regions (central, frontal, occipital, parietal, and temporal) based on complexity values during the Dots and Remember tasks in control (Ctrl) and schizophrenia (Scz) groups.

3. Discussion

In this study, we explored the temporal dynamics of EEG complexity during an attentional and memory task in healthy controls and patients with schizophrenia. Previous research has shown that patients with schizophrenia exhibit reduced task-specific modulation of EEG complexity compared to healthy controls [13]. These findings suggest that patients exhibit a diminished ability of the brain to dynamically adjust its functioning to meet the cognitive demands of the task. In this experimental design, we aimed to further explore brain dynamics by analyzing the temporal evolution of complexity across different cognitive tasks.

Consistent with our hypothesis, patients and controls showed differences in the temporal patterns of EEG complexity. Healthy controls exhibited a highly regular temporal pattern, showing increased complexity during the attentional task consistently over time. However, in the group of patients, differences in complexity between tasks varied inconsistently over time. This variability likely accounts for the reduced complexity differences between tasks reported in previous research; unpredictable fluctuations in patients' EEG complexity levels (sometimes higher, sometimes lower) result in a smaller average difference between tasks.

To our knowledge, this finding of high temporal inconsistency in complexity among patients is entirely novel. It supports the notion that patients with schizophrenia have impaired modulation of brain complexity in response to task demands and struggle to adjust brain dynamics to the task at hand.

Now, we will discuss these differences between patients and controls according to the brain regions of interest (central, frontal, occipital, parietal, and temporal lobes).

3.1. Central, Parietal, and Frontal Lobe

We have grouped the discussion of the results into these regions since the findings are similar across them both in the controls and patients. The CTRL group consistently shows significant positive differences favoring the attentional task. This result may be reflective of efficient somatosensory integration and motor control, as well as executive high-order processes, necessary for attentional tasks [23]. In contrast, the SCZ group exhibits inconsistent patterns of neural complexity differences between tasks, with both positive and negative differences observed. The presence of negative differences in the SCZ group, which are absent in the CTRL group, suggests that individuals with schizophrenia may struggle to maintain stable neural complexity across tasks. This variability could reflect compensatory mechanisms or disruptions in connectivity that impair their ability to dynamically adapt to cognitive demands [9,12]. The presence of negative differences in the SCZ group is particularly intriguing because it indicates greater neural complexity during the memory compared to the attentional task, an atypical pattern not observed in the CTRL group. This may suggest that individuals with schizophrenia require additional neural engagement during memory-based tasks to compensate for deficits in episodic memory retrieval or sensory integration. Such findings align with prior research indicating that schizophrenia is associated with altered connectivity and reduced efficiency in neural networks responsible for attention and memory [9,16]. As we have mentioned before, these findings support the idea that with schizophrenia there is a disruption in neural modulation, leading to both reduced task-specific modulation of EEG complexity and unexpected shifts in cortical complexity. Further investigation into these negative differences could provide valuable insights into the compensatory mechanisms at play in schizophrenia and how they influence cognitive performance.

3.2. Occipital and Temporal Lobes

In the occipital and temporal lobes, the temporal pattern of differences was more similar in the controls and patients. In the control group, negative and positive differences in complexity could be observed, especially in temporal regions. That is, in some temporal steps there was higher complexity in the attentional task, and in others, there was higher complexity in the memory task, possibly reflecting the role of temporal regions in episodic memory retrieval through interactions with the hippocampus and prefrontal cortex [24,25]. Patients with schizophrenia also showed negative and positive differences although again, the pattern was more irregular and variable.

On the whole, our findings support prior research indicating that schizophrenia is associated with deficits in sensory integration and attentional control [9,21,26]. It is reasonable to think that this inability to modulate brain activity on the function of the task contributes to difficulties prioritizing external stimuli versus internal representations, which are key deficits observed in schizophrenia [26].

4. Limitations

The study has certain limitations that should be considered when interpreting the results. The relatively small sample size may limit the generalizability of the results to broader populations. Future studies with larger and more diverse samples could provide additional insights and strengthen the conclusions drawn here. Additionally, the experimental tasks were performed in a fixed order for all participants, which might introduce potential order effects, such as fatigue or learning (although the temporal pattern during each task did not show fatigue or learning effects). While Higuchi's fractal dimension (HFD) is a robust tool for analyzing EEG complexity, it is worth noting that it provides an overall measure of signal complexity without distinguishing between specific neural processes. The division of trials into 40 temporal windows allowed for detailed temporal analysis but may not fully capture rapid fluctuations in neural complexity at finer scales. Exploring alternative methods with even higher temporal resolution could complement these findings in future research. Finally, EEG inherently has limited spatial resolution compared to imaging techniques like fMRI. Although this study focused on cortical regions using well-established electrode groupings, future investigations combining EEG with other neuroimaging modalities could provide a more comprehensive understanding of neural dynamics. Despite these considerations, the study's innovative use of HFD to explore temporal dynamics in schizophrenia offers valuable insights into neural complexity and task-specific modulation. In addition to HFD, future studies could benefit from incorporating other nonlinear and dynamic measures to analyze EEG complexity in schizophrenia. For instance, indices such as approximate entropy, multiscale entropy, or nonlinear correlation coefficients may provide complementary insights into the underlying neural dynamics.

5. Conclusions

Our results highlight significant differences in the dynamics of cortical activity between the CTRL and SCZ groups across various brain regions, emphasizing how schizophrenia disrupts task-specific neural modulation. In the CTRL group, cortical regions, such as the central, frontal, occipital, parietal, and temporal lobes, demonstrate dynamic and efficient neural responses tailored to the demands of attention-focused and memory-based tasks. These findings align with prior research indicating that healthy individuals exhibit robust neural flexibility, allowing them to allocate resources appropriately for sensory integration, attentional control, and episodic memory retrieval [23–25].

In contrast, the SCZ group shows reduced task-specific modulation of EEG complexity across all cortical regions analyzed. This diminished task-specific modulation reflects impairments in sensory integration, attentional control, and memory retrieval—core cognitive deficits associated with schizophrenia. For example, the parietal lobe shows fewer significant differences between tasks in patients with schizophrenia, suggesting difficulties in integrating multisensory information and directing attention effectively [10,27]. The frontal lobe also exhibits reduced modulation in SCZ participants, likely due to disrupted connectivity with other cortical regions such as the hippocampus, key for episodic memory retrieval and contextual integration [24,26].

The correlation analyses also support the hypothesis of a lack of functional specificity in the schizophrenia group. Thus, in the control group, distinct correlation patterns emerge depending on the task, reflecting functional specificity. During the Dots task, stronger correlations are observed between frontal and parietal regions, while the Remember task shows higher synchronization involving occipital and temporal regions. In contrast, the schizophrenia group exhibits consistently high correlations across all regions and tasks, suggesting a generalized hyperconnectivity pattern. This lack of task-dependent variability would indicate impaired functional segregation in the schizophrenia group.

Overall, these findings underscore widespread disruptions in neural networks in schizophrenia, highlighting the critical role of connectivity impairments in cognitive dysfunctions observed in this population.

The findings of this study could have significant clinical implications. Temporal differences in cortical activity may serve as neurophysiological biomarkers for the diagnosis of schizophrenia, the assessment of disease progression, and the evaluation of treatment response [28,29]. Identifying reliable biomarkers is essential for improving early detection and personalized treatment approaches. Therefore, future research should focus on integrating multimodal biomarkers to enhance diagnostic accuracy and optimize therapeutic strategies for schizophrenia. Likewise, these findings could serve as a starting point for future studies on other mental disorders, expanding the understanding of neurophysiological mechanisms beyond schizophrenia.

Author Contributions: Conceptualization, R.M., Y.C.-C., F.J.E., A.V.A., J.R.-Á., M.F.S., A.J.I.-M. and S.I.-P.; methodology, R.M., Y.C.-C., F.J.E., A.V.A., J.R.-Á., M.F.S., A.J.I.-M. and S.I.-P.; software, R.M.,

Y.C.-C., F.J.E., A.V.A., J.R.-Á., M.F.S., A.J.I.-M. and S.I.-P.; validation, R.M., Y.C.-C., F.J.E., A.V.A., J.R.-Á., M.F.S., A.J.I.-M. and S.I.-P.; formal analysis, R.M., Y.C.-C., F.J.E., A.V.A., J.R.-Á., M.F.S., A.J.I.-M. and S.I.-P.; investigation, R.M., Y.C.-C., F.J.E., A.V.A., J.R.-Á., M.F.S., A.J.I.-M. and S.I.-P.; resources, R.M., Y.C.-C., F.J.E., A.V.A., J.R.-Á., M.F.S., A.J.I.-M. and S.I.-P.; data curation, R.M., Y.C.-C., F.J.E., A.V.A., J.R.-Á., M.F.S., A.J.I.-M. and S.I.-P.; writing—original draft preparation: R.M. and Y.C.-C.; writing—review and editing: F.J.E. and A.J.I.-M.; visualization: R.M. and Y.C.-C.; supervision: A.J.I.-M. and S.I.-P.; project administration: A.J.I.-M., M.F.S., S.I-P. and F.J.E. All authors have read and agreed to the published version of the manuscript.

Funding: This research received funding from the University of Jaén (PAIUJA-EI_CTS02_2023), from the Regional Government of Andalusia (BIO-302), from the Ministry of Science and Innovation, the State Research Agency (AEI), and the European Regional Development Fund (ERDF—Ref: PID2021-122991NB-C21 and Ministry of Health and Families of the Regional Government of Andalusia (AP-0033-2020-C1-F2).

Institutional Review Board Statement: The study protocol was approved by the Research Ethics Committee of the Province of Jaén (Protocol number: AP-0033-2020-C1-F2/0032-N-21, approved on 25 February 2021). The study was conducted in accordance with the Declaration of Helsinki and all participants provided written informed consent.

Informed Consent Statement: Not applicable.

Data Availability Statement: The dataset supporting this study has been deposited in FigShare and is publicly available for other researchers. The dataset can be accessed via the following DOI: 10.6084/m9.figshare.28304096 (https://figshare.com/ndownloader/files/52005371, accessed on 10 September 2020).

Conflicts of Interest: The authors declare no conflicts of interest.

Appendix A

Electrodes assigned to each topology

Frontal Topology

Fp1, Fp2, AF7, AF3, AFz, AF4, AF8, F7, F3, Fz, F4, F8, F1, F5, F6, F2.

Central Topology

FC5, FC1, FCz, FC2, FC6, C3, C4, C1, C5, C2, FC3, C6, FC4.

Parietal Topology

CP5, CP1, CPz, CP2, CP6, P3, P4, Pz, P1, P2, P5, P6, P7, P8, CP3, CP4.

Occipital Topology

O1, Oz, O2, PO7, PO3, POz, PO4, PO8.

Temporal Topology

T7, T8, TP9, TP10, FT9, FT10, FT7, FT8, TP7, TP8.

References

1. Kantz, H.; Schreiber, T. *Nonlinear Time Series Analysis*, 2nd ed.; Cambridge University Press: Cambridge, UK, 2003. [CrossRef]
2. Pfurtscheller, G.; Lopes Da Silva, F.H. Event-related EEG/MEG synchronization and desynchronization: Basic principles. *Clin. Neurophysiol.* **1999**, *110*, 1842–1857. [CrossRef] [PubMed]
3. Pincus, S.M. Approximate entropy as a measure of system complexity. *Proc. Natl. Acad. Sci. USA* **1991**, *88*, 2297–2301. [CrossRef] [PubMed]
4. de Miras, J.R.; Ibáñez-Molina, A.J.; Soriano, M.F.; Iglesias-Parro, S. Fractal dimension analysis of resting state functional networks in schizophrenia from EEG signals. *Front. Hum. Neurosci.* **2023**, *17*, 1236832. [CrossRef]
5. Jeong, J. EEG dynamics in patients with Alzheimer's disease. *Clin. Neurophysiol.* **2004**, *115*, 1490–1505. [CrossRef] [PubMed]
6. Babloyantz, A.; Destexhe, A. Low-dimensional chaos in an instance of epilepsy. *Proc. Natl. Acad. Sci. USA* **1986**, *83*, 3513–3517. [CrossRef] [PubMed]
7. Higuchi, T. Approach to an irregular time series on the basis of the fractal theory. *Phys. D Nonlinear Phenom.* **1988**, *31*, 277–283. [CrossRef]

8. Raghavendra, B.S.; Dutt, D.N.; Halahalli, H.N.; John, J.P. Complexity analysis of EEG in patients with schizophrenia using fractal dimension. *Physiol. Meas.* **2009**, *30*, 795–808. [CrossRef] [PubMed]
9. Barch, D.M.; Ceaser, A. Cognition in schizophrenia: Core psychological and neural mechanisms. *Trends Cogn. Sci.* **2012**, *16*, 27–34. [CrossRef]
10. Fornito, A.; Zalesky, A.; Breakspear, M. The connectomics of brain disorders. *Nat. Rev. Neurosci.* **2015**, *16*, 159–172. [CrossRef]
11. Jeong, J.; Chae, J.H.; Kim, S.Y.; Han, S.H. Nonlinear dynamic analysis of the EEG in patients with Alzheimer's disease and vascular dementia. *J. Clin. Neurophysiol.* **2001**, *18*, 58–67. [CrossRef] [PubMed]
12. Kristensen, T.D.; Ambrosen, K.S.; Raghava, J.M.; Syeda, W.T.; Dhollander, T.; Lemvigh, C.K.; Bojesen, K.B.; Barber, A.D.; Nielsen, M.Ø.; Rostrup, E.; et al. Structural and functional connectivity in relation to executive functions in antipsychotic-naïve patients with first episode schizophrenia. *Schizophrenia* **2024**, *10*, 2024. [CrossRef]
13. Iglesias-Parro, S.; Soriano, M.F.; Prieto, M.; Rodríguez, I.; Aznarte, J.I.; Ibáñez-Molina, A.J. Introspective and Neurophysiological Measures of Mind Wandering in Schizophrenia. *Sci. Rep.* **2020**, *10*. [CrossRef] [PubMed]
14. Murray, M.M.; Brunet, D.; Michel, C.M. Topographic ERP analyses: A step-by-step tutorial review. *Brain Topogr.* **2008**, *20*, 249–264. [CrossRef] [PubMed]
15. Miller, E.K.; Cohen, J.D. An integrative theory of prefrontal cortex function. *Annu. Rev. Neurosci.* **2001**, *24*, 167–202. [CrossRef] [PubMed]
16. Fornia, L.; Leonetti, A.; Puglisi, G.; Rossi, M.; Viganò, L.; Della Santa, B.; Simone, L.; Bello, L.; Cerri, G. The parietal architecture binding cognition to sensorimotor integration: A multimodal causal study. *Brain* **2023**, *147*, 297–310. [CrossRef] [PubMed]
17. Culham, J.C.; Kanwisher, N.G. Neuroimaging of cognitive functions in human parietal cortex. *Curr. Opin. Neurobiol.* **2001**, *11*, 157–163. [CrossRef] [PubMed]
18. Grill-Spector, K.; Malach, R. The human visual cortex. *Annu. Rev. Neurosci.* **2004**, *27*, 649–677. [CrossRef]
19. Binder, J.R.; Desai, R.H. The neurobiology of semantic memory. *Trends Cogn. Sci.* **2011**, *15*, 527–536. [CrossRef] [PubMed]
20. Bonilha, L.; Hillis, A.E.; Hickok, G.; Den Ouden, D.B.; Rorden, C.; Fridriksson, J. Temporal lobe networks supporting the comprehension of spoken words. *Brain* **2017**, *140*, 2370–2380. [CrossRef] [PubMed]
21. Koessler, L.; Maillard, L.; Benhadid, A.; Vignal, J.P.; Felblinger, J.; Vespignani, H.; Braun, M. Automated cortical projection of EEG sensors: Anatomical correlation via the international 10-10 system. *NeuroImage* **2009**, *46*, 64–72. [CrossRef] [PubMed]
22. Benjamini, Y.; Hochberg, Y. Controlling The False Discovery Rate—A Practical And Powerful Approach To Multiple Testing. *J. R. Stat. Soc. Ser. B Methodol.* **1995**, *57*, 289–300. [CrossRef]
23. Shanahan, M. Dynamical complexity in small-world networks of spiking neurons. *Phys. Rev. E* **2008**, *78*, 041924. [CrossRef] [PubMed]
24. Gazzaley, A.; Nobre, A.C. Top-down modulation: Bridging selective attention and working memory. *Trends Cogn. Sci.* **2012**, *16*, 129–135. [CrossRef]
25. Squire, L.R.; Wixted, J.T. The cognitive neuroscience of human memory since H.M. *Annu. Rev. Neurosci.* **2011**, *34*, 259–288. [CrossRef] [PubMed]
26. Ragland, J.D.; Ranganath, C.; Harms, M.P.; Barch, D.M.; Gold, J.M.; Layher, E.; Lesh, T.A.; MacDonald, A.W.; Niendam, T.A.; Phillips, J.; et al. Functional and Neuroanatomic Specificity of Episodic Memory Dysfunction in Schizophrenia: A Functional Magnetic Resonance Imaging Study of the Relational and Item-Specific Encoding Task. *JAMA Psychiatry* **2015**, *72*, 909–916. [CrossRef]
27. Bisley, J.W.; Goldberg, M.E. Attention, intention, and priority in the parietal lobe. *Annu. Rev. Neurosci.* **2010**, *33*, 1–21. [CrossRef] [PubMed]
28. Kim, H.K.; Blumberger, D.M.; Daskalakis, Z.J. Neurophysiological Biomarkers in Schizophrenia-P50, Mismatch Negativity, and TMS-EMG and TMS-EEG. *Front. Psychiatry* **2020**, *11*, 795. [CrossRef] [PubMed] [PubMed Central]
29. Light, G.A.; Swerdlow, N.R. Future clinical uses of neurophysiological biomarkers to predict and monitor treatment response for schizophrenia. *Ann. N. Y. Acad. Sci.* **2015**, *1344*, 105–119. [CrossRef] [PubMed] [PubMed Central]

entropy

MDPI

Article

Brain Complexity and Parametrization of Power Spectral Density in Children with Specific Language Impairment

Brenda Y. Angulo-Ruiz [1,*], Elena I. Rodríguez-Martínez [1], Francisco J. Ruiz-Martínez [1], Ana Gómez-Treviño [2], Vanesa Muñoz [1], Sheyla Andalia Crespo [1] and Carlos M. Gómez [1]

1 Human Psychobiology Laboratory, Experimental Psychology Department, University of Seville, 41018 Seville, Spain; elisroma@us.es (E.I.R.-M.); frmartinez@us.es (F.J.R.-M.); lmunnoz@us.es (V.M.); sheandcre@alum.us.es (S.A.C.); cgomez@us.es (C.M.G.)

2 Unidad de Desarrollo Infantil y Atención Temprana (UDIATE) Affiliated with the Hospital Victoria Eugenia, Spanish Red Cross, 41009 Seville, Spain; ana.gomezt@cruzroja.es

* Correspondence: bangulo@us.es

Abstract: This study examined spontaneous activity in children aged 3–11 years with specific language impairment (SLI) using an electroencephalogram (EEG). We compared SLI-diagnosed children with a normo-development group (ND). The signal complexity, multiscale entropy (MSE) and parameterized power spectral density (FOOOF) were analyzed, decomposing the PSD into its aperiodic (AP, proportional to 1/fx) and periodic (P) components. The results showed increases in complexity across scales in both groups. Although the topographic distributions were similar, children with SLI exhibited an increased AP component over a broad frequency range (13–45 Hz) in the medial regions. The P component showed differences in brain activity according to the frequency and region. At 9–12 Hz, ND presented greater central–anterior activity, whereas, in SLI, this was seen for posterior–central. At 33–36 Hz, anterior activity was greater in SLI than in ND. At 37–45 Hz, SLI showed greater activity than ND, with a specific increase in the left, medial and right regions at 41–45 Hz. These findings suggest alterations in the excitatory–inhibitory balance and impaired intra- and interhemispheric connectivity, indicating difficulties in neuronal modulation possibly associated with the cognitive and linguistic characteristics of SLI.

Keywords: multiscale entropy; parameterized PSD; aperiodic component; periodic component; specific language impairment

1. Introduction

Specific language impairment (SLI) is classified within Communication Disorders in the DSM-V. It incorporates expressive language disorder and mixed receptive and expressive language disorder from the DSM-IV edition. Within this classification, SLI would be embedded in the Developmental Language Disorder category in the DSM-V [1,2]. It is defined as an impairment in language acquisition and use, arising from inappropriate social conditions, in the absence of physical, cognitive, sensory or social–affective deprivation [3]. Research suggests that these difficulties are influenced by genetic factors [4–7]. Diagnosis is most common in childhood, with an international prevalence ranging from 1.4% to 16.2% of the school population [3,8]. In Andalusia (Spain), the prevalence ranges from 5.26% to 12.58% [3]. SLI often leads to reading difficulties, such as dyslexia [9], and negatively impacts academic performance [10]. Furthermore, it has been associated with emotional, behavioral and social difficulties during adolescence [11–13], as well as a significant reduction in quality of life and well-being in adulthood [14].

https://doi.org/10.3390/e27060572

Several models have been proposed to understand the underlying mechanisms of SLI. Among these, the speed-of-processing limitation hypothesis [15] suggests that children with SLI have difficulty processing rapid auditory stimuli, which impairs their ability to discriminate phonemes with brief transitions and consequently affects language development. Meanwhile, the statistical learning model [16] proposes that individuals with SLI have difficulty extracting implicit regularities from sequences of stimuli, which may explain their difficulties in acquiring grammatical rules. Finally, the temporal processing hypothesis of language [17] proposes that disruptions in neural synchronization with the rhythmic structures of language may serve as a neurobiological basis for SLI. In other words, difficulties in aligning neural activity with speech rhythms may impair the perception of prosodic and phonological patterns, contributing to deficits in language development. Taken together, these models suggest that SLI possibly is not just a language disorder but also involves impairments in learning and auditory processing.

Neurobiological studies in SLI have shown structural, functional and connectivity changes in key brain regions involved in language processing [18]. At a global level, the findings are inconsistent: while some studies report a reduced total brain volume in SLI [19], others find an increased volume [20]. This discrepancy may be explained by variations in cortical morphology, suggesting that the primary affected variable is the surface area rather than the cortical thickness, with lower values observed in the SLI group [21]. At a regional level, an increase in the volume of the left inferior frontal cortex and a decrease in the right caudate nucleus and superior temporal cortex has been observed in children with SLI [22]. Furthermore, in children under 11 years of age, more extensive growth patterns were identified in regions such as the entorhinal, temporopolar, motor-precentral and precuneus cortex [23], as well as structural enlargement in subcortical regions such as the putamen, nucleus accumbens and right globus pallidus [24]. At the microstructural level, based on diffusion tensor imaging (DTI), an increase in white matter in the frontal and temporal regions has been reported [23], as well as reductions in fractional anisotropy (FA) in the globus pallidus and thalamus [24]. In addition, it has been shown that the dorsal striatum has reduced myelin content, which may reduce the neuronal transmission efficiency [25]. In terms of connectivity, findings suggest an abnormal increase in white matter in children with SLI, indicating impaired connectivity across neural networks [18,20,26].

From a functional perspective, atypical patterns of lateralization and activation have been described in the left hemisphere, with reduced or absent activation in the left inferior frontal gyrus and compensatory activation in the right, according to functional magnetic resonance imaging (fMRI) studies [27,28]. In addition, research using single-photon emission computed tomography (SPECT) has shown the abnormal lateralization of blood flow patterns and/or hypoperfusion in language-related regions [29], while studies using functional transcranial Doppler (fTCD) have shown the right hemisphere lateralization of language or a bilateral distribution in SLI [30]. These studies have linked specific language deficits to the atypical development of both ventral [31] and dorsal [32,33] language streams. Furthermore, a persistent deficit in the maturation of these pathways has been proposed in SLI [33].

Neurophysiological studies using electroencephalograms (EEGs) and computing event-related potentials (ERPs) have identified deficits in children with SLI [34]. These studies highlight limitations in the attentional discrimination of auditory stimuli, evidenced by a reduced or absent mismatch negativity (MMN) component [35–37]. Specific deficits have been observed in lexical access and semantic integration, with a delayed [35] or absent N400 component [38,39]. Impaired speech processing and differences in the allocation of attentional resources to auditory stimuli have been reported, reflected in a delayed P100 component [40]. Difficulties in central auditory processing have also been noted, as indicated by alterations in the T-complex [41], which is a bilateral negative component at temporal sites [42], and the P300 component [43]. Furthermore, deficits in processing and storing visual information in working memory have been evidenced by a decreased P3b amplitude [44].

Resting-state EEG studies have analyzed the signal using both linear (power spectral density [PSD]) and non-linear (multiscale entropy [MSE]) measures in ND subjects and those with neurodevelopmental disorders. Regarding the PSD, during typical brain maturation, the absolute spectral power decreases, while the relative spectral power fluctuates, reflecting a transition from slow (delta and theta) to fast (alpha, beta, and gamma) frequencies. This shift is associated with synaptic pruning, which optimizes neuronal transmission [45–50]. As for MSE, maturation is characterized by increased values at fine scales and decreased values at coarse scales, indicating the strengthening of local connections and the refinement of long-range connections with age [51,52].

In neurodevelopmental disorders such as attention-deficit hyperactivity disorder (ADHD) and autism spectrum disorder (ASD), the typical maturation pattern is altered. In ADHD, the PSD increases in specific frequency bands, including delta and theta [50,53–55], and this occurs in theta, alpha and beta in ASD [56–60]. Regarding MSE, studies indicate a less consistent maturation pattern and lower values in both ADHD [61–64] and ASD [65–68]. However, the literature remains inconsistent for both PSD [53,69,70] and MSE [71,72]. For SLI, although research is more limited, studies report increased delta and theta band activity (relative PSD [73]; absolute PSD [74]; absolute PSD [75]) and reduced alpha and beta band PSD values (relative PSD [73]; absolute PSD [75]; absolute PSD [76]). However, recent findings suggest higher spectral power in the beta band in children with SLI compared to ND children (absolute PSD [75]). As far as we know, complexity measures have not been described for the analysis of resting-state EEG in SLI. This lack of evidence, combined with inconsistencies in the PSD literature, underscores the need for further research to clarify the underlying neurophysiological mechanisms of SLI.

Recently, an innovative approach to PSD analysis has been proposed, offering a more detailed and accurate neurophysiological assessment while minimizing information loss in the interpretation of cognitive and behavioral states [77,78]. This method involves decomposing the PSD into two components: the aperiodic component (AP), characterized by a 1/f-like distribution, which provides insights into the neuronal activation level (offset parameter) and the balance between excitatory and inhibitory synaptic currents (exponent parameter) [79,80], and the periodic component (P), which reflects genuine rhythmic patterns of brain activity [77]. This decomposition is based on the assumption that the EEG power spectrum contains oscillatory activity overlaid on a scale-free background signal. Traditional spectral analyses often combine these components, which can obscure physiologically relevant information. By modeling and subtracting the aperiodic component, the newly proposed method isolates the actual oscillatory peaks, allowing for the more accurate characterization of frequency-specific activity. Technically, this approach is commonly implemented using algorithms such as Fitting Oscillations and One Over F (FOOOF), which iteratively fits the aperiodic background and identifies periodic peaks that exceed the aperiodic estimate [77,78]. These components appear to be altered in neurodevelopmental disorders such as ADHD [81–84] and ASD [85].

The present study aimed to apply this new PSD decomposition methodology to a sample of children with SLI compared to ND children to examine more precisely the patterns of neural activity underlying this disorder. Additionally, given the lack of previous studies on brain signal complexity in SLI, MSE was explored to assess the stability and adaptability of neural activity. We expected to observe an increase in MSE with the order of the scale in both groups, with lower complexity values in the SLI group. Regarding the

PSD components, we hypothesized that there would be group differences in the AP and P components, primarily in language-related regions and at the most relevant frequencies, similarly to those observed in canonical PSD analyses [73,75,76]. The selection of resting-state EEG responds to the need for a neurophysiological assessment that is not dependent on task-specific performance, which is especially important in populations with language difficulties, such as children with SLI. These analyses were computed in a broad age range (3–11 years) given the extension of the SLI impairment across ages and in order to obtain a sufficient number of SLI subjects.

2. Materials and Methods

2.1. Sample

A total of 66 subjects participated in this study, divided into two groups: a normo-development (ND) group and a clinical group diagnosed with specific language impairment (SLI). The SLI group consisted of 30 children aged 3 to 10 years (M = 6.37, SD = 1.75, 19 males). Participants were recruited from the Unidad de Desarrollo Infantil y Atención Temprana (UDIATE: https://hospitalveugenia.com/udiate-atencion-temprana-desarrollo-infantil/ accessed on 1 April 2025), affiliated with the Hospital Victoria Eugenia, part of the Spanish Red Cross, which specializes in the assessment and treatment of neurodevelopmental disorders.

The ND group consisted of a total of 36 participants aged 3 to 11 years (M = 6.69, SD = 2.08, 19 males). This group was recruited from different schools in Seville. Parents did not report any neurological diseases, signs of epileptic discharge, learning difficulties or developmental delays in the children.

The experimental protocol was approved by the biomedical research ethics committee of the Autonomous Community of Andalusia, following the guidelines of the Declaration of Helsinki (code: 0818-N-21). Written informed consent was obtained from the parents, who were provided with written information and an explanation regarding the objectives and characteristics of the study.

2.2. Psychological Test

To be included in the study, SLI children required a clinical report confirming their language deficits, supported by language experts from the clinical center (see above) based on tests such as the *Clinical Evaluation of Language Fundamentals-5* (*CELF-5*; [86]), the *Navarre Oral Language Test—Revised* (*PLON-R*; [87]), the *Illinois Test of Psycholinguistic Abilities* (*ITPA*; [88]), the *Peabody Picture Vocabulary Test* (*PPVT-5*; [89]) or the *Kaufman Brief Intelligence Test* (*KBIT*; [90]) or a structured interview following the *DSM-V* or *CIE-10* criteria, all of which were administered by language therapists. Previously, otorhinolaryngologists did not find basic auditory problems according to audiometry and/or brainstem auditory evoked potentials.

The *KBIT* [90] was used to assess the non-verbal cognitive skills of all children participating in the study.

2.3. EEG Recording

Spontaneous brain electrical activity was recorded using EEG for three minutes while participants had their eyes open. Participants were asked to sit in a comfortable position and look at an hourglass. This element was used as a passive aid for visual fixation, minimizing eye movements and maintaining a stable resting state. Given the young age of many participants, traditional fixation strategies, such as the cross on a blank screen, were not considered optimal for maintaining attention. The hourglass provided a slow and engaging visual stimulus that helped to maintain gaze stability, which promoted consistency in the recording conditions across participants. The recording was obtained from 19 electrodes mounted on an electrode cap (ELECTROCAP), selected from the international 10–20 system (Fp1, Fp2, F3, F4, C3, C4, P3, P4, O1, O2, F7, F8, T3, T4, T5, T6, Cz, Fz, Pz). For all participants, an average reference was used, and impedance was kept below 10 kΩ. Data were recorded in direct current at 1024 Hz, without any filtering. The amplification gain was set to 20,000 using an analog-to-digital acquisition and analysis system (ANT Amplifiers, The Netherlands).

2.4. Data Analysis

2.4.1. EEG Pre-Processing

The EEGLAB software package [91] and Matlab R2021b were used to analyze the raw EEG data. Pre-processing consisted of applying (i) a 47–53 Hz notch filter (EEGLAB function: *eegfiltnew*), (ii) an average reference and (iii) an artifact subspace reconstruction (*ASR*) algorithm (EEGLAB function: *clean_raw_data*). ASR was used to correct segments of the data with a standard deviation greater than 20 times that of the calibration data [92]. The epochs had a duration of 4 s (4000 ms). Eyeblink, muscle and other movement artifacts were removed using independent component analysis (*ICA*; function: *pop_runica*) with the natural gradient function [93,94] and the ICLabel extension classification [95] in EEGLAB. The EEG signal was reconstructed, and all epochs with amplitudes exceeding ±120 μV were rejected using the *eegthresh* function.

2.4.2. Multiscale Entropy

Multiscale entropy (MSE) was calculated for all EEG participants and electrodes using the *Multiscale Sample Entropy* function in MATLAB [96], based on the MSE method proposed by Costa et al. [97]. This method measures the signal complexity by calculating the sample entropy (SE; [98]) on multiple time scales using a coarse-graining procedure. MSE provides an estimate of complexity by dividing the EEG signal into non-overlapping windows of different sample sizes and evaluating the repetition frequency of patterns of length 'm' compared to patterns of length 'm + 1'. To calculate MSE, each time scale is defined by averaging neighboring points (p) in the original time series (length τ). A similarity limit (r) is set to define the tolerance range within which neighboring points are considered similar (k), and this limit is normalized to the standard deviation (SD) of the EEG, according to the relationship $k < r \times SD$ [96]. Sample entropy is calculated for each time scale according to formula (1):

$$SE = \log \frac{p^{m}(r)}{p^{(m+1)}(r)} \tag{1}$$

Following previous recommendations in studies on EEG signal complexity [98–102], the parameters were set to m = 2 and r = 0.5. The MSE was calculated for 136 time scales, corresponding to 30 point windows obtained by collapsing 136 consecutive points sampled in each 4-s trial (0.97656 ms × 136 scales = 132.81 ms). A detailed description of the number of points, sample periods and frequencies covered in MSE is provided in Supplementary Table S1. High MSE values are associated with high complexity, indicating less repetition and greater diversity of patterns at different time scales, suggesting an information-rich signal [103]. In contrast, low MSE values reflect greater regularity or predictability in the signal patterns, which may indicate a reduction in information richness [99,104].

In the MSE calculation, the coarse-graining procedure acts as a filter for higher frequencies as the time scales increase. This implies that, at lower scales, all frequencies present in the EEG signal are included, while, at higher scales, only low frequencies are retained.

Recently, this approach has been linked to power spectrum analysis using methods such as Haar wavelets, which also separate frequencies into different levels [105].

2.4.3. Parameterization of *Fitting Oscillations and One over F (FOOOF)*

A MATLAB-adapted version of the *Fitting Oscillations and One Over F (FOOOF)* function (version 3.8), originally implemented in Python, was used to analyze the power spectral density (PSD) data. The FOOOF *specparam* tool (https://fooof-tools.github.io/fooof/reference.html) [77] allowed us to decompose the spectrum (canonical PSD) into the full model fit (adjusted PSD), the aperiodic component (AP) and the periodic component (P). Given the high similarity between the canonical PSD and the fitted PSD observed in a previous study [106] and confirmed in the present data (Supplementary Figure S1), the present study focused exclusively on the adjusted PSD and the AP and P component analyses. The PSD is provided directly from *specparam*, which implements Welch's method, incorporated into the EEGLAB *spectopo* function, to calculate the power spectral density in the range of 1 to 45 Hz.

The *FOOOF* algorithm was configured using the following recommended parameters: peak_width_limits = [1,8], min_peak_height = 0.05, peak_threshold = 0.5, max_n_peaks = 6 and aperiodic_mode = 'fixed' [77,78]. In this approach, the AP component computed from the offset and exponent parameters models the background activity in log–log space, while the P component represents the oscillations that stand out from this activity, modeled by Gaussian functions with peak power, central frequency and bandwidth parameters (see [77] for details). As *FOOOF* does not directly provide a fitted spectrum for the P component, it was obtained by subtracting the AP component (ap_fit in *specparam*) from the adjusted PSD (2):

$$P = PSD\ adjusted - ap_{fit} \quad (2)$$

To assess the goodness-of-fit of the model across all subjects, the explained variance (R^2) and mean absolute error (MAE) metrics were calculated, comparing the fitted model with the original power spectrum. The results showed the good fit of the algorithm in each data set, as shown in Supplementary Figure S1, following the fit criteria reported by Ostlund et al. [107] (MAE underfit > 0.1, MAE overfit < 0.020).

2.5. Statistical Analysis

The assumptions of normality and homogeneity of variances were tested using the Shapiro–Wilk and Levene tests, respectively. When these assumptions were violated, non-parametric analyses were conducted using the Mann–Whitney U test. This approach was applied to assess group differences in demographic and preprocessing variables such as age, biological sex, the number of remaining ICA components and epochs. For variables that met the parametric assumptions, such as the *KBIT* raw scores, independent-sample *t*-tests were used.

Statistical analyses were performed in SPSS v25 and MATLAB for multiscale entropy (MSE) and power spectrum component (PSD) parametrization using the Parametrization of *Fitting Oscillations and One Over F (FOOOF)* model. For both measurements, the analytical strategy for electrodes was a reduction in dimensionality by creating nine regions covering the whole scalp: left anterior, left central, left posterior, medial anterior, medial central, medial posterior, medial posterior, right anterior, right central and right posterior (Supplementary Table S2).

2.5.1. Multiscale Entropy

The MSE was divided into three types of scales [52] in order to reduce the 136 scales to a lower resolution and dimensionality: (i) fine scales, ranging from scale 1 (0.976 ms,

4095 time points) to scale 25 (24.41 ms, 163 time points); (ii) medium scales, ranging from scale 26 (25.39 ms, 157 time points) to scale 46 (44.92 ms, 89 time points); and (iii) coarse scales, ranging from scale 47 (45.89 ms, 87 time points) to scale 136 (132.81 ms, 30 time points) in each group (ND and SLI). For the parameters of the different scales, see Supplementary Table S1. This classification reflects the underlying neural dynamics: fine scales capture high-frequency activity linked to local processing and tend to increase with age; coarse scales capture low-frequency activity related to long-range communication and tend to decrease with age; and medium scales reflect intermediate dynamics with mixed frequency contributions and typically show minimal age-related changes [52].

A repeated-measures analysis of variance (RM-ANOVA) was applied. Within-subjects factors were (i) scale (fine, medium and coarse); (ii) laterality (left, medial, right); and (iii) anteroposterior orientation (anterior, central, posterior). The intersubject factor was group (ND and SLI), with age in days and biological sex as covariates. The statistical power was also computed, and the effect size was calculated directly using the partial eta-squared (ηp^2). Post hoc analyses were performed using Student's t-test for pairwise comparisons, with Cohen's d as the effect size metric [108] and correction for multiple comparisons using the False Discovery Rate method (FDR; [109]). Only significant results will be discussed when including the group factor, and the FDR-corrected post hoc analyses were significant. Some laterality effects are reported for their possible relation to language lateralization.

2.5.2. Parametrization of Fitting Oscillations and One over F (FOOOF)

The aperiodic component (AP), its offset and exponent parameters and the periodic component (P) were analyzed separately. This decision was based on the recommendations of Donoghue et al. [77], who proposed to model these components independently in order to obtain a more accurate interpretation of the underlying neural dynamics. Although both components are derived from the same decomposition model (*FOOOF*), they reflect distinct neurophysiological processes: the AP component has been linked to the balance between excitation and inhibition, while the P component captures specific oscillatory activity. A topographical analysis was performed for each experimental group (ND and SLI), as well as for the differences between both groups (ND-SLI), on the parameters of both components (AP and P). For the P component, topographic maps were generated for 11 frequency bands (4 Hz windows: 1–4 Hz to 41–45 Hz). It should be noted that the last frequency range included frequencies up to 45 Hz.

Subsequently, independent RM-ANOVAs were performed for the AP component parameters (offset and exponent), as well as for the AP and P components, in each of the 11 frequency bands indicated above. In all cases, the same within-subject factors (laterality and anterior–posterior distribution), the between-subject factor (group: ND vs. SLI) and the covariates (age and biological sex) were included. Post hoc analyses were performed following the same criteria as described for the MSE analysis, including correction for multiple comparisons using FDR and calculation of the effect size with Cohen's d.

In response to possible concerns about the sensitivity to outliers, a complementary analysis was performed using a standard z-score method ($z > 3$) to identify and correct for outliers by replacing them with the mean of the respective variable. This procedure was applied to the periodic and aperiodic components separately.

3. Results

3.1. Demographic, Cognitive and Technical Results of Participants

The normality of the age data was assessed using the Shapiro–Wilk test (appropriate for samples under 50 participants). The ND group showed a normal distribution ($p = 0.191$), whereas the SLI group did not ($p < 0.001$). Homogeneity of variances was confirmed with

Levene's test ($p = 0.947$). No significant age differences were found between groups (Mann–Whitney U = 547.5, Z = 0.097, $p = 0.923$).

As biological sex presented only two categories, they were considered as dummy variables, and no significant statistical differences were found in the subject sample (U = 597, Z = 0.857, $p = 0.391$). Consequently, age and biological sex are not considered relevant factors for the discussion of the results.

The *KBIT* raw scores met the assumptions of normality (ND: $p = 0.132$; SLI: $p = 0.840$) and homogeneity of variances ($p = 0.900$). The statistical analysis revealed a significant difference between the ND and SLI groups in cognitive skills (t(55) = 2.22, $p = 0.030$, d = 0.059) (Table 1).

Table 1. Means (M) and standard deviations of the specific language impairment (SLI) and normo-developmental (ND) groups.

	SLI	ND
Age	M = 6.38, SD = 1.75	M = 6.90, SD = 2.08
Males	M = 6.52, SD = 1.71	M = 7.05, SD = 2.09
Females	M = 6.09, SD = 1.87	M = 6.29, SD = 2.05
KBIT	M = 22.52, SD = 7.49	M = 26.63, SD = 6.49
Components	M = 12.93, SD = 0.254	M = 12.56, SD = 0.652
Epochs	M = 27.87, SD = 2.45	M = 32.58, SD = 7.51

Subjects' recordings in the range of 20–45 epochs and 11–13 remaining components were accepted for analysis. The Shapiro–Wilk test revealed that the number of epochs in both groups did not follow a normal distribution (ND: $p < 0.001$; SLI: $p < 0.001$), which was also observed for the components (ND: $p < 0.001$; SLI: $p < 0.001$). Levene's test indicated that homogeneity of variances was not satisfied for the epochs ($p < 0.001$) or for the components ($p < 0.001$). The Mann–Whitney U-test revealed significant differences in epochs (U = 315, Z = -2.99, $p = 0.003$) and components (U = 702, Z = 2.86, $p = 0.004$) between the groups.

Table 1 shows the means and standard deviations of the specific language impairment (SLI) and normo-developmental (ND) groups.

3.2. Multiscale Entropy (MSE)

Figure 1 presents the results regarding multiscale entropy (MSE) across the nine evaluated areas, comparing participants from the ND and SLI groups. In both groups, we observed a consistent increase in MSE with the order of the scales across all considered areas. The RM-ANOVA analysis did not reveal significant effects between the groups. Some laterality effects are reported in the description of Supplementary Table S3 considering its possible relationship with language lateralization.

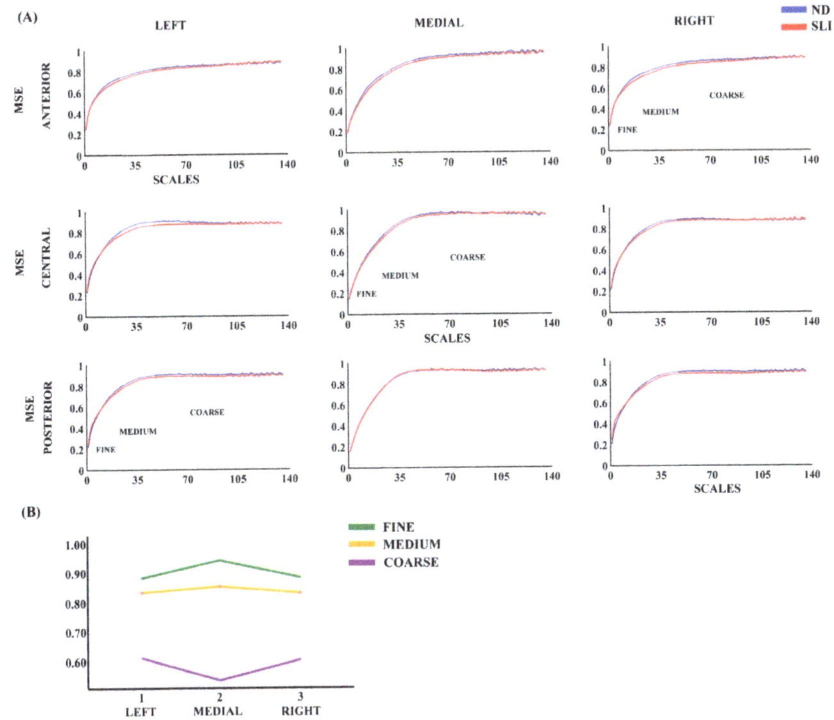

Figure 1. (**A**) Multiscale entropy (MSE) for nine considered areas in each group (normo-development (ND) and specific language impairment (SLI)). (**B**) Marginal means of MSE when collapsing the electrodes across the antero-posterior dimension to observe the laterality effects reported in the RM-ANOVA (Supplementary Table S3).

3.3. *Parametrization of Fitting Oscillations and One over F (FOOOF)*

3.3.1. Aperiodic Component (AP)

Figure 2 shows the topographies of the offset and exponent parameters of the AP component and the differences between the groups (ND-SLI). The offset parameter displays a predominant distribution in the frontocentral and posterior areas, while the exponent is mainly distributed in the frontocentral and parietal areas.

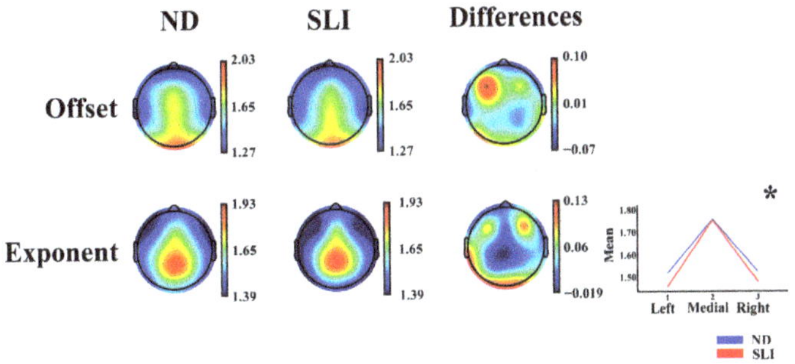

Figure 2. Topographies of the offset and exponent parameters for each group (normo-development (ND) and specific language impairment (SLI)) and the differences between the groups (ND−SLI). The asterisk indicates the interaction of the effects of laterality × group in the RM-ANOVA (marginal means are represented in the inset in the lower-right corner for this significant interaction).

The results of the RM-ANOVA analysis for the offset and exponent parameters of the AP component show a laterality × group interaction for the exponent parameter ($F(1.81,108.498) = 3.48$, $p = 0.039$, np2 = 0.055, power = 0.610). The post hoc results (Supplementary Table S4) did not survive the multiple comparison correction (FDR). Some laterality effects are reported in the description of Supplementary Table S5 considering its possible relationship with language lateralization.

Figure 3 shows the AP component of the PSD in each considered area. Table 2 presents the results of the RM-ANOVA analyzing the group effect for each frequency band of the AP component. Table 3 displays the RM-ANOVA exploring the group effect in the P component. Finally, Table 4 summarizes the significant effects of the post hoc analyses resulting from both analyses after multiple comparison correction (FDR), specifically highlighting the significant laterality × group or anterior–posterior x group interactions, which are described in the following paragraphs.

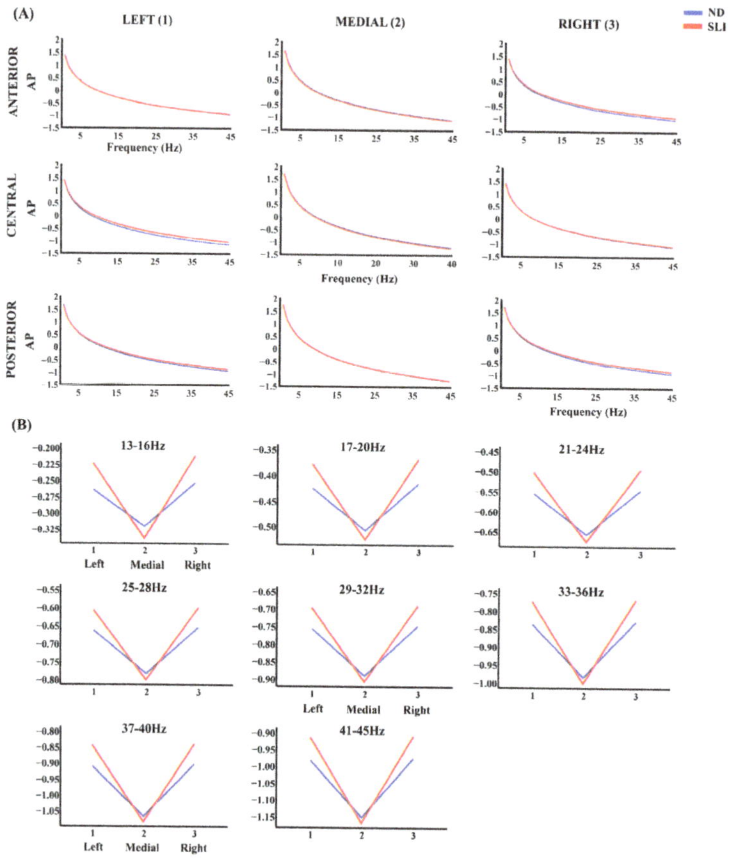

Figure 3. (**A**) Aperiodic component (AP) of the PSD across all frequencies and in each considered area for the ND group (blue line) and the SLI group (red line). (**B**) The marginal means of the AP PSD of the frequencies that showed significant effects of the interaction laterality × group in the RM-ANOVA.

Table 2. Significant group interactions were obtained in the RM-ANOVA of the aperiodic component (AP) of the power spectral density (PSD) with the following factors: subject group (ND and SLI), antero-posterior distribution (anterior, central, posterior) and laterality (left, medial, right). Each RM-ANOVA was performed for each frequency collapse (4 Hz, total 11 frequency ranges). Age in days and biological sex were used as covariates.

Frequency	Within Subjects
13–16 Hz	Laterality × group $p = 0.034$ F(1.89,114.96) = 3.58, np2 = 0.055, power = 0.636
17–20 Hz	Laterality × group $p = 0.025$ F(1.88,114.93) = 4.52, np2 = 0.060, power = 0.676
21–24 Hz	Laterality × group $p = 0.021$ F(1.89,115.11) = 4.12, np2 = 0.063, power = 0.701
25–28 Hz	Laterality × group $p = 0.018$ F(1.89,115.36) = 4.27, np2 = 0.065, power = 0.719
29–32 Hz	Laterality × group $p = 0.016$ F(1.89,115.63) = 4.38, np2 = 0.067, power = 0.731
33–36 Hz	Laterality × group $p = 0.015$ F(1.90,115.90) = 4.47, np2 = 0.068, power = 0.741
37–40 Hz	Laterality × group $p = 0.014$ F(1.90,116.16) = 4.54, np2 = 0.069, power = 0.748
41–45 Hz	Laterality × group $p = 0.013$ F(1.91,116.43) = 4.60, np2 = 0.070, power = 0.755

Table 3. Significant group interactions obtained in the RM-ANOVA of the periodic component (P) of the power spectral density (PSD) with the following factors: subject group (ND and SLI), antero-posterior distribution (anterior, central, posterior) and laterality (left, medial, right). Each RM-ANOVA was performed for each frequency collapse (4 Hz, total 11 frequency ranges). Age in days and biological sex were used as covariates.

	RM-ANOVA	
Frequency	**Within Subjects**	**Between Subjects**
1–4 Hz	Laterality × group $p = 0.050$ F(1.98,120.93) = 3.07, np2 = 0.048, power = 0.581	-
9–12 Hz	Laterality × group $p = 0.017$ F(1.57,95.48) = 4.78, np2 = 0.073, power = 0.710 Antero-posterior × group $p = 0.031$ F(1.97,120.13) = 3.59, np2 = 0.056, power = 0.650	-
13–16 Hz	Antero-posterior × group $p = 0.030$ F(1.89,115.23) = 3.69, np2 = 0.057, power = 0.650	-
33–36 Hz	Antero-posterior × group $p = 0.036$ F(1.53,93.23) = 3.84, np2 = 0.059, power = 0.603	-
37–40 Hz	-	Group $p = 0.005$ F(1,61) = 8.61, np2 = 0.124, power = 0.823
41–45 Hz	Laterality × group $p = 0.037$ F(1.97,120.09) = 3.40, np2 = 0.053, power = 0.626	Group $p = 0.006$ F(1,61) = 8.03, np2 = 0.116, power = 0.796

Table 4. Summary of post hoc analysis for aperiodic and periodic components across selected frequency ranges (4 Hz). Only significant group effect remains after false discovery rate (FDR) correction.

	Aperiodic	
Frequency Range	**Between Subjects**	**Within Subjects**
13–16 Hz	-	Left–Medial (SLI > ND) Right–Medial (SLI > ND)
17–20 Hz	-	Left–Medial (SLI > ND) Right–Medial (SLI > ND)
21–24 Hz		Left–Medial (SLI > ND) Right–Medial (SLI > ND)
25–28 Hz	-	Left–Medial (SLI > ND) Right–Medial (SLI > ND)
29–32 Hz	-	Left–Medial (SLI > ND) Right–Medial (SLI > ND)
33–36 Hz	-	Left–Medial (SLI > ND) Right–Medial (SLI > ND))
37–40 Hz	-	Left–Medial (SLI > ND) Right–Medial (SLI > ND)
41–45 Hz	-	Left–Medial (SLI > ND) Right–Medial (SLI > ND)
	Periodic	
9–12 Hz	-	Central–Anterior (ND>SLI) Posterior–Central (SLI > ND)
33–36 Hz	-	Anterior (SLI > ND)
37–40 Hz	SLI > ND	-
41–45 Hz	SLI > ND	Left (SLI > ND) Medial (SLI > ND) Right (SLI > ND)

For the AP component, post hoc analyses show that, for the 13–16 Hz range, after FDR correction, the left–medial area difference (t(64) = 2.44, p = 0.047, d = 0.60) is higher for the SLI group (M = 0.114, SD = 0.099) compared to the ND group (M = 0.059, SD = 0.084). In addition, the right–medial area difference (t(64) = 2.19, p = 0.047, d = 0.054) shows higher values in the SLI group (M = 0.124, SD = 0.098) than the ND group (M = 0.069, SD = 0.105). For the 17–20 Hz range, interactions between group and laterality were also found. The left–medial area difference (t(64) = 2.53, p = 0.037, d = 0.62) is higher in the SLI group (M = 0.145, SD = 0.105) compared to the ND group (M = 0.085, SD = 0.088). Moreover, the right–medial area difference (t(64) = 2.30, p = 0.037, d = 0.06) is greater in the SLI group (M = 0.153, SD = 0.103) compared to the ND group (M = 0.094, SD = 0.107). Similar patterns are observed at higher frequencies. For 21–24 Hz, the left–medial area difference (t(64) = 2.58, p = 0.031, d = 0.63) is higher in the SLI group (M = 0.169, SD = 0.111) vs. the ND group (M = 0.105, SD = 0.093), as well as the right–medial area difference (t(64) = 2.37, p = 0.031, d = 0.058; SLI: M = 0.177, SD = 0.107; ND: M = 0.113, SD = 0.109). For 25–28 Hz, the left–medial area difference (t(64) = 2.61, p = 0.031, d = 0.638) is higher in the SLI group (M = 0.190, SD = 0.115) compared to the ND group (M = 0.122, SD = 0.097), as well as the right–medial area difference (t(64) = 2.41, p = 0.031, d = 0.060; SLI: M = 0.196, SD = 0.111; ND: M = 0.130, SD = 0.112). For 29–32 Hz, significant laterality × group interactions were observed. The difference between the left and medial areas (t(64) = 2.62, p = 0.026, d = 0.643) shows higher values in the SLI group (M = 0.207, SD = 0.119) compared to the ND group (M = 0.137, SD = 0.100), similarly to the difference between the right and medial areas (t(64) = 2.45, p = 0.026, d = 0.061; SLI: M = 0.213, SD = 0.114; ND: M = 0.144, SD = 0.114). For 33–36 Hz, laterality × group interactions were found regarding the left–medial area difference (t(64) = 2.63, p = 0.024, d = 0.646) and right–medial area difference; (t(64) = 2.48, p = 0.024, d = 0.061). For both, the SLI group had consistently higher scores (left vs. medial: M = 0.223, SD = 0.124; right vs. medial: M = 0.227, SD = 0.117) compared to the ND group (left vs. medial: M = 0.149, SD = 0.104; right vs. medial: M = 0.156, SD = 0.116). For 37–40 Hz, interactions were also observed in the left–medial area difference (t(64) = 2.64, p = 0.023, d = 0.648) and the right–medial area difference (t(64) = 2.50, p = 0.023, d = 0.062), being higher in the SLI group (left–medial: M = 0.237, SD = 0.127; right–medial: SLI: M = 0.240, SD = 0.119) compared to the ND group (left–medial: M = 0.161, SD = 0.107; right vs. medial: M = 0.167, SD = 0.118). Finally, significant interactions of Laterality × group were found for 41–45 Hz. The left–medial area difference (t(64) = 2.65, p = 0.022, d = 0.649) is higher in the SLI group (M = 0.251, SD = 0.131) compared to the ND group (M = 0.172, SD = 0.110), and the right–medial area difference (t(64) = 2.51, p = 0.022, d = 0.062) is higher in the SLI group (M = 0.255, SD = 0.122) compared to the ND group (M = 0.178, SD = 0.120). Supplementary Table S6 shows the significant interactions without the group effect.

3.3.2. Periodic Component (P)

Figure 4 shows the topographies of the periodic component in both groups (ND and SLI) and their differences (ND-SLI). A well-defined regional distribution is observed for the different frequencies, with a higher intensity in the SLI group in high frequencies.

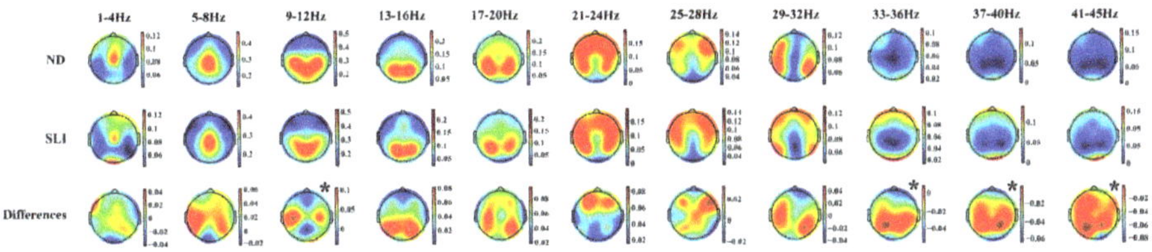

Figure 4. Topographies of the periodic component in the 11 collapsed frequency bands (4 Hz) for each group (ND and SLI) and the differences between them. The asterisk indicates significant differences between groups in the RM-ANOVA.

Figure 5 shows the periodic component's power values (P) in each defined area. The RM-ANOVA results for the P component of the PSD revealed significant group differences across specific frequency ranges (Table 3). Post hoc analyses (summary in Table 4) indicated that, in the 37–40 Hz range (t(64) = 3.27, p = 0.003, d = 0.787), the SLI group (M = 0.047, SD = 0.039) had significantly higher values compared to the ND group (M = 0.022, SD = 0.020). Similarly, in the 41–45 Hz range (t(64) = 3.23, p = 0.004, d = 0.774), the SLI group (M = 0.048, SD = 0.040) had higher values than the ND group (M = 0.023, SD = 0.019). Interactions of the effect of the group with laterality were also found. Additionally, for the same frequency range (9–12 Hz), a significant antero-posterior x group interaction was found. In this case, the difference between the central and anterior areas (t(64) = 2.29, p = 0.038, d = 0.057) is higher in the ND group (M = 0.194, SD = 0.101) compared to the SLI group (M = 0.142, SD = 0.076). Moreover, the difference between the posterior and central areas (t(64) = 2.37, p = 0.038, d = 0.058) shows higher values in the SLI group (M = 0.066, SD = 0.078) than in the ND group (M = 0.017, SD = 0.086). In the 33–36 Hz range, an antero-posterior × group interaction was found. Post hoc analyses revealed that the anterior area (t(64) = 2.97, p = 0.013, d = 0.72) had significantly higher values in the SLI group (M = 0.069, SD = 0.058) compared to the ND group (M = 0.035, SD = 0.034). For the 41–45 Hz range, a significant Laterality × group interaction was observed, with differences across all three areas. In the left area (t(64) = 2.46, p = 0.035, d = 0.591), the SLI group (M = 0.054, SD = 0.048) showed significantly higher values compared to the ND group (M = 0.031, SD = 0.026). A similar pattern was found in the medial area (t(64) = 2.29, p = 0.035, d = 0.551), where the SLI group (M = 0.029, SD = 0.038) had higher values than the ND group (M = 0.013, SD = 0.020). Finally, the right area (t(64) = 3.86, p = 0.002, d = 0.927) exhibited the most pronounced difference, with the SLI group (M = 0.060, SD = 0.047) showing significantly higher values than the ND group (M = 0.026, SD = 0.024). In Supplementary Table S7, we show the significant interactions without the group effect. Significant post hoc results that did not survive multiple-comparisons correction are shown in Supplementary Table S8.

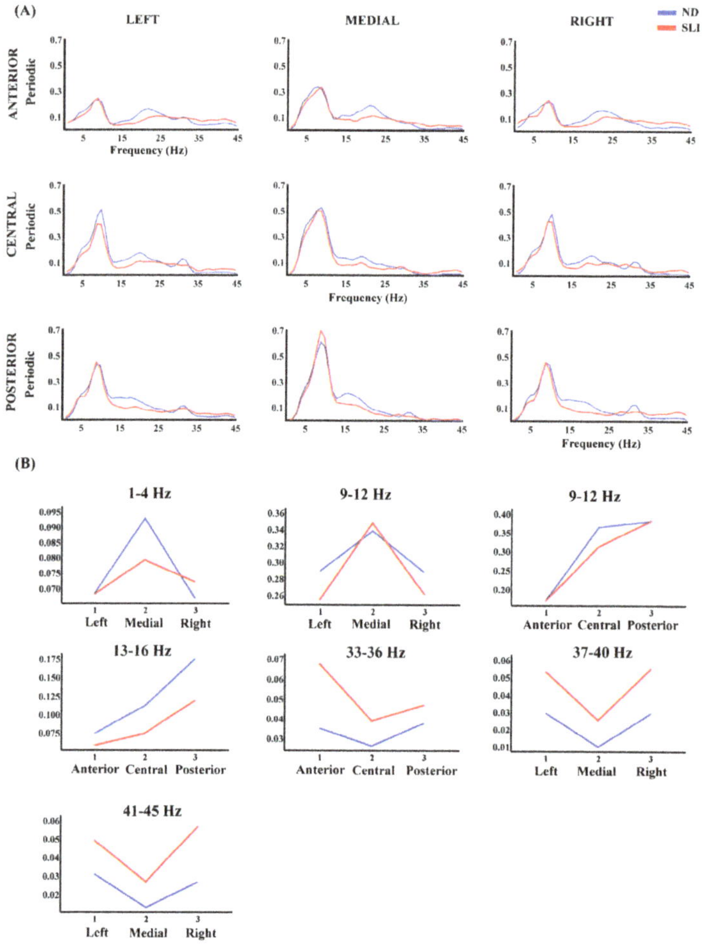

Figure 5. (**A**) Periodic component of the PSD across all frequencies and in each considered area for the ND group (blue line) and SLI group (red line). (**B**) Marginal means of the significant interaction effects of the factors Laterality × group and antero-posterior x group in the RM-ANOVA.

As a complementary robustness check, an additional RM-ANOVA analysis was performed after correcting for possible outliers (z-score > 3) in the periodic and aperiodic components. This analysis revealed a slight improvement in the stability and statistical robustness of the previously reported effects, as well as some additional significant interactions (see Supplementary Table S9). Notably, the proportion of corrected data remained low across all groups and components (ND: aperiodic = 0.79%, periodic = 1.77%; SLI: aperiodic = 0.07%, periodic = 1.25%).

4. Discussion

This study analyzed brain activity at rest in a group of children diagnosed with SLI compared to an ND group. The objective was, on one hand, to assess the stability and adaptability of neuronal activity measured through MSE. On the other hand, the study aimed to examine more precisely the patterns of neuronal activity underlying this disorder, measured with an innovative approach that decomposes the PSD into its AP and P components.

4.1. *Multiscale Entropy (MSE)*

Our findings show an increase in MSE values across the scales in both groups, but with a non-significant tendency toward lower values in the SLI group. These findings are consistent with previous studies on complexity in neurodevelopmental disorders, such as ADHD [61–64] and ASD [65–68], which have reported lower variability in neural network dynamics. However, the absence of significant differences between the groups suggests that the variability in the neural dynamics of children with SLI would be more similar to that of typically developing children, indicating that their adaptability, although slightly reduced, is not as compromised as in other neurodevelopmental disorders.

4.2. *Parametrization of Fitting Oscillations and One over F (FOOOF)*

4.2.1. Topographies of the Exponent and Offset Parameters of the Aperiodic Component

The topographies of the AP component parameters (offset and exponent) show independence in the regional distributions for each parameter, consistent with previous studies [110,111]. The offset parameter presents a predominant distribution in the frontocentral and posterior regions, with higher values in posterior areas, and the exponent parameter is primarily distributed in frontocentral and parietal areas, being consistent with previous results [106]. These findings suggest that the offset and exponent reflect distinct neurophysiological mechanisms. While the offset is related to the overall cortical excitability and global electroencephalogram amplitude [112–114], its distribution suggests that it may be influenced by the default mode network [115,116]. On the other hand, the exponent is linked to the spectral organization of neuronal activity and the balance between excitatory and inhibitory processes [79,80], and it seems to focus on regions crucial for cognitive control and sensory integration. The lower value of the exponent in the SLI group could indicate alterations in cortical dynamics, with the impaired maturation of the balance of excitatory and inhibitory synaptic currents (E/I, balance), as observed in other neurodevelopmental disorders [81,83–85], suggesting alterations in neuronal communication efficiency and sensory and cognitive processing in children with SLI. This change in the E/I balance, indexed by the decrease in the exponent in the SLI group, could have been due to a reduced inhibitory or some increased excitatory synaptic currents in the SLI group during development.

4.2.2. Aperiodic Component (AP)

The analysis of the aperiodic component (AP) shows significant interactions between the laterality and group, with higher aperiodic values in the SLI group in the 13–45 Hz frequency range. These alterations in the AP component suggest the atypical modulation of background activity and increased cortical excitability in the lateral areas in the SLI group. The elevated aperiodic component values in this same region for the SLI group, if due to a lower exponent—as suggested by the statistical trend for a lower exponent in SLI—would suggest an atypical E/I balance with a bias for excitatory activity [79,80].

These spectral differences in lateralized regions could be related to dysfunctions in neural connectivity, which have been associated in the literature with language development [117,118]. However, as this study did not include direct measures of functional connectivity, this interpretation should be viewed with caution and considered as a hypothesis for future studies integrating measures of connectivity and language performance.

4.2.3. Topographies of the Periodic Component

Moreover, the topographies of the P component exhibit a well-defined regional distribution across the different frequency ranges analyzed. This regional organization aligns with the canonical frequency bands described in the literature, including delta (1–4 Hz), theta (4–8 Hz), alpha (8–12 Hz), beta (12–30 Hz) and gamma (>30 Hz) [49,50,77,119–121]. The low P beta topography (13–20 Hz) is slightly different from the canonical PSD beta topography given that the posterior frequency in P beta is prolonged to higher frequencies than in the canonical beta. The distribution patterns observed in this study are consistent with those reported in a previous parametrization analysis conducted on a sample of 240 neurotypical participants [106]. The increased intensity at higher frequencies (33–45 Hz) in the SLI group suggests that, while the spectral organization of low-frequency (1–8 Hz) oscillations remains preserved in these children, differences emerge in the magnitude of this organization at higher frequencies. This finding may indicate that, although the fundamental architecture of rhythmic neural activity is not disrupted in SLI, the deficits observed in language processing and other cognitive functions, such as attention [40,122], memory [123] and executive function [124,125], could be associated with reduced neuronal synchronization efficiency, impairing the modulation of high-frequency oscillatory activity in these children.

4.2.4. Periodic Component (P)

The results of the group comparison for the P component show an increase in power at high frequencies (37–45 Hz) in the SLI group compared to the ND group. Previous studies have reported an increase in activity in slower bands (delta and theta) and a decrease in faster bands (alpha and beta) in children with SLI [73–76]. However, our results indicate a different pattern in the high-frequency gamma band (33–45 Hz), which has been linked to processes of sensory integration, neuronal synchronization and cognitive modulation [126]. An increase in the gamma P band in SLI could reflect an imbalance, once again, in the regulation of cortical excitability, possibly related to the reduced efficiency of inhibitory mechanisms, as suggested by the AP PSD component, increasing the neuronal capacity for the generation of synchronization for high frequencies [126] in children diagnosed with SLI.

These results are not homogeneous across all regions or high frequencies but instead vary based on the laterality (41–45 Hz) and antero-posterior distribution (33–36 Hz). Thus, we found (i) greater high-frequency power (41–45 Hz) in the SLI group in the left, medial and right areas, suggesting a disruption in the synchronization of interhemispheric networks, and (ii) greater power in the anterior area for the SLI group (33–36 Hz), suggesting an impairment in intra-hemispheric synchronization. This could be related to alterations in executive control and attentional modulation—functions that are strongly associated with activity in the prefrontal cortex [127]. These findings, on the one hand, reinforce previous evidence of structural alterations in lateralized regions (left and right) in children with SLI [22,27]. Additionally, they suggest that these differences are not solely explained by an increase or decrease in certain frequency bands but by an atypical pattern of oscillatory modulation between regions, which could impact the intra- and inter-hemispheric functional dynamics of the SLI brain.

On the other hand, the results for the 9–12 Hz range, corresponding to the canonical alpha band [128], show that the difference between the central area and the anterior area is greater in ND children, which is similar to the results of Stanojević et al. [76], while the difference between the posterior area and the central area is greater in children with an SLI diagnosis. This suggests an antero-posterior alteration of these frequencies and the modulation of oscillatory activity. Previous studies have indicated that alpha rhythm modulation is key for corticocortical and intracortical coordination [129], as well as for the integration of sensory and linguistic information [130,131]. The differences found in this study, regarding the distribution of alpha oscillatory activity, could be related to difficulties in neuronal synchronization between regions associated with processes such as cortical inhibition, attention regulation and language [130], as well as with the modulation of sensory perception [131] in children with SLI.

4.3. Limitations

While these findings highlight the importance of research on how brain activity in neurodevelopmental disorders such as SLI could reflect alterations in the underlying neuronal dynamics, this study presents some limitations that must be considered. Firstly, the small sample size could have led to the lack of significance in some of our results, such as the MSE, although the observed trend suggests similarities with other disorders, such as ADHD [64] and ASD [68]. Moreover, the different tests used for the delimitation of the diagnostic threshold do not permit the computation of correlations between behavioral tests and EEG-derived parameters. A methodological limitation of this study is that no correction for multiple comparisons was applied to the p-values obtained in the RM-ANOVA models by frequency band. Although FDR correction was applied to the post hoc analyses, the band omnibus models were not corrected. This decision was based on the fact that the main objective was to identify general patterns of interaction and differences between groups, rather than to derive independent statistical inferences for each frequency band. However, we recognize that this approach may increase the risk of type I errors, so the results should be interpreted with caution. Furthermore, although our main analyses were conducted without modifying the outliers, a complementary robustness analysis using z-score correction revealed consistent and even strengthened effects in some cases. These results support the robustness of our findings despite the presence of a small number of outliers. However, we recognize that the definition of outliers in physiological data is context-dependent, and future studies could further explore this aspect using robust statistical methods. Despite these limitations, the findings of this study provide a better understanding of the differences in connectivity, measured indirectly, and the modulation of brain activity, which could have implications for diagnosis and the development of interventions to improve cognitive and linguistic performance in these children.

5. Conclusions

This study analyzed brain activity in children with specific language impairment (SLI) compared to typically developing children (ND), using measures of neural complexity (MSE) and the decomposition of the power spectrum (PSD) into its periodic (P) and aperiodic (AP) components. Although no significant differences were observed in the multiscale entropy (MSE) or in the overall topographical organization of the AP and P components, statistically significant differences were identified in specific frequency bands and specific brain regions, after correction for multiple comparisons. These alterations could reflect changes in spectral dynamics linked to neurophysiological processes such as cortical excitation–inhibition balance or neuronal synchronization, particularly at high frequencies and in the alpha band. It is important to note that no direct functional connectivity analyses were performed, nor was the involvement of networks such as the default mode network (DMN) assessed, so such interpretations should be viewed with caution.

From a clinical perspective, these findings underline the value of analyzing the AP and P components of the PSD separately to identify atypical neuroelectrical patterns in children with language difficulties. In the future, the more precise characterization of these differences could contribute to the development of EEG biomarkers that support early diagnosis and the personalization of interventions in SLI. However, further studies should include direct measures of functional connectivity and correlations with linguistic and cognitive performance to confirm and extend these results.

Supplementary Materials: The following supporting information can be downloaded at: https://www.mdpi.com/article/10.3390/e27060572/s1. Supplementary Figure S1: Goodness-of-fit histograms; Supplementary Table S1: Description of the types of MSE scales; Supplementary Table S2: Areas of interest; Supplementary Table S3: RM-ANOVA of the MSE; Supplementary Table S4: Post-hoc analysis of exponent parameter; Supplementary Table S5: RM-ANOVA of exponent and offset parameters; Supplementary Table S6: RM-ANOVA of aperiodic component; Supplementary Table S7: RM-ANOVA of periodic component; Supplementary Table S8: Post-hoc analysis of the periodic component; Supplementary Table S9: Additional significant effects observed in the RM-ANOVA of Aperiodic and Periodic components

Author Contributions: Conceptualización, B.Y.A.-R., E.I.R.-M., A.G.-T. and C.M.G.; Curación de datos, B.Y.A.-R.; Análisis formal, B.Y.A.-R. and C.M.G.; Adquisición de financiación, E.I.R.-M. and C.M.G.; Investigación, B.Y.A.-R. and C.M.G.; Metodología, B.Y.A.-R., E.I.R.-M., V.M. and C.M.G.; Administración de proyectos, F.J.R.-M. and C.M.G.; Recursos, E.I.R.-M., F.J.R.-M., A.G.-T. and S.A.C.; Software, B.Y.A.-R.; Supervisión, E.I.R.-M., F.J.R.-M. and C.M.G.; Validación, V.M. and C.M.G.; Visualización, B.Y.A.-R. and A.G.-T.; Escritura—borrador original, B.Y.A.-R., V.M. and C.M.G.; Redacción, revisión y edición, B.Y.A.-R., E.I.R.-M., F.J.R.-M., V.M., S.A.C. and C.M.G. All authors have read and agreed to the published version of the manuscript.

Funding: This research was funded by Agencia Estatal de Investigación (AEI), grant number PID2022–139151OB-I00 (FEDER funds).

Institutional Review Board Statement: The experimental protocol was approved by the biomedical research ethics committee of the Autonomous Community of Andalusia, following the guidelines of the Declaration of Helsinki (code: 0818-N-21).

Data Availability Statement: The data and code related to the present study are available under reasonable request to the corresponding author (bangulo@us.es).

Acknowledgments: We are grateful to the children, adolescents and young adults who participated in the present study. We also wish to express our gratitude to the Unidad de Desarrollo Infantil y Atención Temprana (UDIATE), affiliated with the Victoria Eugenia Hospital of the Spanish Red Cross, for their valuable assistance in referring children for registration. The authors have reviewed and edited the output and take full responsibility for the content of this publication.

Conflicts of Interest: The authors declare no conflicts of interest.

Abbreviations

The following abbreviations are used in this manuscript:

ADHD	Attention-Deficit Hyperactivity Disorder
AP	Aperiodic
ASD	Autism Spectrum Disorder
ASR	Artifact Subspace Reconstruction
CELF	Clinical Evaluation of Language Fundamentals
DTI	Diffusion Tensor Imaging
EEG	Electroencephalogram
ERPs	Event-Related Potentials
FDR	False Discovery Rate

fMRI	Functional Magnetic Resonance Imaging
FOOOF	Fitting Oscillations and One Over F
fTCD	Functional Transcranial Doppler
ICA	Independent Component Analysis
ITPA	Illinois Test of Psycholinguistic Abilities
KBIT	Kaufman Brief Intelligence Test
M	Mean
MAE	Mean Absolute Error
MMN	Mismatch Negativity
MSE	Multiscale Entropy
ND	Normo-Development
P	Periodic
PLON-R	Navarre Oral Language Test—Revised
PSD	Power Spectral Density
PPVT-5	Peabody Picture Vocabulary Test
RM-ANOVA	Repeated-Measures Analysis of Variance
SD	Standard Deviation
SE	Sample Entropy
SLI	Specific Language Impairment
SPECT	Single-Photon Emission Computed Tomography
UDIATE	Unidad de Desarrollo Infantil y Atención Temprana

References

1. American Psychiatric Association. *Diagnostics and Statistical Manual of Mental Disorders, IV Text Revision*; American Psychiatric Association: Arlington, VA, USA, 2000.
2. American Psychiatric Association. DSM-V Development. 2012. Available online: https://www.federaciocatalanatdah.org/wp-content/uploads/2018/12/dsm5-manualdiagnsticoyestadisticodelostrastornosmentales-161006005112.pdf (accessed on 1 December 2024).
3. Villegas, L.F. Specific language impairment in Andalusia, Spain: Prevelance by subtype and gender. *Rev. Logop. Foniatría Audiol.* **2022**, *42*, 147–157. [CrossRef]
4. Bishop, D.V. The role of genes in the etiology of specific language impairment. *J. Commun. Disord.* **2002**, *35*, 311–328. [CrossRef] [PubMed]
5. Bishop, D.V. Genes, cognition, and communication: Insights from neurodevelopmental disorders. *Ann. N. Y. Acad. Sci.* **2009**, *1156*, 1–18. [CrossRef]
6. Bishop, D.V. Cerebral asymmetry and language development: Cause, correlate, or consequence? *Science* **2013**, *340*, 1230531. [CrossRef]
7. Li, N.; Bartlett, C.W. Defining the genetic architecture of human developmental language impairment. *Life Sci.* **2012**, *90*, 469–475. [CrossRef]
8. Tomblin, J.B.; Records, N.L.; Buckwalter, P.; Zhang, X.; Smith, E.; O'Brien, M. Prevalence of specific language impairment in kindergarten children. *J. Speech Lang. Hear. Res.* **1997**, *40*, 1245–1260. [CrossRef]
9. Pennington, B.F.; Bishop, D.V. Relations among speech, language, and reading disorders. *Annu. Rev. Psychol.* **2009**, *60*, 283–306. [CrossRef] [PubMed]
10. Young, A.R.; Beitchman, J.H.; Johnson, C.; Douglas, L.; Atkinson, L.; Escobar, M.; Wilson, B. Young adult academic outcomes in a longitudinal sample of early identified language impaired and control children. *J. Child Psychol. Psychiatry Allied Discip.* **2002**, *43*, 635–645. [CrossRef]
11. Wadman, R.; Botting, N.; Durkin, K.; Conti-Ramsden, G. Changes in emotional health symptoms in adolescents with specific language impairment. *Int. J. Lang. Commun. Disord.* **2011**, *46*, 641–656. [CrossRef]
12. Wadman, R.; Durkin, K.; Conti-Ramsden, G. Close relationships in adolescents with and without a history of specific language impairment. *Lang. Speech Hear. Serv. Sch.* **2011**, *42*, 41–51. [CrossRef]
13. Wadman, R.; Durkin, K.; Conti-Ramsden, G. Social stress in young people with specific language impairment. *J. Adolesc.* **2011**, *34*, 421–431. [CrossRef] [PubMed]
14. Arkkila, E.; Rasanen, P.; Roine, R.P.; Vilkman, E. Specific language impairment in childhood is associated with impaired mental and social well-being in adulthood. *Logop. Phoniatr. Vocol.* **2008**, *33*, 179–189. [CrossRef] [PubMed]
15. Tallal, P. Improving language and literacy is a matter of time. *Nat. Rev. Neurosci.* **2004**, *5*, 721–728. [CrossRef]

16. Hsu, H.J.; Tomblin, J.B.; Christiansen, M.H. Impaired statistical learning of non-adjacent dependencies in adolescents with specific language impairment. *Front. Psychol.* **2014**, *5*, 175. [CrossRef]
17. Goswami, U. Speech rhythm and language acquisition: An amplitude modulation phase hierarchy perspective. *Ann. N. Y. Acad. Sci.* **2019**, *1453*, 67–78. [CrossRef]
18. Abbott, N.; Love, T. Bridging the Divide: Brain and Behavior in Developmental Language Disorder. *Brain Sci.* **2023**, *13*, 1606. [CrossRef]
19. Gauger, L.M.; Lombardino, L.J.; Leonard, C.M. Brain morphology in children with specific language impairment. *J. Speech Lang. Hear. Res.* **1997**, *40*, 1272–1284. [CrossRef]
20. Herbert, M.R.; Ziegler, D.A.; Makris, N.; Filipek, P.A.; Kemper, T.L.; Normandin, J.J.; Sanders, H.A.; Kennedy, D.N.; Caviness, V.S., Jr. Localization of white matter volume increase in autism and developmental language disorder. *Ann. Neurol.* **2004**, *55*, 530–540. [CrossRef] [PubMed]
21. Bahar, N.; Cler, G.J.; Krishnan, S.; Asaridou, S.S.; Smith, H.J.; Willis, H.E.; Healy, M.P.; Watkins, K.E. Differences in Cortical Surface Area in Developmental Language Disorder. *Neurobiol. Lang.* **2024**, *5*, 288–314. [CrossRef]
22. Badcock, N.A.; Bishop, D.V.; Hardiman, M.J.; Barry, J.G.; Watkins, K.E. Co-localisation of abnormal brain structure and function in specific language impairment. *Brain Lang.* **2012**, *120*, 310–320. [CrossRef]
23. Soriano-Mas, C.; Pujol, J.; Ortiz, H.; Deus, J.; López-Sala, A.; Sans, A. Age-related brain structural alterations in children with specific language impairment. *Hum. Brain Mapp.* **2009**, *30*, 1626–1636. [CrossRef] [PubMed]
24. Lee, J.C.; Nopoulos, P.C.; Bruce Tomblin, J. Abnormal subcortical components of the corticostriatal system in young adults with DLI: A combined structural MRI and DTI study. *Neuropsychologia* **2013**, *51*, 2154–2161. [CrossRef] [PubMed]
25. Krishnan, S.; Cler, G.J.; Smith, H.J.; Willis, H.E.; Asaridou, S.S.; Healy, M.P.; Papp, D.; Watkins, K.E. Quantitative MRI reveals differences in striatal myelin in children with DLD. *eLife* **2022**, *11*, e74242. [CrossRef] [PubMed]
26. Herbert, M.R.; Ziegler, D.A.; Makris, N.; Bakardjiev, A.; Hodgson, J.; Adrien, K.T.; Kennedy, D.N.; Filipek, P.A.; Caviness, V.S., Jr. Largerbrain and white matter volumes in children with developmental language disorder. *Dev. Sci.* **2003**, *6*, F11–F22. [CrossRef]
27. Hugdahl, K.; Gundersen, H.; Brekke, C.; Thomsen, T.; Rimol, L.M.; Ersland, L.; Niemi, J. FMRI brain activation in a finnish family with specific language impairment compared with a normal control group. *J. Speech Lang. Hear. Res.* **2004**, *47*, 162–172. [CrossRef]
28. de Guibert, C.; Maumet, C.; Jannin, P.; Ferré, J.C.; Tréguier, C.; Barillot, C.; Le Rumeur, E.; Allaire, C.; Biraben, A. Abnormal functional lateralization and activity of language brain areas in typical specific language impairment (developmental dysphasia). *Brain* **2011**, *134*, 3044–3058. [CrossRef]
29. Ors, M.; Ryding, E.; Lindgren, M.; Gustafsson, P.; Blennow, G.; Rosén, I. SPECT findings in children with specific language impairment. *Cortex J. Devoted Study Nerv. Syst. Behav.* **2005**, *41*, 316–326. [CrossRef]
30. Whitehouse, A.J.; Bishop, D.V. Cerebral dominance for language function in adults with specific language impairment or autism. *Brain J. Neurol.* **2008**, *131*, 3193–3200. [CrossRef]
31. Vydrova, R.; Komarek, V.; Sanda, J.; Sterbova, K.; Jahodova, A.; Maulisova, A.; Zackova, J.; Reissigova, J.; Krsek, P.; Kyncl, M. Structural alterations of the language connectome in children with specific language impairment. *Brain Lang.* **2015**, *151*, 35–41. [CrossRef]
32. Roberts, T.P.; Heiken, K.; Zarnow, D.; Dell, J.; Nagae, L.; Blaskey, L.; Solot, C.; Levy, S.E.; Berman, J.I.; Edgar, J.C. Left hemisphere diffusivity of the arcuate fasciculus: Influences of autism spectrum disorder and language impairment. *Am. J. Neuroradiol.* **2014**, *35*, 587–592. [CrossRef]
33. Lee, J.C.; Dick, A.S.; Tomblin, J.B. Altered brain structures in the dorsal and ventral language pathways in individuals with and without developmental language disorder (DLD). *Brain Imaging Behav.* **2020**, *14*, 2569–2586. [CrossRef] [PubMed]
34. Bishop, D.V.; McArthur, G.M. Individual differences in auditory processing in specific language impairment: A follow-up study using event-related potentials and behavioural thresholds. *Cortex J. Devoted Study Nerv. Syst. Behav.* **2005**, *41*, 327–341. [CrossRef] [PubMed]
35. Shafer, V.L.; Morr, M.L.; Datta, H.; Kurtzberg, D.; Schwartz, R.G. Neurophysiological indexes of speech processing deficits in children with specific language impairment. *J. Cogn. Neurosci.* **2005**, *17*, 1168–1180. [CrossRef] [PubMed]
36. Datta, H.; Shafer, V.L.; Morr, M.L.; Kurtzberg, D.; Schwartz, R.G. Electrophysiological indices of discrimination of long-duration, phonetically similar vowels in children with typical and atypical language development. *J. Speech Lang. Hear. Res.* **2010**, *53*, 757–777. [CrossRef]
37. Kujala, T.; Leminen, M. Low-level neural auditory discrimination dysfunctions in specific language impairment-A review on mismatch negativity findings. *Dev. Cogn. Neurosci.* **2017**, *28*, 65–75. [CrossRef]
38. Sabisch, B.; Hahne, A.; Glass, E.; von Suchodoletz, W.; Friederici, A.D. Lexical–semantic processes in children with specific language impairment. *NeuroReport* **2006**, *17*, 1511–1514. [CrossRef]
39. Haebig, E.; Leonard, L.; Usler, E.; Deevy, P.; Weber, C. An Initial Investigation of the Neural Correlates of Word Processing in Preschoolers with Specific Language Impairment. *J. Speech Lang. Hear. Res.* **2018**, *61*, 729–739. [CrossRef]

40. Shafer, V.L.; Ponton, C.; Datta, H.; Morr, M.L.; Schwartz, R.G. Neurophysiological indices of attention to speech in children with specific language impairment. *Clin. Neurophysiol. Off. J. Int. Fed. Clin. Neurophysiol.* **2007**, *118*, 1230–1243. [CrossRef]
41. Shafer, V.L.; Schwartz, R.G.; Martin, B. Evidence of deficient central speech processing in children with specific language impairment: The T-complex. *Clin. Neurophysiol. Off. J. Int. Fed. Clin. Neurophysiol.* **2011**, *122*, 1137–1155. [CrossRef]
42. Albrecht, R.; Suchodoletz, W.; Uwer, R. The development of auditory evoked dipole source activity from childhood to adulthood. *Clin. Neurophysiol. Off. J. Int. Fed. Clin. Neurophysiol.* **2000**, *111*, 2268–2276. [CrossRef]
43. Shaheen, E.A.; Shohdy, S.S.; Abd Al Raouf, M.; Mohamed El Abd, S.; Abd Elhamid, A. Relation between language, audio-vocal psycholinguistic abilities and P300 in children having specific language impairment. *Int. J. Pediatr. Otorhinolaryngol.* **2011**, *75*, 1117–1122. [CrossRef]
44. Evans, J.L.; Selinger, C.; Pollak, S.D. P300 as a measure of processing capacity in auditory and visual domains in specific language impairment. *Brain Res.* **2011**, *1389*, 93–102. [CrossRef]
45. Gasser, T.; Verleger, R.; Bächer, P.; Sroka, L. Development of the EEG of school-age children and adolescents. I. Analysis of band power. *Electroencephalogr. Clin. Neurophysiol.* **1988**, *69*, 91–99. [CrossRef]
46. Whitford, T.J.; Rennie, C.J.; Grieve, S.M.; Clark, C.R.; Gordon, E.; Williams, L.M. Brain maturation in adolescence: Concurrent changes in neuroanatomy and neurophysiology. *Hum. Brain Mapp.* **2007**, *28*, 228–237. [CrossRef]
47. Segalowitz, S.J.; Santesso, D.L.; Jetha, M.K. Electrophysiological changes during adolescence: A review. *Brain Cogn.* **2010**, *72*, 86–100. [CrossRef] [PubMed]
48. Miskovic, V.; Ma, X.; Chou, C.A.; Fan, M.; Owens, M.; Sayama, H.; Gibb, B.E. Developmental changes in spontaneous electrocortical activity and network organization from early to late childhood. *NeuroImage* **2015**, *118*, 237–247. [CrossRef]
49. Rodríguez-Martínez, E.I.; Ruiz-Martínez, F.J.; Barriga Paulino, C.I.; Gómez, C.M. Frequency shift in topography of spontaneous brain rhythms from childhood to adulthood. *Cogn. Neurodyn.* **2017**, *11*, 23–33. [CrossRef]
50. Rodríguez-Martínez, E.I.; Angulo-Ruiz, B.Y.; Arjona-Valladares, A.; Rufo, M.; Gómez-González, J.; Gómez, C.M. Frequency coupling of low and high frequencies in the EEG of ADHD children and adolescents in closed and open eyes conditions. *Res. Dev. Disabil.* **2020**, *96*, 103520. [CrossRef] [PubMed]
51. Lea-Carnall, C.A.; Montemurro, M.A.; Trujillo-Barreto, N.J.; Parkes, L.M.; El-Deredy, W. Cortical Resonance Frequencies Emerge from Network Size and Connectivity. *PLoS Comput. Biol.* **2016**, *12*, e1004740. [CrossRef] [PubMed]
52. Szostakiwskyj, J.M.H.; Willatt, S.E.; Cortese, F.; Protzner, A.B. The modulation of EEG variability between internally- and externally-driven cognitive states varies with maturation and task performance. *PLoS ONE* **2017**, *12*, e0181894. [CrossRef]
53. Barry, R.J.; Clarke, A.R.; Johnstone, S.J.; McCarthy, R.; Selikowitz, M. Electroencephalogram theta/beta ratio and arousal in attention-deficit/hyperactivity disorder: Evidence of independent processes. *Biol. Psychiatry* **2009**, *66*, 398–401. [CrossRef]
54. Newson, J.J.; Thiagarajan, T.C. EEG Frequency Bands in Psychiatric Disorders: A Review of Resting State Studies. *Front. Hum. Neurosci.* **2019**, *12*, 521. [CrossRef]
55. Clarke, A.R.; Barry, R.J.; Johnstone, S. Resting state EEG power research in Attention-Deficit/Hyperactivity Disorder: A review update. *Clin. Neurophysiol. Off. J. Int. Fed. Clin. Neurophysiol.* **2020**, *131*, 1463–1479. [CrossRef]
56. Dawson, G.; Klinger, L.G.; Panagiotides, H.; Lewy, A.; Castelloe, P. Subgroups of autistic children based on social behavior display distinct patterns of brain activity. *J. Abnorm. Child Psychol.* **1995**, *23*, 569–583. [CrossRef]
57. Daoust, A.M.; Limoges, E.; Bolduc, C.; Mottron, L.; Godbout, R. EEG spectral analysis of wakefulness and REM sleep in high functioning autistic spectrum disorders. *Clin. Neurophysiol. Off. J. Int. Fed. Clin. Neurophysiol.* **2004**, *115*, 1368–1373. [CrossRef]
58. Chan, A.S.; Sze, S.L.; Cheung, M.C. Quantitative electroencephalographic profiles for children with autistic spectrum disorder. *Neuropsychology* **2007**, *21*, 74–81. [CrossRef]
59. Pop-Jordanova, N.; Zorcec, T.; Demerdzieva, A.; Gucev, Z. QEEG characteristics and spectrum weighted frequency for children diagnosed as autistic spectrum disorder. *Nonlinear Biomed. Phys.* **2010**, *4*, 4. [CrossRef]
60. Wang, J.; Barstein, J.; Ethridge, L.E.; Mosconi, M.W.; Takarae, Y.; Sweeney, J.A. Resting state EEG abnormalities in autism spectrum disorders. *J. Neurodev. Disord.* **2013**, *5*, 24. [CrossRef]
61. Sohn, H.; Kim, I.; Lee, W.; Peterson, B.S.; Hong, H.; Chae, J.H.; Hong, S.; Jeong, J. Linear and non-linear EEG analysis of adolescents with attention-deficit/hyperactivity disorder during a cognitive task. *Clin. Neurophysiol. Off. J. Int. Fed. Clin. Neurophysiol.* **2010**, *121*, 1863–1870. [CrossRef]
62. Sokunbi, M.O.; Fung, W.; Sawlani, V.; Choppin, S.; Linden, D.E.; Thome, J. Resting state fMRI entropy probes complexity of brain activity in adults with ADHD. *Psychiatry Res.* **2013**, *214*, 341–348. [CrossRef] [PubMed]
63. Rezaeezadeh, M.; Shamekhi, S.; Shamsi, M. Attention Deficit Hyperactivity Disorder Diagnosis using non-linear univariate and multivariate EEG measurements: A preliminary study. *Phys. Eng. Sci. Med.* **2020**, *43*, 577–592. [CrossRef]
64. Angulo-Ruiz, B.Y.; Muñoz, V.; Rodríguez-Martínez, E.I.; Cabello-Navarro, C.; Gómez, C.M. Multiscale entropy of ADHD children during resting state condition. *Cogn. Neurodyn.* **2022**, *17*, 869–891. [CrossRef]
65. Chu, Y.J.; Chang, C.F.; Shieh, J.S.; Lee, W.T. The Potential Application of Multiscale Entropy Analysis of Electroencephalography in Children with Neurological and Neuropsychiatric Disorders. *Entropy* **2017**, *19*, 428. [CrossRef]

66. Liu, T.; Chen, Y.; Chen, D.; Li, C.; Qiu, Y.; Wang, J. Altered electroencephalogram complexity in autistic children shown by the multiscale entropy approach. *NeuroReport* **2017**, *28*, 169–173. [CrossRef]
67. Milne, E.; Gomez, R.; Giannadou, A.; Jones, M. Atypical EEG in autism spectrum disorder: Comparing a dimensional and a categorical approach. *J. Abnorm. Psychol.* **2019**, *128*, 442–452. [CrossRef]
68. Angulo-Ruiz, B.Y.; Ruiz-Martínez, F.J.; Rodríguez-Martínez, E.I.; Ionescu, A.; Saldaña, D.; Gómez, C.M. Linear and Non-linear Analyses of EEG in a Group of ASD Children During Resting State Condition. *Brain Topogr.* **2023**, *36*, 736–749. [CrossRef]
69. DiStefano, C.; Dickinson, A.; Baker, E.; Jeste, S.S. EEG Data Collection in Children with ASD: The Role of State in Data Quality and Spectral Power. *Res. Autism Spectr. Disord.* **2019**, *57*, 132–144. [CrossRef]
70. Pierce, S.; Kadlaskar, G.; Edmondson, D.A.; McNally Keehn, R.; Dydak, U.; Keehn, B. Associations between sensory processing and electrophysiological and neurochemical measures in children with ASD: An EEG-MRS study. *J. Neurodev. Disord.* **2021**, *13*, 5. [CrossRef]
71. Takahashi, T.; Yoshimura, Y.; Hiraishi, H.; Hasegawa, C.; Munesue, T.; Higashida, H.; Minabe, Y.; Kikuchi, M. Enhanced brain signal variability in children with autism spectrum disorder during early childhood. *Hum. Brain Mapp.* **2016**, *37*, 1038–1050. [CrossRef]
72. Bosl, W.J.; Loddenkemper, T.; Nelson, C.A. Nonlinear EEG biomarker profiles for autism and absence epilepsy. *Neuropsychiatr. Electrophysiol.* **2017**, *3*, 1. [CrossRef]
73. Nenadović, V.; Stokić, M.; Vuković, M.; Đoković, S.; Subotić, M. Cognitive and electrophysiological characteristics of children with specific language impairment and subclinical epileptiform electroencephalogram. *J. Clin. Exp. Neuropsychol.* **2014**, *36*, 981–991. [CrossRef]
74. Chutko, L.S.; Surushkina, S.Y.; Yakovenko, E.A.; Sergeev, A.V.; Rozhkova, A.V.; Anosova, L.V.; Chistyakova, N.P. Clinical and electroencephalographic characteristics of specific language impairment in children and an evaluation of the efficacy of cerebrolysin. *Zhurnal Nevrologii i Psikhiatrii Imeni SS Korsakova* **2015**, *115*, 98–102. [CrossRef]
75. Fatić, S.; Stanojević, N.; Stokić, M.; Nenadović, V.; Jeličić, L.; Bilibajkić, R.; Gavrilović, A.; Maksimović, S.; Adamović, T.; Subotić, M. Electroencephalography correlates of word and non-word listening in children with specific language impairment: An observational study20F0. *Medicine* **2022**, *101*, e31840. [CrossRef]
76. Stanojevic, N.; Fatic, S.; Jelicic, L.; Nenadovic, V.; Stokic, M.; Bilibajkic, R.; Subotic, M.; Boskovic Matic, T.; Konstantinovic, L.; Cirovic, D. Resting-state EEG alpha rhythm spectral power in children with specific language impairment: A cross-sectional study. *J. Appl. Biomed.* **2023**, *21*, 113–120. [CrossRef]
77. Donoghue, T.; Haller, M.; Peterson, E.J.; Varma, P.; Sebastian, P.; Gao, R.; Noto, T.; Lara, A.H.; Wallis, J.D.; Knight, R.T.; et al. Parameterizing neural power spectra into periodic and aperiodic components. *Nat. Neurosci.* **2020**, *23*, 1655–1665. [CrossRef]
78. McSweeney, M.; Morales, S.; Valadez, E.A.; Buzzell, G.A.; Yoder, L.; Fifer, W.P.; Pini, N.; Shuffrey, L.C.; Elliott, A.J.; Isler, J.R.; et al. Age-related trends in aperiodic EEG activity and alpha oscillations during early- to middle-childhood. *NeuroImage* **2023**, *269*, 119925. [CrossRef]
79. Buzsáki, G.; Anastassiou, C.A.; Koch, C. The origin of extracellular fields and currents--EEG, ECoG, LFP and spikes. Nature reviews. *Neuroscience* **2012**, *13*, 407–420. [CrossRef]
80. Gao, R.; Peterson, E.J.; Voytek, B. Inferring synaptic excitation/inhibition balance from field potentials. *NeuroImage* **2017**, *158*, 70–78. [CrossRef]
81. Robertson, M.M.; Furlong, S.; Voytek, B.; Donoghue, T.; Boettiger, C.A.; Sheridan, M.A. EEG power spectral slope differs by ADHD status and stimulant medication exposure in early childhood. *J. Neurophysiol.* **2019**, *122*, 2427–2437. [CrossRef]
82. Pertermann, M.; Bluschke, A.; Roessner, V.; Beste, C. The Modulation of Neural Noise Underlies the Effectiveness of Methylphenidate Treatment in Attention-Deficit/Hyperactivity Disorder. *Biol. Psychiatry Cogn. Neurosci. Neuroimaging* **2019**, *4*, 743–750. [CrossRef]
83. Mamiya, P.C.; Arnett, A.B.; Stein, M.A. Precision Medicine Care in ADHD: The Case for Neural Excitation and Inhibition. *Brain Sci.* **2021**, *11*, 91. [CrossRef]
84. Ostlund, B.D.; Alperin, B.R.; Drew, T.; Karalunas, S.L. Behavioral and cognitive correlates of the aperiodic (1/f-like) exponent of the EEG power spectrum in adolescents with and without ADHD. *Dev. Cogn. Neurosci.* **2021**, *48*, 100931. [CrossRef]
85. Levin, A.R.; Naples, A.J.; Scheffler, A.W.; Webb, S.J.; Shic, F.; Sugar, C.A.; Murias, M.; Bernier, R.A.; Chawarska, K.; Dawson, G.; et al. Day-to-Day Test-Retest Reliability of EEG Profiles in Children With Autism Spectrum Disorder and Typical Development. *Front. Integr. Neurosci.* **2020**, *14*, 21. [CrossRef]
86. Wiig, E.H.; Semel, E.; Secord, W.A. *Clinical Evaluation of Language Fundamentals—Fifth Edition (CELF-5)*; NCS Pearson: Bloomington, MN, USA, 2013.
87. Aguinaga Ayerra, G. *PLON-R: Prueba de Lenguaje Oral Navarra Revisada*, 2nd ed.; TEA Ediciones, S.A.U.: Madrid, España, 2005.
88. Kirk, S.A.; McCarthy, J.J. The Illinois test of psycholinguistic abilities--an approach to differential diagnosis. *Am. J. Ment. Defic.* **1961**, *66*, 399–412.
89. Dunn, L.M. *Peabody Picture Vocabulary Test*; American Guidance Service: Miami, FL, USA, 1959.

90. Kaufman, A.S.; Kaufman, N.L. *KBIT: Kaufman Brief Intelligence Test (KBIT Spanish Version)*; ASD Editions: Madrid, Spain, 2004.
91. Delorme, A.; Makeig, S. EEGLAB: An open source toolbox for analysis of single-trial EEG dynamics including independent component analysis. *J. Neurosci. Methods* **2004**, *134*, 9–21. [CrossRef]
92. Mullen, T.R.; Kothe, C.A.; Chi, Y.M.; Ojeda, A.; Kerth, T.; Makeig, S.; Jung, T.P.; Cauwenberghs, G. Real-Time Neuroimaging and Cognitive Monitoring Using Wearable Dry EEG. *IEEE Trans. Bio-Med. Eng.* **2015**, *62*, 2553–2567. [CrossRef]
93. Bell, A.J.; Sejnowski, T.J. An information-maximization approach to blind separation and blind deconvolution. *Neural Comput.* **1995**, *7*, 1129–1159. [CrossRef]
94. Amari, S.; Cichocki, A.; Yang, H.H. *A New Learning Algorithm for Blind Signal Separation*; NIPS: San Diego, CA, USA, 1995.
95. Pion-Tonachini, L.; Kreutz-Delgado, K.; Makeig, S. ICLabel: An automated electroencephalographic independent component classifier, dataset, and website. *NeuroImage* **2019**, *198*, 181–197. [CrossRef]
96. Malik, J. Multiscale Sample Entropy. 2022. Available online: https://www.mathworks.com/matlabcentral/fileexchange/62706-multiscale-sample-entropy (accessed on 1 December 2024).
97. Costa, M.; Goldberger, A.L.; Peng, C.K. Multiscale entropy analysis of biological signals. *Phys. Rev. E* **2005**, *71*, 021906. [CrossRef]
98. Richman, J.S.; Moorman, J.R. Physiological time-series analysis using approximate entropy and sample entropy. *Am. J. physiology. Heart Circ. Physiol.* **2000**, *278*, H2039–H2049. [CrossRef]
99. McIntosh, A.R.; Kovacevic, N.; Itier, R.J. Increased brain signal variability accompanies lower behavioral variability in development. *PLoS Comput. Biol.* **2008**, *4*, e1000106. [CrossRef]
100. Miskovic, V.; Owens, M.; Kuntzelman, K.; Gibb, B.E. Charting moment-to-moment brain signal variability from early to late childhood. *Cortex J. Devoted Study Nerv. Syst. Behav.* **2016**, *83*, 51–61. [CrossRef]
101. Kloosterman, N.A.; Kosciessa, J.Q.; Lindenberger, U.; Fahrenfort, J.J.; Garrett, D.D. Boosting brain signal variability underlies liberal shifts in decision bias. *bioRxiv* **2019**. [CrossRef]
102. Kosciessa, J.Q.; Kloosterman, N.A.; Garrett, D.D. Standard multiscale entropy reflects neural dynamics at mismatched temporal scales: What's signal irregularity got to do with it? *PLoS Comput. Biol.* **2020**, *16*, e1007885. [CrossRef]
103. Papaioannou, A.G.; Kalantzi, E.; Papageorgiou, C.C.; Korombili, K.; Bokou, A.; Pehlivanidis, A.; Papageorgiou, C.C.; Papaioannou, G. Complexity analysis of the brain activity in Autism Spectrum Disorder (ASD) and Attention Deficit Hyperactivity Disorder (ADHD) due to cognitive loads/demands induced by Aristotle's type of syllogism/reasoning. A Power Spectral Density and multiscale entropy (MSE) analysis. *Heliyon* **2021**, *7*, e07984. [CrossRef]
104. Garrett, D.D.; Samanez-Larkin, G.R.; MacDonald, S.W.; Lindenberger, U.; McIntosh, A.R.; Grady, C.L. Moment-to-moment brain signal variability: A next frontier in human brain mapping? *Neurosci. Biobehav. Rev.* **2013**, *37*, 610–624. [CrossRef]
105. Bosl, W.J.; Loddenkemper, T.; Vieluf, S. Coarse-graining and the Haar wavelet transform for multiscale analysis. *Bioelectron. Med.* **2022**, *8*, 3. [CrossRef]
106. Angulo-Ruiz, B.Y.; Rodríguez-Martínez, E.I.; Muñoz, V.; Gómez, C.M. Unveiling the hidden electroencephalographical rhythms during development: Aperiodic and Periodic activity in healthy subjects. *Clin. Neurophysiol. Off. J. Int. Fed. Clin. Neurophysiol.* **2024**, *169*, 53–64. [CrossRef]
107. Ostlund, B.; Donoghue, T.; Anaya, B.; Gunther, K.E.; Karalunas, S.L.; Voytek, B.; Pérez-Edgar, K.E. Spectral parameterization for studying neurodevelopment: How and why. *Dev. Cogn. Neurosci.* **2022**, *54*, 101073. [CrossRef]
108. Cohen, J. *Statistical Power Analysis for the Behavioral Sciences*, 2nd ed.; Lawrence Erlbaum Associates: Hillsdale, MI, USA, 1988.
109. Benjamini, Y.; Hochberg, Y. Controlling the false discovery rate: A practical and powerful approach to multiple testing. *J. R. Stat. Soc.* **1995**, *57*, 289–300. [CrossRef]
110. Jacob, M.S.; Roach, B.J.; Sargent, K.; Mathalon, D.H.; Ford, J.M. Aperiodic measures of neural excitability are associated with anticorrelated hemodynamic networks at rest: A combined EEG-fMRI study. *NeuroImage* **2021**, *245*, 1187052021. [CrossRef]
111. Hill, A.T.; Clark, G.M.; Bigelow, F.J.; Lum, J.A.G.; Enticott, P.G. Periodic and aperiodic neural activity displays age-dependent changes across early-to-middle childhood. *Dev. Cogn. Neurosci.* **2022**, *54*, 101076. [CrossRef]
112. Manning, J.R.; Jacobs, J.; Fried, I.; Kahana, M.J. Broadband shifts in local field potential power spectra are correlated with single-neuron spiking in humans. *J. Neurosci. Off. J. Soc. Neurosci.* **2009**, *29*, 13613–13620. [CrossRef]
113. Miller, K.J.; Honey, C.J.; Hermes, D.; Rao, R.P.; denNijs, M.; Ojemann, J.G. Broadband changes in the cortical surface potential track activation of functionally diverse neuronal populations. *NeuroImage* **2014**, *85 Pt 2*, 711–720. [CrossRef]
114. Voytek, B.; Knight, R.T. Dynamic network communication as a unifying neural basis for cognition, development, aging, and disease. *Biol. Psychiatry* **2015**, *77*, 1089–1097. [CrossRef]
115. Buckner, R.L.; Andrews-Hanna, J.R.; Schacter, D.L. The brain's default network: Anatomy, function, and relevance to disease. *Ann. N. Y. Acad. Sci.* **2008**, *1124*, 1–38. [CrossRef]
116. Proal, E.; Álvarez-Segura, M.; De la Iglesia-Vayá, M.; Martí-Bonmatí, L.; Castellanos, F.X.; Spanish Resting State Network. Actividad funcional cerebral en estado de reposo: Redes en conexion [Functional cerebral activity in a state of rest: Connectivity networks]. *Rev. Neurol.* **2011**, *52* (Suppl. S1), S3–S10. [CrossRef]

117. Wilke, M.; Hauser, T.K.; Krägeloh-Mann, I.; Lidzba, K. Specific impairment of functional connectivity between language regions in former early preterms. *Hum. Brain Mapp.* **2014**, *35*, 3372–3384. [CrossRef]
118. Barnes-Davis, M.E.; Williamson, B.J.; Merhar, S.L.; Holland, S.K.; Kadis, D.S. Rewiring the extremely preterm brain: Altered structural connectivity relates to language function. *NeuroImage Clin.* **2020**, *25*, 102194. [CrossRef]
119. Chen, A.C.; Feng, W.; Zhao, H.; Yin, Y.; Wang, P. EEG default mode network in the human brain: Spectral regional field powers. *NeuroImage* **2008**, *41*, 561–574. [CrossRef]
120. Gómez, C.M.; Marco-Pallarés, J.; Grau, C. Location of brain rhythms and their modulation by preparatory attention estimated by current density. *Brain Res.* **2006**, *1107*, 151–160. [CrossRef]
121. Rodríguez-Martinez, E.I.; Barriga-Paulino, C.I.; Zapata, M.I.; Chinchilla, C.; López-Jiménez, A.M.; Gómez, C.M. Narrow band quantitative and multivariate electroencephalogram analysis of peri-adolescent period. *BMC Neurosci.* **2012**, *13*, 104. [CrossRef]
122. Finneran, D.A.; Francis, A.L.; Leonard, L.B. Sustained attention in children with specific language impairment (SLI). *J. Speech Lang. Hear. Res.* **2009**, *52*, 915–929. [CrossRef]
123. Verche, B.E.; Hernández, E.S.; Quintero, F.I.; Acosta, R.V.M. Alteraciones en la memoria en el Trastorno Específico del Lenguaje. Una perspectiva neuropsicológica. *Rev. Logop. Foniatría Audiol.* **2013**, *33*, 179–185. [CrossRef]
124. Quintero, I.; Hernández, S.; Verche, E.; Acosta, V.; Hernández, A. Disfunción ejecutiva en el Trastorno Específico del Lenguaje. *Rev. Logop. Foniatría Audiol.* **2013**, *33*, 172–178. [CrossRef]
125. Rodríguez, A.; Ramírez, S.G.M.; Expósito, H. Funciones ejecutivas y lenguaje en subtipos de niños con trastorno específico del lenguaje. *Neurología* **2017**, *32*, 355–362. [CrossRef]
126. Bosman, C.A.; Lansink, C.S.; Pennartz, C.M. Functions of gamma-band synchronization in cognition: From single circuits to functional diversity across cortical and subcortical systems. *Eur. J. Neurosci.* **2014**, *39*, 1982–1999. [CrossRef]
127. Friedman, N.P.; Robbins, T.W. The role of prefrontal cortex in cognitive control and executive function. *Neuropsychopharmacol. Off. Publ. Am. Coll. Neuropsychopharmacol.* **2022**, *47*, 72–89. [CrossRef]
128. Mazza, A.; Dal Monte, O.; Schintu, S.; Colombo, S.; Michielli, N.; Sarasso, P.; Törlind, P.; Cantamessa, M.; Montagna, F.; Ricci, R. Beyond alpha-band: The neural correlate of creative thinking. *Neuropsychologia* **2023**, *179*, 108446. [CrossRef]
129. Hindriks, R.; van Putten, M.J.A.M.; Deco, G. Intra-cortical propagation of EEG alpha oscillations. *NeuroImage* **2014**, *103*, 444–453. [CrossRef]
130. Klimesch, W. α-band oscillations, attention, and controlled access to stored information. *Trends Cogn. Sci.* **2012**, *16*, 606–617. [CrossRef] [PubMed]
131. London, R.E.; Benwell, C.S.Y.; Cecere, R.; Quak, M.; Thut, G.; Talsma, D. EEG alpha power predicts the temporal sensitivity of multisensory perception. *Eur. J. Neurosci.* **2022**, *55*, 3241–3255. [CrossRef] [PubMed]

entropy

Article

Time-Resolved Information-Theoretic and Spectral Analysis of fNIRS Signals from Multi-Channel Prototypal Device

Irene Franzone [1], Yuri Antonacci [1,*], Fabrizio Giuliano [1], Riccardo Pernice [1], Alessandro Busacca [1], Luca Faes [1,2] and Giuseppe Costantino Giaconia [1]

1 Department of Engineering, University of Palermo, 90128 Palermo, Italy; irene.franzone@unipa.it (I.F.); fabrizio.giuliano@unipa.it (F.G.); riccardo.pernice@unipa.it (R.P.); alessandro.busacca@unipa.it (A.B.); luca.faes@unipa.it (L.F.); costantino.giaconia@unipa.it (G.C.G.)
2 Faculty of Technical Sciences, University of Novi Sad, Trg Dositeja Obradovića 6, 21102 Novi Sad, Serbia
* Correspondence: yuri.antonacci@unipa.it

Abstract

Functional near-infrared spectroscopy (fNIRS) is a non-invasive imaging technique that measures brain hemodynamic activity by detecting changes in oxyhemoglobin and deoxyhemoglobin concentrations using light in the near-infrared spectrum. This study aims to provide a comprehensive characterization of fNIRS signals acquired with a prototypal continuous-wave fNIRS device during a breath-holding task, to evaluate the impact of respiratory activity on scalp hemodynamics within the framework of Network Physiology. To this end, information-theoretic and spectral analysis methods were applied to characterize the dynamics of fNIRS signals. In the time domain, time-resolved information-theoretic measures, including entropy, conditional entropy and, information storage, were employed to assess the complexity and predictability of the fNIRS signals. These measures highlighted distinct informational dynamics across the breathing and apnea phases, with conditional entropy showing a significant modulation driven by respiratory activity. In the frequency domain, power spectral density was estimated using a parametric method, allowing the identification of distinct frequency bands related to vascular and respiratory components. The analysis revealed significant modulations in both the amplitude and frequency of oscillations during the task, particularly in the high-frequency band associated with respiratory activity. Our observations demonstrate that the proposed analysis provides novel insights into the characterization of fNIRS signals, enhancing the understanding of the impact of task-induced peripheral cardiovascular responses on NIRS hemodynamics.

Keywords: network physiology; functional near-infrared spectroscopy; power spectral density; Information Dynamics; time-resolved analysis; time-varying autoregressive modeling; recursive least-squares

1. Introduction

Functional near-infrared spectroscopy (fNIRS) is a non-invasive imaging technique which relies on the use of light at specific wavelengths for measuring the hemodynamic activity of brain tissue [1]. Specifically, the near-infrared region (650–950 nm) enables the estimation of the concentration of oxyhemoglobin (HbO_2) and deoxyhemoglobin (HHb) by exploiting their different absorption spectra at different wavelengths [2,3]. Among the different fNIRS techniques available on the market, the continuous-wave (CW) modality employs constant tissue illumination to quantify light attenuation through the head. Given its reliance on low-cost electronic components, as well as its high portability and ease of use, the CW modality has become the most commonly used technique. Indeed, only recently novel multichannel fNIRS systems were developed to answer the need of a portable system that can monitor hemodynamic activity in ecological and clinical settings [4,5].

In a recent study, an integrated system was developed that combines a novel continuous-wave fNIRS device with a modified commercial electroencephalographic (EEG) system [4]. This hybrid system has been validated for monitoring Event-Related Potentials across three distinct experimental paradigms designed to elicit the following: (i) motor cortex activity via a finger-tapping task [6]; (ii) visual cortex activity using a flickering black-and-white checkerboard at a frequency of 2 Hz [7]; and (iii) frontal cortex activity through a Stroop task [8].

However, it is well established that cognitive and emotional tasks can influence peripheral physiology by altering heart rate, respiration, blood pressure, and skin conductance [9,10]. These physiological changes affect both scalp and brain hemodynamics, thereby contributing to the fNIRS signal in ways that extend beyond neural activity [11]. In particular, systemic physiological variations—especially those driven by autonomic or cardiorespiratory activity—can impact both superficial and intracerebral hemodynamics. Previous research has shown that signals recorded from a given scalp location do not exclusively reflect neurovascular coupling but may also result from hemodynamic changes in extracerebral tissues (e.g., skin, muscles, skull) as well as in deeper brain regions influenced by systemic processes [12]. This issue is especially pronounced in CW fNIRS systems with short source–detector separations (typically < 4 cm), where the sensitivity to extracerebral tissues often exceeds that to cortical areas. Recent studies confirm that the extracerebral compartment—including scalp, skull, and cerebrospinal fluid—absorbs a substantial portion of the incident light, often leading to signals dominated by superficial layers [13–15]. This limited depth sensitivity remains a major limitation of CW fNIRS, particularly in adult populations, and must be considered when interpreting signal origin.

As a result, fNIRS measurements are affected by various background physiological components that are often considered "noise" in traditional analyses of brain connectivity, as they can obscure the detection of evoked neural responses. Yet, the influence of these systemic signals on both brain and scalp hemodynamics remains underexplored. Most analysis techniques aim to remove such components from the fNIRS signal [16,17]. Recent developments in the field of network physiology, however, propose a paradigm shift: rather than viewing these fluctuations as artifacts, they can be interpreted as meaningful indicators of physiological interactions [18,19]. According to this perspective, the human body can be seen as an integrated network, where each physiological system, despite having its own regulatory mechanisms, continuously interacts with others to coordinate functions and generate distinct physiological states in both health and disease [20]. In this framework, cardiorespiratory dynamics arise from the coordination of respiratory, cardiovagal, and sympathetic functions, which operate across various levels of the nervous system. These interactions play a critical role in maintaining physiological equilibrium and have only recently been shown to have a direct impact on both scalp and neural signals in specific clinical conditions [21,22]. To investigate these complex interactions, we designed a protocol based on the breath-holding task, in which the presence of respiratory signals is systematically modulated, allowing for a detailed assessment of the influence of respiratory activity on the local concentration of oxy- and deoxyhemoglobin measured over the scalp [23]. While previous studies have demonstrated strong responses in all brain chromophores due to respiratory activity using well-established methods [17], traditional approaches do not fully characterize the regularity, complexity, or transient dynamics of these signals.

Therefore, the aim of this work is to characterize the effects of respiratory activity on scalp hemodynamics as measured by a prototypal fNIRS system. We employ a multi-domain analysis that includes state-of-the-art signal processing techniques, which have not yet been extensively applied in the context of fNIRS signal analysis. These techniques are designed to assess physiological oscillations and their modulation during breath-holding, as well as the transient dynamics during the intermittent presence of respiratory activity. Specifically, to investigate the complexity and predictability of the fNIRS signals dynamics, we leverage information-theoretic measures computed in a time-resolved fashion such as conditional entropy (CE) [24] and information storage (IS) [25]. Moreover, given the presence of oscillatory activity that can be measured over the scalp, we perform a frequency-domain analysis to explore the distribution of power across frequency bands, providing an indirect measure of both the intensity and temporal evolution of neural activity [26].

2. Materials and Methods

2.1. Dataset Description and Pre-Processing

A total of 6 healthy subjects (4 males, 2 females; mean age: 27 ± 2.5 years) were enrolled in this study and provided informed consent prior to participation. The experimental procedure was approved by the ethical committee of the University of Palermo and was conducted in accordance with the ethical standards of the Helsinki Declaration. The subjects were seated in a chair in an isolated room and instructed to breathe normally, avoiding other actions or excessive movements. The experimental protocol consisted of a five-minute recording alternating two different phases: (i) apnea episodes, during which subjects were instructed to hold their breath, starting at predefined time points (8, 50, 110, 170, and 230 s), and (ii) a breathing phase, during which the subject resumed normal breathing before the onset of the subsequent apnea.

The signal acquisitions were performed using a multi-channel CW-fNIRS system. This system includes 16 bicolor light-emitting diodes (LEDs) operating at wavelengths of 735 nm and 850 nm, alongside 16 silicon photomultipliers (SiPMs). Each SiPM is positioned approximately 3 cm away from its corresponding LED source and acquires signals at a sampling frequency (f_s) of 130 Hz. This is in line with previous studies with a similar setup [22,27], and with a work demonstrating that, under optimal conditions, optode spacings up to 5 cm are usable with NIRS equipment [28]. According to a "banana-shaped" model, which assumes that the penetration is similar for all wavelengths, the resulting wavepath had a maximum penetration depth of about 1.5 cm [29,30]. Thus, these fNIRS measurements are a superposition of signals from multiple tissue layers, with dominant contributions from extracerebral tissues [14,15]. Further details about the prototype electronic components and the realization procedure can be found in [4]. The selected configuration of the optodes is shown in Figure 1, where each optode is labeled with the letter 'S' (source) if it is an LED, or 'D' (detector) if it is a SiPM. Each optode is also assigned a number to identify its position on the scalp. In particular, the positions of the 6 LEDs are marked in red, while the 6 SiPMs are marked in blue. Together, they provide 16 different channels, which are listed in the table of the same figure.

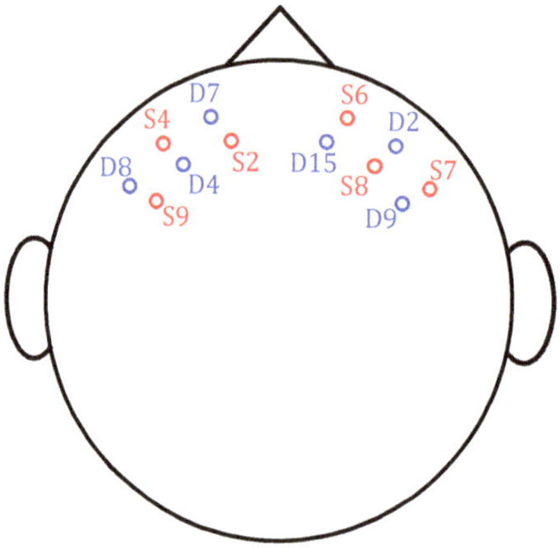

CHANNELS	
1) S6-D7	9) S4-D4
2) S6-D2	10) S4-D8
3) S4-D7	11) S7-D2
4) S2-D7	12) S8-D2
5) S6-D15	13) S7-D9
6) S8-D15	14) S8-D9
7) S2-D15	15) S9-D4
8) S2-D4	16) S9-D8

Figure 1. fNIRS optodes configuration and selected channels.

The pre-processing pipeline employed in this study included the following steps: (i) the raw intensity signals were converted to optical density (OD) signals using the equation $OD = -\ln(I(t)/I_0)$, where $I(t)$ is the time-dependent recorded signal intensity and I_0 is its initial value [4]; (ii) the Temporal Derivative Distribution Repair (TDDR) method was applied to the OD signals as implemented in the fNIRS Brain AnalyzIR Python toolbox [31]. TDDR is a motion correction algorithm designed to eliminate the two most common motion artifacts in fNIRS data: spikes and baseline shifts; (iii) the corrected OD signals were converted to hemoglobin concentration signals (HbO_2 and HHb) using the modified Beer–Lambert law [32], following the procedure described in [4]. The resulting signals were then low-pass-filtered with a cutoff frequency of 0.5 Hz to prevent aliasing and subsequently downsampled to 1 Hz; (iv) finally, a high-pass autoregressive (AR) filter with infinite impulse response (IIR) and zero phase [33], with a cutoff frequency of 0.018 Hz, was applied to emphasize variations in the respiratory frequency band induced by the task. The resulting time series, each consisting of $N = 300$ samples, were normalized to zero mean and unit variance. Representative trends for one channel and one subject are shown in Figure 2a.1 for HbO_2 and Figure 2a.2 for HHb, highlighting the variations observed during the breath phases (red windows) and the apnea phases (blue windows).

2.2. Time-Resolved Information Measures

Information-theoretic measures provide powerful approaches to assess the regularity and predictability of the dynamics of a given system. While traditional analyses provide overall measures of complexity and predictability, recent advancements extend these measures to enable time-resolved characterization [24,34,35]. Here, we apply these methods to study the transient evolution of scalp hemodynamics during breath-holding tasks—revealing dynamic responses in extracerebral oxygenation that conventional fNIRS analyses cannot resolve.

The analysis of any given dynamical system can be performed by mapping the system activity with a set of random variables and then studying the statistical dependencies among the observed realizations of the variables collected in the form of time series. In the general field of information theory, starting from the basic concept of entropy introduced by Shannon [36], it is possible to dissect the information processed in a dynamical system into meaningful elements of computation to quantify the complexity and predictability of the aforementioned system.

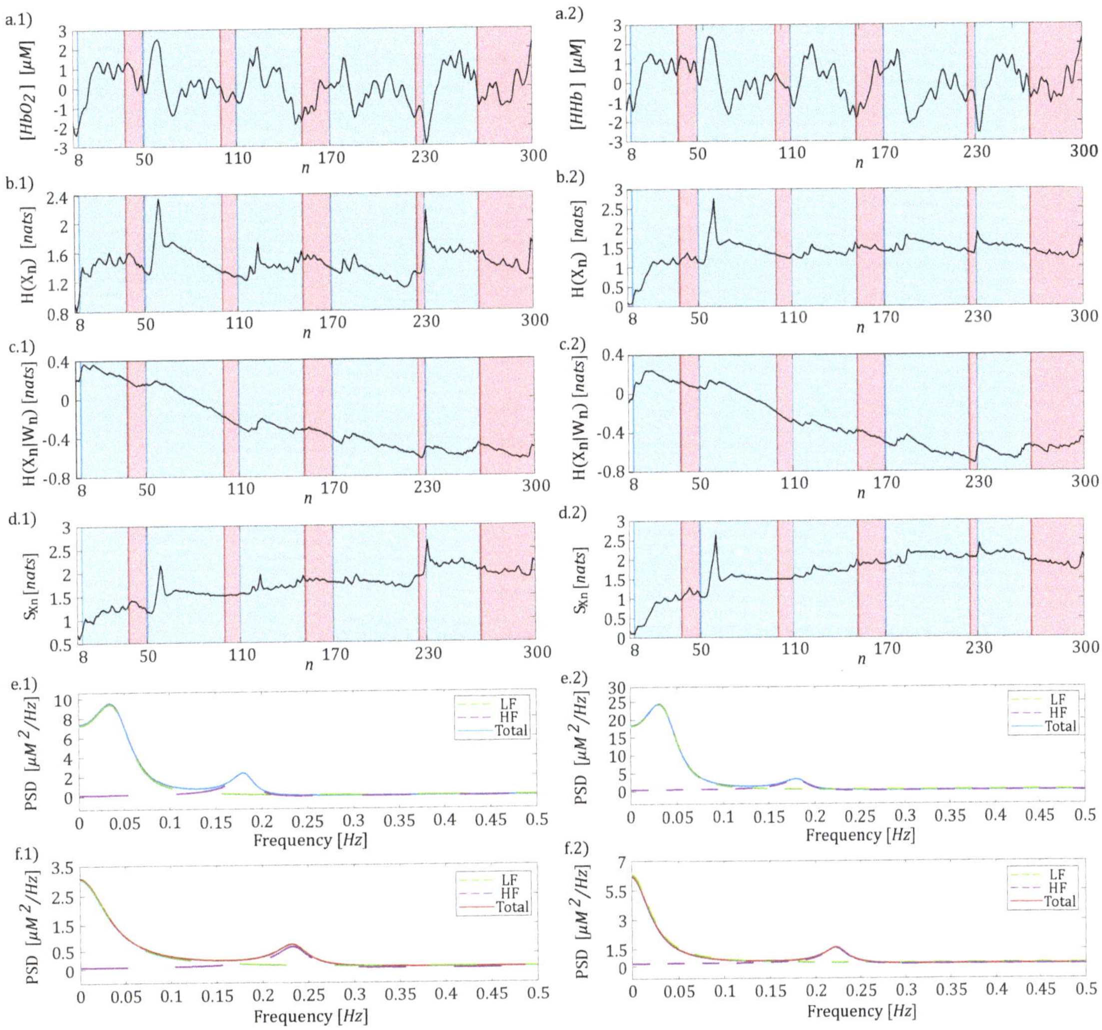

Figure 2. Time series showing the variation in hemoglobin concentration over a period of 5 min (panels (**a.1**,**a.2**)) obtained for a representative channel for HbO_2 (panel (**a.1**–**f.1**)) and HHb (panel (**a.2**–**f.2**)). The blue and red windows correspond to the time intervals associated with the apnea (blue area) and breathing (red area) phases, respectively. Time-resolved analysis of entropy (panels (**b.1**,**b.2**)), conditional entropy (panels (**c.1**,**c.2**)), and information storage (panels (**d.1**,**d.2**)) obtained for HbO_2 and HHb, respectively. PSD trends along with their spectral decompositions obtained for the apnea (panels (**e.1**,**e.2**)) and breathing phases (panels (**f.1**,**f.2**)) computed separately for HbO_2 and HHb, respectively.

The HbO_2 or HHb signal, acquired for each subject and channel, was modeled as a realization of a zero-mean stochastic process X, with X_n representing the random variable sampling the process at time n (temporal counter). Then, the Shannon entropy (H) can be used to quantify the amount of information needed to describe X_n as follows: $H(X_n) = -\mathbb{E}[\log p(x_n)]$, where x_n refers to a realization of X_n and $p(x_n)$ is the probability density function of X_n measured for the outcome x_n, and $\mathbb{E}[\cdot]$ is the expectation operator computing the statistical average over all possible values of x_n. Assuming X as a Markov process, its past history can be truncated up to a lag p to obtain the p-dimensional vector $W_n = [X_{n-1}, \ldots, X_{n-p}]^\top$. To account for the dynamical evolution of the system X, it is possible to quantify the new information carried by the present state X_n of the process which cannot be inferred from its past W_n, i.e., the Conditional Entropy (CE) [37]: $H(X_n|W_n) = -\mathbb{E}[\log p(x_n|w_n)]$ where w_n refers to a realization of W_n and $p(\cdot|\cdot)$ is the conditional probability. Specifically, the higher the CE, the more complex the dynamics generated by the system. Entropy and CE serve as measures to introduce the Information Storage (IS), which quantifies the amount of information contained in the present state that can be predicted by the knowledge of its past state [38] and can be defined as

$$S_{X,n} = I(X_n; W_n) = \mathbb{E}\Big[\log \frac{p(x_n|w_n)}{p(x_n)}\Big]. \tag{1}$$

The IS measures the predictability of the process at time n by quantifying the average level of uncertainty about the current state of the process X_n that can be resolved by knowledge of its past states W_n.

To describe the time-resolved behavior of the fNIRS signal (HbO_2, HHb) during breathing and apnea phases, a time-varying analysis of the measures of H, CE, and IS was performed, through the adoption of a time-varying version of an autoregressive (TV-AR) model [24,34]:

$$X_n = \sum_{k=1}^{p} a_{k,n} X_{n-k} + U_n, \tag{2}$$

where U_n represents the prediction error, while $a_{k,n}$ denotes the AR coefficient describing the interaction from X_{n-k} to X_n at lag k, relevant to the time instant n. Then, under Gaussian assumption of X_n, the time-resolved H can be expressed as follows [24]: $H(X_n) = \frac{1}{2}log(2\pi e \cdot \sigma^2_{X_n})$, where $\sigma^2_{X_n}$ represents the variance of X_n. Moreover, if X_n and W_n are jointly Gaussian, the time-resolved CE of X_n given W_n can be expressed as follows [39]: $H(X_n|W_n) = \frac{1}{2}log(2\pi e \cdot \sigma^2_{U_n})$, where $\sigma^2_{U_n}$ is the variance of the prediction error U_n at the time step n. Thus, the equation given in (1) can be rewritten as

$$S_{X,n} = H(X_n) - H(X_n|W_n) = \frac{1}{2}log\frac{\sigma^2_{X_n}}{\sigma^2_{U_n}}. \tag{3}$$

The identification procedure of the TV-AR model (2) can be performed through the recursive least-squares (RLS) as described in [40]. Briefly, the RLS consists of the following steps: (i) choose a value for the adaption factor $c \in (0,1)$ and an order p of the AR model; (ii) define proper initial conditions for the vector of coefficients at time p, $A_p = [a_{1,p}, \ldots, a_{p,p}] \in \mathbb{R}^{1\times p}$ and for the correlation matrix of the past state of X stored in W_n,$\mathbf{\Sigma}_{W_n} = \mathbb{E}[W_n W_n^\top] \in \mathbb{R}^{p\times p}$; (iii) considering N-p consecutive time steps, for $n = p+1$ to N, repeat the following steps:

$$\mathbf{\Sigma}_{W_n} = (1-c)\mathbf{\Sigma}_{W_{n-1}} + W_n W_n^\top, \tag{4a}$$

$$K_n = (\mathbf{\Sigma}_{W_n})^{-1} W_n, \tag{4b}$$

$$Z_n = X_n - A_{n-1} W_n, \tag{4c}$$

$$A_n = A_{n-1} + Z_n K_n^\top, \tag{4d}$$

where $K_n \in \mathbb{R}^{p\times 1}$ is the so-called gain vector and $Z_n \in \mathbb{R}^{1\times 1}$ is intended as the a priori estimation error before updating the AR coefficients vector. The parameter $(1-c)$ controls the memory of the algorithm allowing it to follow possible statistical variations in the property of X in non-stationary conditions. To complete the identification procedure, a recursive estimation of the time-varying innovation variance can be obtained as follows: $\sigma^2_{U_n} = \sigma^2_{U_{n-1}} + c(Z_n^2 - \sigma^2_{U_{n-1}})$ [24,34]. The required recursive estimation of the process variance $\sigma^2_{X_n}$ can be directly derived from the structure of the linear TV-AR representation of the process X as described in [24].

2.3. *Spectral Analysis of fNIRS Signals*

Biomedical signals are often rich with oscillatory content, and therefore naturally lend themselves to spectral representation. Classical approaches integrating the power spectral density profile within the spectral bands of interest attempt to obtain band-specific time-domain powers, while the method of spectral decomposition used here allows for focusing only on the spectral components with frequencies within those bands, thus avoiding spurious contributions due to broadband oscillations [41].

To characterize the oscillatory content of each fNIRS signal (HbO_2, HHb), a linear model governed by the following difference equation was employed:

$$X_n = \sum_{k=1}^{p} a_k X_{n-k} + U_n, \tag{5}$$

where U_n is the prediction error with a variance σ^2_U and a_k is the AR coefficient describing the interaction from X_{n-k} to X_n at lag k. The linear model (5) can be represented in the Z-domain through its Z-transform yielding $X(z) = H(z)U(z)$, where $H(z) = [1 - \sum_{k=1}^{p} a_k z^{-k}]^{-1}$ is the transfer function relating the Fourier Transform (FT) of U to the FT of the process X. Applying the residue theorem, $H(z)$ can be expressed as follows [41]:

$$H(z) = \frac{z^p}{\prod_{k=1}^{p}(z - p_k)} = \prod_{k=1}^{p} H^{(k)}(z), \tag{6}$$

where p_k, $k = 1, \ldots, p$, are the p poles of the AR process, while the terms $H^{(k)}(z) = \frac{z}{z-p_k} \cdot \frac{1/z^*}{1/z^* - p_k}$ are pole-specific factors associated each with a given pole p_k, with * indicating the Hermitian transpose. The power spectral density (PSD) of the process X can be expressed in the Z-domain as $P(z) = H(z)\sigma^2_U H^*(\frac{1}{z^*})$. Then, by using the Heaviside decomposition, it can be expanded into simple fractions corresponding to the poles of the system. These include the poles inside the unit circle, p_k, and their reciprocals outside the unit circle, $\bar{p}_k = p_k^{-1}$, for $k = 1, \ldots, p$, weighted by the residuals of $P(z)$, specifically $r_k p_k$ and $-r_k p_k^{-1}$. This results in the following expression [41]:

$$P(z) = \sum_{k=1}^{p} P^{(k)}(z) = \sum_{k=1}^{p} \left[\frac{r_k p_k}{z - p_k} - \frac{r_k p_k^{-1}}{z - p_k^{-1}} \right], \tag{7}$$

where the residuals are given by $r_k = \left. \frac{\sigma^2_U}{z \prod_{h \neq k}(z - p_h) \cdot \prod(z^{-1} - p_h)} \right|_{z = p_k}$, $k = 1, \ldots, p$. Evaluating $P(z)$ on the unit circle of the complex plane, specifically $P(f) = P(z)|_{z = e^{j2\pi f / f_s}}$, where $f \in [-f_s/2, f_s/2]$ and f_s is the sampling frequency (1 Hz), allows for the derivation of the spectral profile, $P(f)$, along with its kth component, $P^{(k)}(f)$. Each spectral component is characterized by a unique profile, which is determined by the central frequency of the oscillation, derived from the argument of the pole ($f_k = \frac{\arg(p_k)}{2\pi}$), and by the power related to the residual of the pole. For real poles, the power is $\sigma_k^2 = r_k$, and for complex conjugate poles, it is $\sigma_k^2 = r_k + r_k^*$. Notably, the total variance σ^2_X of the process is equal to the sum of the variances of all poles, σ_k^2, for $k = 1, \ldots, p$.

The computation of the PSD of X relies on the identification procedure of the AR model (5) which herein was performed through the ordinary least-squares (OLS) method [42]. By recalling the p-dimensional vector W_n containing the past states of X and considering N consecutive time steps, a compact representation of the VAR model (5) can be defined as $\mathbf{X} = \mathbf{AW} + \mathbf{U}$, where $\mathbf{A} = [a_1, \ldots, a_p]$ is the $1 \times p$ vector of unknown coefficients, $\mathbf{X} = [X_{p+1}, \ldots, X_N]$ and $\mathbf{U} = [U_{p+1}, \ldots, U_N]$ are $1 \times (N-p)$ vectors collecting the present states and the residuals, and $\mathbf{W} = [W_{p+1}, \ldots, W_N]$ is a $p \times (N-p)$ matrix collecting the regressor terms. Then by using the OLS formula an estimate of the AR coefficient can be obtained as follows: $\hat{\mathbf{A}} = \mathbf{X}(\mathbf{W})^\top[\mathbf{W}(\mathbf{W})^\top]^{-1}$. The innovation process can be estimated as the residual time-series $\hat{\mathbf{U}} = \mathbf{X} - \hat{\mathbf{A}}\mathbf{W}$, whose $\hat{\sigma}_U^2$ is an estimate of the innovation variance.

2.4. Data Analysis and Statistical Validation

All the analyses were performed by considering separately the concentrations of HbO_2 and HHb for each channel and for each subject. Regarding their spectral analysis, two distinct identification procedures for the model (5) were performed, separately analyzing the breathing phase and the apnea task. Specifically, given $O = [O_1, \ldots, O_6] = [8, 50, 110, 170, 230, 300]$ as a vector representing the time instants of the apnea task onsets, with the last element indicating the total length of the time series, we defined l_i and m_i, $i \in \{1, \ldots, 5\}$, as the i-th time window during which the subjects breathed normally (red windows in Figure 2) or performed the apnea task, respectively (blue windows in Figure 2). Let B_i denote a generic instant when the subject begins to breathe normally after an apnea task (which may differ for each apnea event and subject). Using this notation, we can define the following time windows: $m_i = [O_i + (p+1), \ldots, B_i - 1]$, and $l_i = [B_i + (p+1), \ldots, O_{i+1} - 1]$. Then when the breath task was considered, we used $\mathbf{X} = [X_{l_1}, \ldots, X_{l_5}]$ as the present state, while the matrix of regressors $\mathbf{W}$ was defined by simply switching the content of each window l_i of one temporal unit in the past, up to p lags. The same rationale was used for the apnea task where $\mathbf{X} = [X_{m_1}, \ldots, X_{m_5}]$. This procedure was implemented to prevent potential discontinuities that could arise from simply concatenating different time windows and to increase the number of observations, thereby reducing the estimation bias [43]. Spectral profiles were computed using (7), and the low-frequency (LF) and high-frequency (HF) components were identified based on poles with central frequencies in the ranges [0.04–0.15] Hz and [0.15–0.4] Hz [44]. The PSD profile relevant to each pole, $P^{(k)}(f)$, was then integrated within the whole frequency spectrum to obtain the power content associated with the LF and HF components, respectively. A representative example of the spectral decomposition performed on one representative channel for HbO_2 and HHb is reported in Figure 2 (panel e: apnea; panel f: breath). The code used for performing the Power Spectral Density Analysis can be found here: https://github.com/YuriAntonacci/LSP_toolbox/tree/main, accessed on 12 December 2024. To test whether the spectral content of fNIRS signals was modulated during the breath-holding task, a statistical comparison between the distributions of central frequency and spectral power obtained across subjects during the breath and apnea phases was performed using the paired non-parametric Wilcoxon signed-rank test ($\alpha < 0.05$).

As for the time-resolved analysis of the different information-theoretic measures, the identification procedure of the TV-AR model (2) was performed setting $p = 4$ and $(1-c) = 0.975$. The model order p was selected using the Akaike Information Criterion (AIC) [42] as a guide. Analysis across the entire scalp and for all subjects returned values ranging from four to six. To ensure consistency and avoid duplicate or negative peaks in the subsequent spectral analysis, the model order was fixed at four [44]. As regard for the forgetting factor $(1-c)$, the selection was performed according with previous studies, which highlighted the interval $[0.97, 0.98]$ as an ideal compromise for the bias–variance trade-off [24,34].

The time-resolved measures of entropy ($H(X_n)$), conditional entropy ($H(X_n|W_n)$), and information storage ($S_{X,n}$) (panels b–d of Figure 2) were then computed and averaged within each i-th time window corresponding to the analyzed experimental condition (breath and apnea). The obtained values were further averaged across the five available windows for each experimental condition to obtain a representative value for each of the six subjects. Statistically significant differences between the apnea and breath conditions were evaluated for each information-theoretic measure using the Wilcoxon test for paired data ($\alpha < 0.05$). The code necessary to compute time-resolved information measures can be found here: https://github.com/YuriAntonacci/Time-VaryingIS, accessed on 12 December 2024.

Therefore, a measure of the effect size was also determined to assess the magnitude of the differences observed among experimental conditions. Specifically, denoting with μ_{S_1}, μ_{S_2}, $\sigma^2_{S_1}$, and $\sigma^2_{S_2}$, the mean and the variance of two distributions S_1 and S_2 obtained measuring the mean and the SD of the distribution across subjects of each information theoretic measure, we computed the Cohen's d measure defined for equally sized groups as follows [45]:

$$d = \frac{\mu_{S_1} - \mu_{S_2}}{\sqrt{(\sigma^2_{S_1} + \sigma^2_{S_2})/2}}. \tag{8}$$

Typically, a small effect size arises for $d = 0.2$, a medium effect size between 0.2 and 0.8, and large was considered when $d = 0.8$ [45]. Finally, we assessed the statistical power of each test to evaluate how the sample size (N = 6) would influence the results. Statistical power ($1 - \beta$) represents the probability of correctly rejecting a false null hypothesis (i.e., the likelihood of detecting a true physiological effect if one exists).

3. Results

3.1. Frequency Specific Analysis of fNIRS Signals

Panels e,f of Figure 2 present the spectral decomposition of the PSD for HbO_2 (panel 1) and HHb (panel 2) signals from one representative subject during the apnea (panel e) and breath (panel f) phases. Regardless of the analyzed time series (HbO_2 or HHb), the PSD trends reveal two distinct peaks centered at approximately 0.05 Hz and 0.2 Hz with the latter showing a modulation when the respiratory activity is considered.

Figures 3 and 4 report the boxplot distributions illustrating the frequency locations of the LF and HF spectral peaks for the apnea (blue) and breath (red) phases, computed separately for HbO_2 (panel a) and HHb (panel b). Figure 3 displays the distribution of spectral peak frequencies within the [0.04–0.15] Hz frequency band, which reveals negligible modulation between apnea and breathing phases, indicating that respiratory activity has limited influence on low-frequency oscillations.

Figure 4 displays the distribution of spectral peak frequencies within the [0.15–0.4] Hz frequency band, revealing significant task-dependent shifts. Overall, regardless of the time series analyzed (i.e., HbO_2 or HHb), there is a consistent trend of increased oscillation frequency observed during the breathing task across all fNIRS channels. This increase becomes statistically significant for HHb, occurring in four specific channels (Figure 4b). Remarkably, the corresponding Cohen's d values consistently exceed 1.2 ($1 - \beta$ >0.6), indicating a strong difference between the two experimental conditions.

Figures 5 and 6 display the boxplot distributions obtained for the spectral power of the fNIRS signal relevant to the LF and HF frequency bands, respectively. The results highlight an increase in both LF and HF power associated with breathing activity. Though this increase is statistically significant only for a few channels, when HHb is considered, the corresponding values of Cohen's d are always greater than 1 ($1 - \beta > 0.55$), highlighting the presence of a strong modulation of the spectral content of fNIRS signals as an effect of respiratory activity.

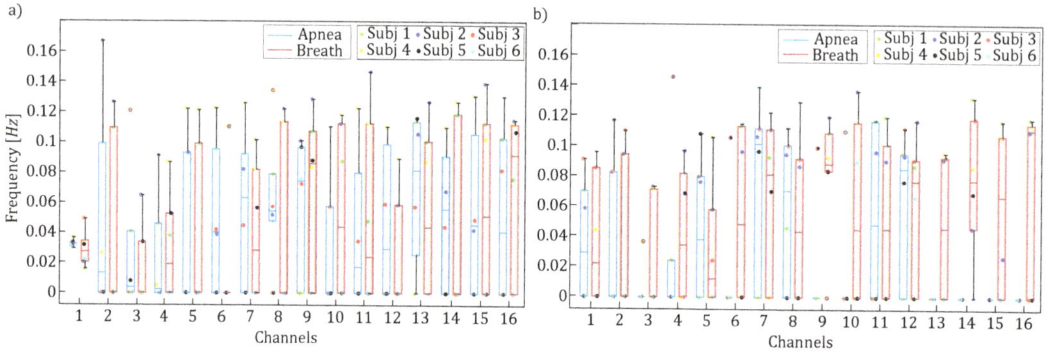

Figure 3. Boxplot distributions and individual values of the frequency locations of the LF spectral peaks for HbO_2 (panel (**a**)) and HHb (panel (**b**)) are shown for each channel, separately for apnea (blue) and breath (red) phases.

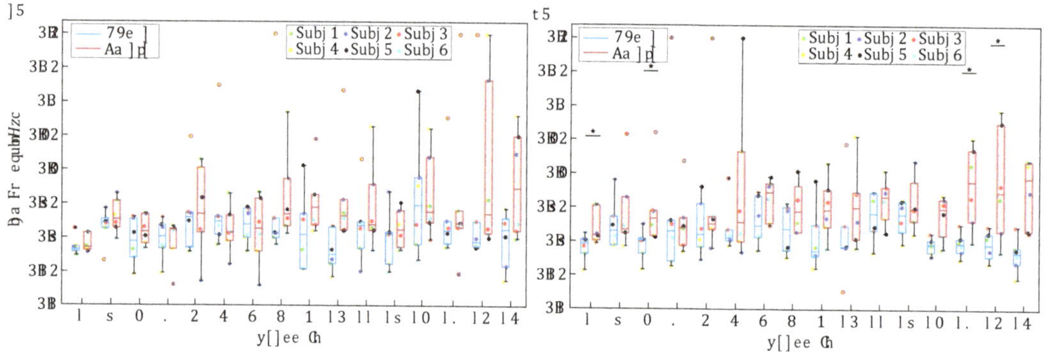

Figure 4. Boxplot distributions and individual values of the frequency locations of the HF spectral peaks for HbO_2 (panel (**a**)) and HHb (panel (**b**)) are shown for each channel, separately for apnea (blue) and breath (red) phases. Statistically significant differences between breath and apnea windows are indicated with * ($p < 0.05$, Wilcoxon signed rank test).

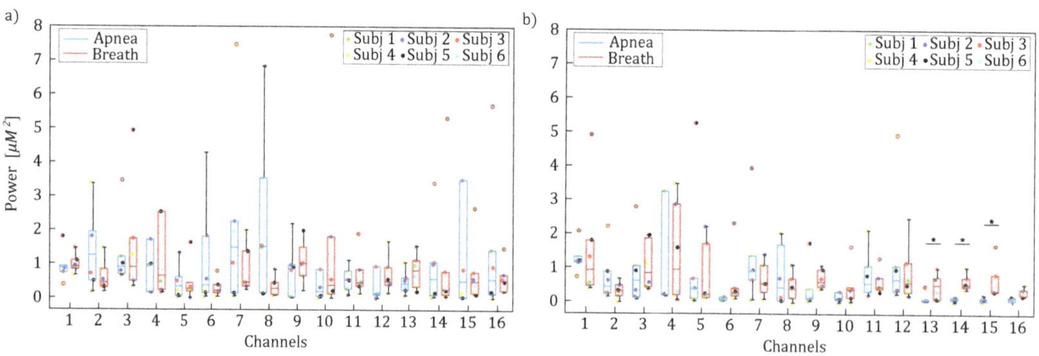

Figure 5. Boxplot distributions and individual values of the spectral power averaged over the LF ([0.04–0.15] Hz) frequency band for HbO_2 (panel (**a**)) and HHb (panel (**b**)), computed for each channel and displayed separately for the apnea (blue) and breath (red) phases. Statistically significant differences between breath and apnea windows are marked with * ($p < 0.05$, Wilcoxon signed rank test).

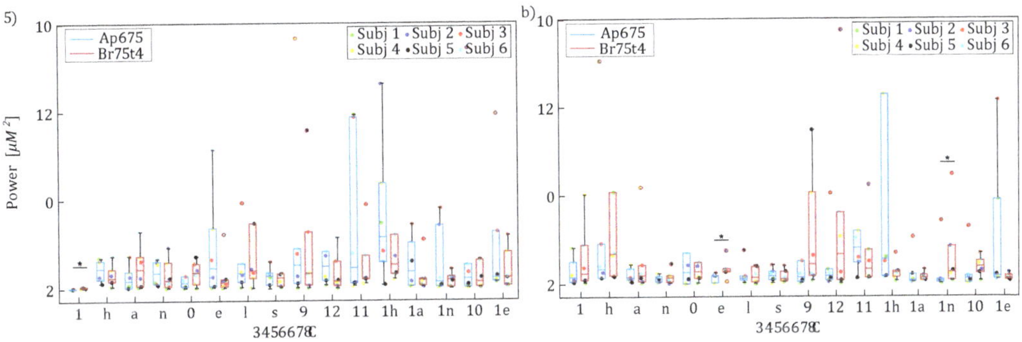

Figure 6. Boxplot distributions and individual values of the spectral power averaged over the HF ([0.15–0.4] Hz) frequency band for HbO_2 (panel (**a**)) and HHb (panel (**b**)), computed for each channel and displayed separately for the apnea (blue) and breath (red) phases. Statistically significant differences between breath and apnea windows are marked with * ($p < 0.05$, Wilcoxon signed rank test).

3.2. Time-Resolved Information-Theoretic Measures

Figures 7–9 display the boxplot distributions of the time-resolved information measures of entropy, conditional entropy, and information storage, averaged across five different time windows relevant to the apnea (blue color) and breath (red color) phases, computed separately for HbO_2 (panel a) and HHb (panel b). The results in Figure 7 suggest a modulation of uncertainty in the oxyhemoglobin and deoxyhemoglobin time series during breathing compared to the apnea condition. However, this difference is not statistically significant, indicating that the amount of information contained in the concentrations of oxyhemoglobin and deoxyhemoglobin remained largely unchanged despite the breath-holding task. Conversely, Figures 8 and 9 reveal statistically significant modulations of the time-resolved conditional entropy and the information storage measures, respectively. In detail, the analysis of time-resolved conditional entropy reveals a statistically significant decrease in the complexity of scalp hemodynamics during breathing periods, regardless of the hemoglobin time series considered, and across nearly all analyzed channels. In contrast, time-resolved information storage shows a statistically significant increase in the predictability of scalp hemodynamics during the same periods. A variability can be observed in these measures across different channels even though Figures 8 and 9 consistently show Cohen's d values > 0.6 with statistical power $(1-\beta) > 0.8$.

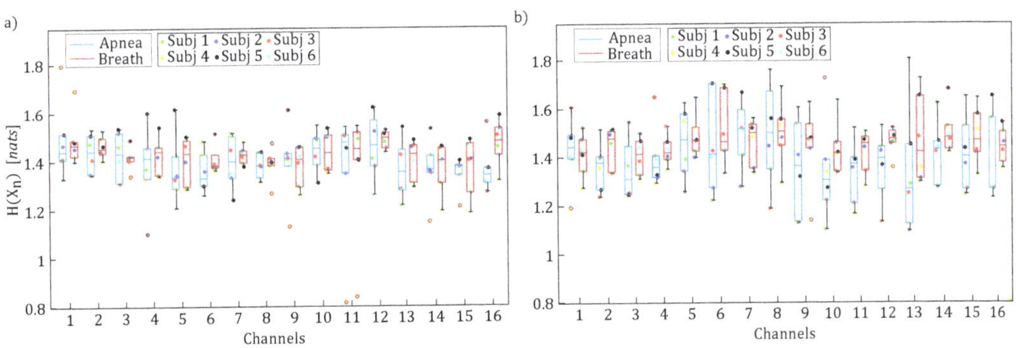

Figure 7. Boxplot distributions and individual values of the time-resolved entropy ($H(X_n)$) for HbO_2 (panel (**a**)) and HHb (panel (**b**)), computed for each channel and displayed separately for the apnea (blue) and breath (red) phases.

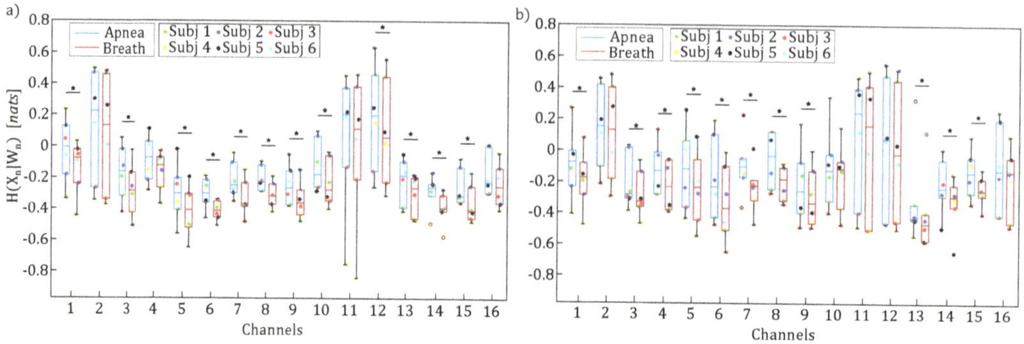

Figure 8. Boxplot distributions and individual values of the time-resolved conditional entropy ($H(X_n|W_n)$) for HbO_2 (panel (**a**)) and HHb (panel (**b**)), computed for each channel and displayed separately for the apnea (blue) and breath (red) phases. Statistically significant differences between the breath and apnea windows are marked with * ($p < 0.05$, Wilcoxon signed-rank test).

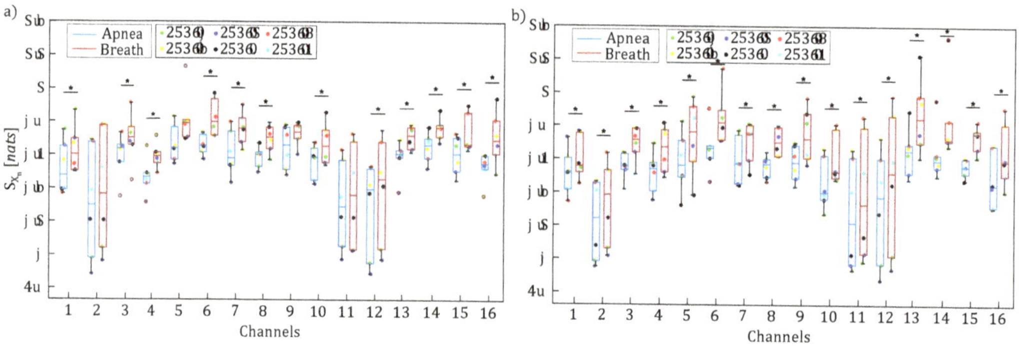

Figure 9. Boxplot distributions and individual values of the time-resolved information storage ($S_{X,n}$) for HbO_2 (panel (**a**)) and HHb (panel (**b**)), computed for each channel and displayed separately for the apnea (blue) and breath (red) phases. Statistically significant differences between the breath and apnea windows are marked with * ($p < 0.05$, Wilcoxon signed-rank test).

4. Discussion

The main results of this work can be summarized as follows: (a) the presence of two different peaks in the PSD of fNIRS signals centered approximately at 0.05 Hz and 0.2 Hz and mainly related to vascular and respiratory components [46–49]; (b) the modulation in both the amplitude and frequency of oscillations in the time series of oxyhemoglobin and deoxyhemoglobin concentrations during the breath-holding task, which may be sustained by the strong coupling between blood pressure and the hemodynamic responses measured in extracerebral tissues [22,50]; (c) the presence of a transition in the information processing characteristics of fNIRS dynamics associated with the shift between breathing and apnea phases, which is marked by an increase in signal predictability during respiratory activity—an effect that cannot be captured using simple measures such as Shannon entropy.

The results reported in Figures 3 and 5 highlighted the presence of a vascular component, known as Mayer waves, consisting of rhythmic fluctuations in arterial pressure caused by changes in vasomotor tone, typically occurring within frequencies $\leq$0.1 Hz, which are reflected in the hemodynamic oscillations of superficial scalp tissues, thereby modulating the optical fNIRS signals [47]. The reduced amplitude of low-frequency components (<0.1 Hz) observed during apnea, even if limited to a few channels, may be attributed to the induction of hypercapnia, i.e., an increase in CO_2 levels during breath-holding tasks in humans [51]. Indeed, the temporary suspension of breathing activity leads to a rise in carbon dioxide levels, triggering a vasodilation response. This vasodilation increases cerebral blood flow, which directly modulates the amplitude of low-frequency oscillations associated with Mayer waves [52].

The results displayed in Figures 4 and 6 revealed the presence of a spectral component in the 0.15–0.4 Hz range, which can be attributed to respiratory activity. This component reflects respiratory-driven variations in arterial pressure as well as in cerebral and extracerebral blood flow. The observed shift in frequency location, occurring exclusively within the HF band, has been previously reported and is physiologically linked to increased respiratory rates during the breathing phase compared to the apnea phase [53]. Interestingly, the peak relevant to the HF range was also observed during apnea and reflects fundamental respiratory physiology: while voluntary breath-holding suppresses overt chest movements, it cannot interrupt the brainstem's central respiratory rhythm [54]. We speculate that these HF oscillations can be related to respiratory sinus arrhythmia—a well-documented phenomenon where autonomic nervous system activity causes a person's heart rate to oscillate in phase with the residual respiratory drive, even during apparent apnea [55]. However, the expected increase in the spectral content within HF components during the breathing phase, as reported in previous studies, was not clearly observed in our analysis—possibly due to the limited duration of the experimental period, which should ideally last at least 28–30 s to reliably capture such changes [56].

Overall, our results revealed dynamic changes in both vascular and respiratory components for oxygenated and deoxygenated hemoglobin, consistent with established mechanisms of respiratory–cardiovascular coupling [54]. Previous studies show that spontaneous hemodynamic oscillations, although not directly elicited by external stimuli or tasks, can still exhibit variations in amplitude and frequency. These fluctuations significantly affect frequency-domain analyses, as they shape the spectral profile of the recorded signals and underscore the influence of underlying physiological rhythms on hemodynamic activity [11,57–59].

The results shown in Figures 7–9 indicated less-complex fNIRS signal dynamics associated with the presence of respiratory activity. Although not directly comparable with previous studies characterizing brain dynamics and physiological systems using information-theoretic approaches [24,34,38,60], these works consistently show that the emergence of a predominant oscillation—often driven by synchronization phenomena—is associated with increased predictability of brain dynamics across different contexts [61]. These findings are consistent with the PSD analysis of fNIRS signals dynamics herein performed, where the breathing task was associated with an increase in the PSD within both the LF and HF bands. Moreover, since a transition occurs after the onset of apnea, we can link the observed decrease in predictability to previous studies, which have highlighted that such transitions are associated with a reduction of the time-resolved information storage [34]. Lastly, we observed a variability in the information-theoretic measures across fNIRS channels, which may originate from the following: (i) spatial heterogeneity in fNIRS sensitivity (superficial vs. deep tissue contributions) [14]; (ii) regional variations in neurovascular coupling [62]; and (iii) task-specific activation patterns [63]. Despite this channel-wise variability, the measures of time-resolved information storage and conditional entropy consistently showed Cohen's d values > 0.6 with statistical power $(1 - \beta) > 0.8$, indicating robust effects of breath-holding on both time-resolved conditional entropy and information storage.

Final Remarks and Limitations

Our results demonstrated that integrating time-resolved information-theoretic measures (entropy, conditional entropy, information storage) with parametric spectral analysis can yield novel insights into fNIRS signal dynamics. Power spectral density analysis revealed distinct spectral signatures: Mayer waves (0.04–0.15 Hz) exhibited minimal modulation during breath-holding, suggesting autonomic stability, while high-frequency oscillations (0.15–0.4 Hz) persisted during apnea, reflecting the ongoing influence of the central respiratory rhythm and respiratory sinus arrhythmia. Time-resolved measures further showed that the presence of respiratory activity reduces hemodynamic complexity and increases signal predictability compared to apnea, underscoring the role of respiration in shaping scalp-measured hemodynamic activity.

The main limitation of our study is that the presented results cannot be directly attributed to cortical activity, as CW single-distance NIRS is known to be more sensitive to superficial tissues than to brain tissue. The measured signals likely reflect a superposition of responses from multiple depths, with a substantial contribution from scalp hemodynamics [14]. Indeed, both incident and backscattered light traverses highly vascularized extracerebral layers such as the skin, subcutaneous fat, and scalp muscles, making the signals particularly sensitive to hemodynamic fluctuations in these regions [30].

Finally, the generally low statistical power and small effect sizes observed in our analyses suggest that some nonsignificant results may be attributed to insufficient statistical sensitivity, primarily due to the limited sample size. This represents a key limitation of the study and highlights the need for larger datasets to enable more robust statistical assessments of the observed effects.

5. Conclusions

This study highlights the significance of fNIRS monitoring and advanced signal processing techniques in exploring scalp hemodynamics during physiological tasks, specifically the breath-holding task. As a non-invasive and portable modality, fNIRS offers a unique capability to monitor real-time hemodynamic responses, complementing other neuroimaging methods [4]. Our findings support the value of a multi-domain analytical approach for reinterpreting physiological "noise" as evidence of functional interactions within the framework of network physiology, and demonstrate the feasibility of applying advanced signal processing techniques even with prototype CW fNIRS systems. Indeed, combining fNIRS with information-theoretic measures allows for the identification of subtle dynamics in both cerebral and extracerebral functions, thereby deepening the physiological insights that can be gained from fNIRS data.

Our results reveal that novel aspects of respiratory–scalp coupling could enable practical applications in the following: (i) clinical monitoring of patients with impaired respiratory–cerebral coupling, such as those with sleep apnea or cerebrovascular disorders; (ii) passive brain–computer interfaces, where entropy-based measures could provide real-time assessment of cognitive state [64,65]. Future work should focus on the validation of these measures in larger cohorts ($N > 30$ subjects), comparing performance across different fNIRS systems (including time-domain devices), and developing hybrid fNIRS-EEG protocols to better separate neural and vascular contributions.

Author Contributions: Conceptualization, L.F. and Y.A.; methodology, L.F. and Y.A.; software, Y.A. and L.F.; validation, I.F., G.C.G., A.B. and F.G.; formal analysis, I.F. and Y.A.; investigation, I.F., Y.A., R.P., L.F. and F.G.; resources, G.C.G., L.F. and A.B.; data curation, G.C.G. and F.G.; writing—original draft preparation, I.F., R.P. and Y.A.; writing—review and editing, F.G., L.F., A.B., R.P. and G.C.G.

visualization, I.F.; supervision, L.F., Y.A. and G.C.G.; project administration, G.C.G., L.F. and A.B.; funding acquisition, L.F., A.B. and G.C.G. All authors have read and agreed to the published version of the manuscript.

Funding: This work was supported by the SiciliAn MicronanOTecH Research And Innovation CEnter "SAMOTHRACE" (MUR, PNRR-M4C2, ECS 00000022), spoke 3—Università degli Studi di Palermo S2-COMMs—Micro and Nanotechnologies for Smart & Sustainable Communities, by the PRIN 2022 project "HONEST-High Order Dynamical Networks in Computational Neuroscience and Physiology: an Information-Theoretic Framework" (funded by MUR, code 2022YMHNPY, CUP B53D23003020006), and by the European Union under the Italian National Recovery and Resilience Plan (NRRP) of NextGenerationEU: partnership on "Telecommunications of the Future" (PE00000001—program "RESTART", projects S10—SEXTET, CUP E63C22002070006).

Institutional Review Board Statement: The study was conducted in accordance with the Declaration of Helsinki, and approved by the Institutional Review Board of the University of Palermo (13925-30 January 2023).

Informed Consent Statement: Informed consent was obtained from all subjects involved in the study.

Data Availability Statement: The data presented in this study are available on request from the corresponding author.

Conflicts of Interest: The authors declare no conflicts of interest.

References

1. Nosrati, R.; Lin, S.; Ramadeen, A.; Monjazebi, D.; Dorian, P.; Toronov, V. Cerebral hemodynamics and metabolism during cardiac arrest and cardiopulmonary resuscitation using hyperspectral near infrared spectroscopy. *Circ. J.* **2017**, *81*, 879–887. [CrossRef] [PubMed]
2. Ferrari, M.; Quaresima, V. A brief review on the history of human functional near-infrared spectroscopy (fNIRS) development and fields of application. *Neuroimage* **2012**, *63*, 921–935. [CrossRef] [PubMed]
3. Ferrari, M.; Mottola, L.; Quaresima, V. Principles, techniques, and limitations of near infrared spectroscopy. *Can. J. Appl. Physiol.* **2004**, *29*, 463–487. [CrossRef] [PubMed]
4. Chiarelli, A.M.; Perpetuini, D.; Croce, P.; Greco, G.; Mistretta, L.; Rizzo, R.; Vinciguerra, V.; Romeo, M.F.; Zappasodi, F.; Merla, A.; et al. Fiberless, multi-channel fNIRS-EEG system based on silicon photomultipliers: Towards sensitive and ecological mapping of brain activity and neurovascular coupling. *Sensors* **2020**, *20*, 2831. [CrossRef]
5. Kassab, A.; Le Lan, J.; Tremblay, J.; Vannasing, P.; Dehbozorgi, M.; Pouliot, P.; Gallagher, A.; Lesage, F.; Sawan, M.; Nguyen, D.K. *Multichannel Wearable f NIRS-EEG System for Long-Term Clinical Monitoring*; Technical Report; Wiley Online Library: New York, NY, USA , 2018.
6. Barut, Ç.; Kızıltan, E.; Gelir, E.; Köktürk, F. Advanced analysis of finger-tapping performance: A preliminary study. *Balk. Med. J.* **2013**, *2013*, 167–171. [CrossRef]
7. Eggebrecht, A.T.; White, B.R.; Ferradal, S.L.; Chen, C.; Zhan, Y.; Snyder, A.Z.; Dehghani, H.; Culver, J.P. A quantitative spatial comparison of high-density diffuse optical tomography and fMRI cortical mapping. *Neuroimage* **2012**, *61*, 1120–1128. [CrossRef]
8. Vendrell, P.; Junqué, C.; Pujol, J.; Jurado, M.A.; Molet, J.; Grafman, J. The role of prefrontal regions in the Stroop task. *Neuropsychologia* **1995**, *33*, 341–352. [CrossRef]
9. Lloyd-Fox, S.; Blasi, A.; Elwell, C. Illuminating the developing brain: The past, present and future of functional near infrared spectroscopy. *Neurosci. Biobehav. Rev.* **2010**, *34*, 269–284. [CrossRef]
10. Wolf, M.; Ferrari, M.; Quaresima, V. Progress of near-infrared spectroscopy and topography for brain and muscle clinical applications. *J. Biomed. Opt.* **2007**, *12*, 062104.
11. Scholkmann, F.; Kleiser, S.; Metz, A.J.; Zimmermann, R.; Pavia, J.M.; Wolf, U.; Wolf, M. A review on continuous wave functional near-infrared spectroscopy and imaging instrumentation and methodology. *Neuroimage* **2014**, *85*, 6–27. [CrossRef]
12. Tachtsidis, I.; Scholkmann, F. False positives and false negatives in functional near-infrared spectroscopy: Issues, challenges, and the way forward. *Neurophotonics* **2016**, *3*, 031405. [CrossRef] [PubMed]
13. Gerouah, Z.; Lin, S.; Toronov, V. Measurement of adult human brain responses to breath-holding by multi-distance hyperspectral near-infrared spectroscopy. *Appl. Sci.* **2021**, *12*, 371. [CrossRef]
14. Strangman, G.E.; Li, Z.; Zhang, Q. Depth sensitivity and source-detector separations for near infrared spectroscopy based on the Colin27 brain template. *PLoS ONE* **2013**, *8*, e66319. [CrossRef]

15. Haeussinger, F.B.; Heinzel, S.; Hahn, T.; Schecklmann, M.; Ehlis, A.C.; Fallgatter, A.J. Simulation of near-infrared light absorption considering individual head and prefrontal cortex anatomy: Implications for optical neuroimaging. *PLoS ONE* **2011**, *6*, e26377. [CrossRef]
16. Scarpa, F.; Cutini, S.; Scatturin, P.; Dell'Acqua, R.; Sparacino, G. Bayesian filtering of human brain hemodynamic activity elicited by visual short-term maintenance recorded through functional near-infrared spectroscopy (fNIRS). *Opt. Express* **2010**, *18*, 26550–26568. [CrossRef]
17. Pinti, P.; Tachtsidis, I.; Hamilton, A.; Hirsch, J.; Aichelburg, C.; Gilbert, S.; Burgess, P.W. The present and future use of functional near-infrared spectroscopy (fNIRS) for cognitive neuroscience. *Ann. N. Y. Acad. Sci.* **2020**, *1464*, 5–29. [CrossRef]
18. Ivanov, P.C. The new field of network physiology: Building the human physiolome. *Front. Netw. Physiol.* **2021**, *1*, 711778. [CrossRef]
19. Bröhl, T.; von Wrede, R.; Lehnertz, K. Impact of biological rhythms on the importance hierarchy of constituents in time-dependent functional brain networks. *Front. Netw. Physiol.* **2023**, *3*, 1237004. [CrossRef]
20. Bashan, A.; Bartsch, R.P.; Kantelhardt, J.W.; Havlin, S.; Ivanov, P.C. Network physiology reveals relations between network topology and physiological function. *Nat. Commun.* **2012**, *3*, 702. [CrossRef]
21. Kandimalla, M.; Lim, S.; Thakkar, J.; Dewan, S.; Kang, D.; In, M.H.; Jo, H.J.; Jang, D.P.; Nedelska, Z.; Lapid, M.I.; et al. Cardiorespiratory Dynamics in the Brain: Review on the Significance of Cardiovascular and Respiratory Correlates in functional MRI signal. *NeuroImage* **2025**, *306*, 121000. [CrossRef]
22. Minati, L.; Kress, I.U.; Visani, E.; Medford, N.; Critchley, H.D. Intra-and extra-cranial effects of transient blood pressure changes on brain near-infrared spectroscopy (NIRS) measurements. *J. Neurosci. Methods* **2011**, *197*, 283–288. [CrossRef] [PubMed]
23. Sütterlin, S.; Schroijen, M.; Constantinou, E.; Smets, E.; Van den Bergh, O.; Van Diest, I. Breath holding duration as a measure of distress tolerance: Examining its relation to measures of executive control. *Front. Psychol.* **2013**, *4*, 483. [CrossRef] [PubMed]
24. Antonacci, Y.; Barà, C.; Zaccaro, A.; Ferri, F.; Pernice, R.; Faes, L. Time-varying information measures: An adaptive estimation of information storage with application to brain-heart interactions. *Front. Netw. Physiol.* **2023**, *3*, 1242505. [CrossRef] [PubMed]
25. Azami, H.; Faes, L.; Escudero, J.; Humeau-Heurtier, A.; Silva, L.E. Entropy analysis of univariate biomedical signals: Review and comparison of methods. In *Frontiers in Entropy Across the Disciplines: Panorama of Entropy: Theory, Computation, and Applications*; World Scientific Publishing Co. Pte. Ltd.: Singapore, 2023; pp. 233–286.
26. Zhang, J.; Chen, H.; Fang, F.; Liao, W. Quantitative analysis of asymmetrical cortical activity based on power spectrum changes. *Brain Topogr.* **2010**, *23*, 257–268. [CrossRef]
27. Minati, L.; Jones, C.L.; Gray, M.A.; Medford, N.; Harrison, N.A.; Critchley, H.D. Emotional modulation of visual cortex activity: A functional near-infrared spectroscopy study. *Neuroreport* **2009**, *20*, 1344–1350. [CrossRef]
28. Tachtsidis, I.; Tisdall, M.; Delpy, D.T.; Smith, M.; Elwell, C.E. Measurement of cerebral tissue oxygenation in young healthy volunteers during acetazolamide provocation: A transcranial Doppler and near-infrared spectroscopy investigation. In *Proceedings of the Oxygen Transport to Tissue XXIX*; Springer: Berlin/Heidelberg, Germany, 2008; pp. 389–396.
29. Cui, W.; Kumar, C.; Chance, B. Experimental study of migration depth for the photons measured at sample surface. In *Proceedings of the Time-Resolved Spectroscopy and Imaging of Tissues*; SPIE: Bellingham, WA, USA, 1991; Volume 1431, pp. 180–191.
30. Strangman, G.; Boas, D.A.; Sutton, J.P. Non-invasive neuroimaging using near-infrared light. *Biol. Psychiatry* **2002**, *52*, 679–693. [CrossRef]
31. Fishburn, F.A.; Ludlum, R.S.; Vaidya, C.J.; Medvedev, A.V. Temporal derivative distribution repair (TDDR): A motion correction method for fNIRS. *Neuroimage* **2019**, *184*, 171–179. [CrossRef]
32. Delpy, D.T.; Cope, M.; van der Zee, P.; Arridge, S.; Wray, S.; Wyatt, J. Estimation of optical pathlength through tissue from direct time of flight measurement. *Phys. Med. Biol.* **1988**, *33*, 1433. [CrossRef]
33. Nollo, G.; Faes, L.; Pellegrini, B.; Porta, A.; Antolini, R. Synchronization index for quantifying nonlinear causal coupling between RR interval and systolic arterial pressure after myocardial infarction. In Proceedings of the Computers in Cardiology 2000. Vol.27 (Cat. 00CH37163), Cambridge, MA, USA, 24–27 September 2000; IEEE: Piscataway, NJ, USA; pp. 143–146.
34. Antonacci, Y.; Barà, C.; de Felice, G.; Sferlazza, A.; Pernice, R.; Faes, L. Exploring transient neurophysiological states through local and time-varying measures of Information Dynamics. *Appl. Math. Comput.* **2025**, *500*, 129437. [CrossRef]
35. Antonacci, Y.; Bara, C.; Sparacino, L.; Mijatovic, G.; Minati, L.; Faes, L. A Method for the Time-Frequency Analysis of High-Order Interactions in Non-Stationary Physiological Networks. *arXiv* **2025**, arXiv:2503.12421.
36. Shannon, C.E. Communication theory of secrecy systems. *Bell Syst. Tech. J.* **1949**, *28*, 656–715. [CrossRef]
37. Barà, C.; Pernice, R.; Catania, C.A.; Hilal, M.; Porta, A.; Humeau-Heurtier, A.; Faes, L. Comparison of entropy rate measures for the evaluation of time series complexity: Simulations and application to heart rate and respiratory variability. *Biocybern. Biomed. Eng.* **2024**, *44*, 380–392. [CrossRef]
38. Lizier, J.T.; Prokopenko, M.; Zomaya, A.Y. A framework for the local information dynamics of distributed computation in complex systems. In *Guided Self-Organization: Inception*; Springer: Berlin/Heidelberg, Germany, 2014; pp. 115–158.

39. Barnett, L.; Barrett, A.B.; Seth, A.K. Granger causality and transfer entropy are equivalent for Gaussian variables. *Phys. Rev. Lett.* **2009**, *103*, 238701. [CrossRef]
40. Haykin, S.S. *Adaptive Filter Theory*; Pearson Education: Noida, India, 2002.
41. Baselli, G.; Porta, A.; Rimoldi, O.; Pagani, M.; Cerutti, S. Spectral decomposition in multichannel recordings based on multivariate parametric identification. *IEEE Trans. Biomed. Eng.* **1997**, *44*, 1092–1101. [CrossRef]
42. Lütkepohl, H. *New Introduction to Multiple Time Series Analysis*; Springer Science & Business Media: New York, NY, USA, 2005.
43. Antonacci, Y.; Toppi, J.; Pietrabissa, A.; Anzolin, A.; Astolfi, L. Measuring connectivity in linear multivariate processes with penalized regression techniques. *IEEE Access* **2024**, *12*, 30638–30652. [CrossRef]
44. Sparacino, L.; Antonacci, Y.; Barà, C.; Švec, D.; Javorka, M.; Faes, L. A method to assess linear self-predictability of physiologic processes in the frequency domain: Application to beat-to-beat variability of arterial compliance. *Front. Netw. Physiol.* **2024**, *4*, 1346424. [CrossRef]
45. Sullivan, G.M.; Feinn, R. Using effect size—Or why the P value is not enough. *J. Grad. Med. Educ.* **2012**, *4*, 279–282. [CrossRef]
46. Chance, B.; Zhuang, Z.; UnAh, C.; Alter, C.; Lipton, L. Cognition-activated low-frequency modulation of light absorption in human brain. *Proc. Natl. Acad. Sci. USA* **1993**, *90*, 3770–3774. [CrossRef]
47. Elwell, C.; Springett, R.; Hillman, E.; Delpy, D.T. Oscillations in cerebral haemodynamics: Implications for functional activation studies. In *Oxygen Transport to Tissue XXI*; Springer: Berlin/Heidelberg, Germany, 1999; pp. 57–65.
48. Emir, U.E.; Ozturk, C.; Akin, A. Multimodal investigation of fMRI and fNIRS derived breath hold BOLD signals with an expanded balloon model. *Physiol. Meas.* **2007**, *29*, 49. [CrossRef]
49. Akin, A.; Emir, U.E.; Kalsin, S.; Sayli, O. Frequency components in breath holding experiments. In *Proceedings of the Optical Tomography and Spectroscopy of Tissue VI*; SPIE: Bellingham, WA, USA, 2005; Volume 5693, pp. 172–178.
50. Tachtsidis, I.; Leung, T.S.; Chopra, A.; Koh, P.H.; Reid, C.B.; Elwell, C.E. False positives in functional nearinfrared topography. In *Oxygen Transport to Tissue XXX*; Springer: Berlin/Heidelberg, Germany, 2009; pp. 307–314.
51. Kastrup, A.; Krüger, G.; Neumann-Haefelin, T.; Moseley, M.E. Assessment of cerebrovascular reactivity with functional magnetic resonance imaging: Comparison of CO_2 and breath holding. *Magn. Reson. Imaging* **2001**, *19*, 13–20. [CrossRef]
52. Zhang, R.; Zuckerman, J.H.; Levine, B.D. Spontaneous fluctuations in cerebral blood flow: Insights from extended-duration recordings in humans. *Am. J. -Physiol.-Heart Circ. Physiol.* **2000**, *278*, H1848–H1855. [CrossRef] [PubMed]
53. Morelli, M.S.; Vanello, N.; Callara, A.L.; Hartwig, V.; Maestri, M.; Bonanni, E.; Emdin, M.; Passino, C.; Giannoni, A. Breath-hold task induces temporal heterogeneity in electroencephalographic regional field power in healthy subjects. *J. Appl. Physiol.* **2021**, *130*, 298–307. [CrossRef] [PubMed]
54. Parkes, M. Breath-holding and its breakpoint. *Exp. Physiol.* **2006**, *91*, 1–15. [CrossRef]
55. Eckberg, D.L. Human sinus arrhythmia as an index of vagal cardiac outflow. *J. Appl. Physiol.* **1983**, *54*, 961–966. [CrossRef]
56. Hakimi, N.; Shahbakhti, M.; Sappia, S.; Horschig, J.M.; Bronkhorst, M.; Floor-Westerdijk, M.; Valenza, G.; Dudink, J.; Colier, W.N. Estimation of respiratory rate from functional near-infrared spectroscopy (fNIRS): A new perspective on respiratory interference. *Biosensors* **2022**, *12*, 1170. [CrossRef]
57. Obrig, H.; Neufang, M.; Wenzel, R.; Kohl, M.; Steinbrink, J.; Einhäupl, K.; Villringer, A. Spontaneous low frequency oscillations of cerebral hemodynamics and metabolism in human adults. *Neuroimage* **2000**, *12*, 623–639. [CrossRef]
58. Näsi, T.; Virtanen, J.; Noponen, T.; Toppila, J.; Salmi, T.; Ilmoniemi, R.J. Spontaneous hemodynamic oscillations during human sleep and sleep stage transitions characterized with near-infrared spectroscopy. *PLoS ONE* **2011**, *6*, e25415. [CrossRef]
59. Tachtsidis, I.; Elwell, C.E.; Leung, T.S.; Lee, C.W.; Smith, M.; Delpy, D.T. Investigation of cerebral haemodynamics by near-infrared spectroscopy in young healthy volunteers reveals posture-dependent spontaneous oscillations. *Physiol. Meas.* **2004**, *25*, 437. [CrossRef]
60. Barà, C.; Zaccaro, A.; Antonacci, Y.; Dalla Riva, M.; Busacca, A.; Ferri, F.; Faes, L.; Pernice, R. Local and global measures of information storage for the assessment of heartbeat-evoked cortical responses. *Biomed. Signal Process. Control* **2023**, *86*, 105315. [CrossRef]
61. Antonacci, Y.; Barà, C.; Sparacino, L.; Pirovano, I.; Mastropietro, A.; Rizzo, G.; Faes, L. Spectral Information Dynamics of Cortical Signals Uncover the Hierarchical Organization of the Human Brain's Motor Network. *IEEE Trans. Biomed. Eng.* **2024**, *72*, 1655–1664. [CrossRef]
62. Tak, S.; Polimeni, J.R.; Wang, D.J.; Yan, L.; Chen, J.J. Associations of resting-state fMRI functional connectivity with flow-BOLD coupling and regional vasculature. *Brain Connect.* **2015**, *5*, 137–146. [CrossRef] [PubMed]
63. Holton, P.; Huang, Y.; Bahuri, N.F.A.; Boccard, S.; Hyam, J.A.; Paterson, D.J.; Dorrington, K.L.; Aziz, T.Z.; Moosavi, S.H.; Green, A.L. Differential responses to breath-holding, voluntary deep breathing and hypercapnia in left and right dorsal anterior cingulate. *Exp. Physiol.* **2021**, *106*, 726–735. [CrossRef] [PubMed]

64. Ronca, V.; Babiloni, F.; Aricò, P. A Novel Mutual Information-based Approach for Neurophysiological Characterization of Sense of Presence in Virtual Reality. *IEEE Trans. Biomed. Eng.* **2025** . [CrossRef] [PubMed]
65. Giorgi, A.; Ronca, V.; Vozzi, A.; Aricò, P.; Borghini, G.; Capotorto, R.; Tamborra, L.; Simonetti, I.; Sportiello, S.; Petrelli, M.; et al. Neurophysiological mental fatigue assessment for developing user-centered Artificial Intelligence as a solution for autonomous driving. *Front. Neurorobot.* **2023**, *17*, 1240933. [CrossRef]

entropy MDPI

Article

Multifractal Multiscale Analysis of Human Movements during Cognitive Tasks

Andrea Faini [1,2], Laurent M. Arsac [3], Veronique Deschodt-Arsac [3] and Paolo Castiglioni [4,5,*]

1 Department of Cardiovascular, Neural and Metabolic Sciences, Istituto Auxologico Italiano, IRCCS, 20149 Milan, Italy; a.faini@auxologico.it
2 Department of Electronics Information and Bioengineering, Politecnico di Milano, 20156 Milan, Italy
3 University of Bordeaux, CNRS, Laboratoire IMS, UMR 5218, 33405 Talence, France; laurent.arsac@u-bordeaux.fr (L.M.A.); veronique.arsac@u-bordeaux.fr (V.D.-A.)
4 Department of Biotechnology and Life Sciences (DBSV), University of Insubria, 21100 Varese, Italy
5 IRCCS Fondazione Don Carlo Gnocchi ONLUS, 20148 Milan, Italy
* Correspondence: pcastiglioni@dongnocchi.it

Abstract: Continuous adaptations of the movement system to changing environments or task demands rely on superposed fractal processes exhibiting power laws, that is, multifractality. The estimators of the multifractal spectrum potentially reflect the adaptive use of perception, cognition, and action. To observe time-specific behavior in multifractal dynamics, a multiscale multifractal analysis based on DFA (MFMS-DFA) has been recently proposed and applied to cardiovascular dynamics. Here we aimed at evaluating whether MFMS-DFA allows identifying multiscale structures in the dynamics of human movements. Thirty-six (12 females) participants pedaled freely, after a metronomic initiation of the cadence at 60 rpm, against a light workload for 10 min: in reference to cycling (C), cycling while playing "Tetris" on a computer, alone (CT) or collaboratively (CTC) with another pedaling participant. Pedal revolution periods (PRP) series were examined with MFMS-DFA and compared to linearized surrogates, which attested to a presence of multifractality at almost all scales. A marked alteration in multifractality when playing Tetris was evidenced at two scales, $\tau \approx 16$ and $\tau \approx 64$ s, yet less marked at $\tau \approx 16$ s when playing collaboratively. Playing Tetris in collaboration attenuated these alterations, especially in the best Tetris players. This observation suggests the high sensitivity to cognitive demand of MFMS-DFA estimators, extending to the assessment of skill/demand interplay from individual behavior. So, by identifying scale-dependent multifractal structures in movement dynamics, MFMS-DFA has obvious potential for examining brain-movement coordinative structures, likely with sufficient sensitivity to find echo in diagnosing disorders and monitoring the progress of diseases that affect cognition and movement control.

Keywords: detrended fluctuation analysis; multifractal; multiscale analysis; cycling; Tetris; multifractal cumulative function; Legendre spectrum

1. Introduction

The complex dynamics of human movements result from the interactions among perceptive, cognitive, and motor systems aimed at responding to changes in the environment or achieving specific tasks. Such interactions are responsible for continuous motor adaptations, and the resulting variability reflects the superposition of fractal processes with different power laws. For this reason, the intrinsic variability of human movements is characterized by multifractal dynamics [1–4].

Among the estimators of the multifractal spectrum, the detrended fluctuation analysis (DFA) [5] for multifractal series (MF-DFA) [6] has gained popularity in the study of motor control due to the statistical performance of the recently proposed focus-based approach [7]. In fact, a limit of most studies is to find a reasonable trade-off between two contrasting requirements: On the one hand, recording long time series to improve the statistical consistency of the regressions that estimate the scale exponents; on the other hand, restricting the duration of motor or cognitive tasks to avoid fatigue, learning or habituation phenomena. The focus-based approach assumes that the DFA regressions of each multifractal exponent converge to a theoretical focus, and it improves the estimates by calculating the focus empirically and by forcing the passage of the regression lines through the calculated focus. Since it provides statistically stable estimates for relatively short series, the focus-based MF-DFA has been used to assess human movements during concurrent visuomotor tasks [8] or complexity-matching phenomena between coupled neural networks [9].

Often, however, the fractal dynamics of physiological systems exhibit different power laws at different temporal scales, that is, a multiscale spectrum. This was soon recognized in the heart-rate series for which the DFA coefficients are traditionally estimated at short and long scales separately [5]. Multiscale analyses have also been used for the monofractal DFA of human movements [10,11]. By contrast, the focus-based approach implicitly assumes that the multifractal exponents are the same at all scales; thus, it cannot separate "short-term" from "long-term" exponents. Therefore, the possible presence of multiscale structures can only be inferred by progressively removing the shorter scales from the multifractal analysis [9].

In this regard, a method for quantifying the multifractal DFA coefficients at each scale separately has been proposed recently [12]. This method improves the statistical consistency of the estimates by splitting the series into maximally overlapped blocks and does not require any assumption on the linear convergence of the DFA variability functions into focus. Based on this approach, a scale-by-scale measure of multifractality that avoids the inapplicability of the Legendre transform when the convexity hypothesis is not satisfied has been also proposed [13]. This multifractal and multiscale approach (MFMS-DFA) has been applied so far to cardiovascular series only [14], but it could be a valuable tool for studying the temporal structures of the multifractal spectrum of human movements as well. By investigating the multifractal dynamics at different scales, such a tool could allow for better assessing the interactions of cognitive- and motor-neural networks in patients after a trauma, or more accurately monitoring the progression of degenerative diseases, as well as the effectiveness of rehabilitation treatments, over time.

Therefore, the present methodological work aims to evaluate whether the recently proposed MFMS-DFA allows effective identification of multiscale structures also in the multifractal dynamics of human movements and quantifying their possible alterations. For this aim, we will describe the multifractal multiscale structure of cycling by maintaining a target pedal rate as the reference multifractal motor task. Furthermore, we will quantify the possible alterations in the multifractal structure of the pedal revolution intervals when a cognitive task is performed simultaneously. We expect that the performance of the reference motor task may be altered due to the greater dual-task cognitive demand and because of the limitations in simultaneously processing the associated interfering streams of information [15,16]. Thus, we aim to describe the intrinsic multifractal dynamics that should characterize the time series of pedal revolution periods and the possible alterations induced in such structure by a dual-task cognitive load, which in our study consists of playing Tetris while cycling.

Therefore, if the recently proposed MFMS-DFA can properly assess the complex dynamics of human movements, we expect to identify possible alterations induced by the cognitive tasks in the multifractal structure at specific scales; and that the presence of multifractality, which is expected to characterize the intrinsic variability of human movements, should be detected also within this temporal structure of multifractal scale coefficients.

2. Materials and Methods

2.1. Subjects and Data Collection

Thirty-six healthy participants (12 females) aged 31.1 ± 12.5 years, all students or university members, gave their written informed consent to participate in this study, which was approved and authorized by the Institutional Review Board Faculte des STAPS and followed the rules of the Declaration of Helsinki.

In order to get temporal variability in movement repetitions emitted by a free-running neurophysiological system, the participants followed a synchronization-continuation paradigm [17]. They were asked to pedal on a friction-loaded cycle ergometer (Monark 818E, Monark, Vansbro, Sweden), imposing the cadence of 60 revolutions per minute (rpm) by a metronome for the first 60 s period (no recordings). The metronome was subsequently stopped, and the participants were asked to keep the same pedaling cadence for the next 10 min against a friction load amounting to 10N. A previous study already reported that such a cycling protocol generates fractal fluctuations in the pedaling cadence [18].

The duration of each pedal revolution period (PRP) was obtained thanks to a Light Meter Pod connected to a PowerLab acquisition system (ADInstruments, Sydney, Australia) at a 1 kHz sampling rate, detecting the changes in light when the pedal passed by the sensor. Each participant reiterated the above procedure in three conditions. The first is the reference cycling (C) condition with the participant's arm resting on an elevated table in front of the bicycle. The second is the dual-task condition, cycling and playing "Tetris" [19] on a computer placed on the table (CT). In Tetris, the player observes a field on the screen in which pieces of different geometric shapes descend from the top. The player can rotate and move the pieces laterally and accelerate them as they fall to create the greatest number of complete horizontal lines of blocks; when a line is completed, it disappears granting points. Thus, Tetris can be considered a progressively demanding cognitive task requiring visuospatial functions. The third condition is again a dual-task performance but cycling simultaneously playing Tetris collaboratively (CTC) on shared screens with the one cycling concomitantly on their side (hidden by a board). In the latter condition, one subject could turn the pieces, the other one could shift them horizontally while they drop and accelerate the drop. Since in CTC the tasks of turning the pieces and moving them horizontally/vertically are no longer performed by a single player, as in CT, but each of the two players takes care of a single task, we expect a somehow lower cognitive load of playing Tetris collaboratively. Tetris collaborations were matched by gender and age.

We recorded the Tetris best score achieved by each participant during CT. At the end of each condition, participants reported the workload they perceived by compiling the NASA Task Load Index (NASA-TLX) questionnaire [20]. The score of the overall workload may range between the minimum value of 0 up to the maximum value equal to 100. Figure 1 shows an example of a PRP series recorded in one participant.

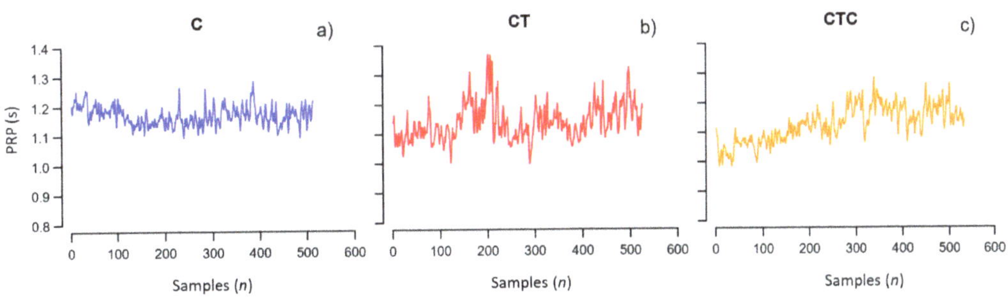

Figure 1. Example of pedal revolution periods, PRP. Recording in the same participant (**a**) during cycling, C; (**b**) cycling playing Tetris alone, CT; (**c**) cycling playing Tetris collaboratively, CTC.

2.2. *Multifractal Multiscale DFA*

2.2.1. Estimation of Multifractal Multiscale Coefficients

The PRP fluctuations around the desired set point of 60 rpm result from the superimposition and interaction of several factors. If these factors act independently without any underlying feedback control, we may expect that they produce uncorrelated deviations from the set point. On the other hand, if they interact with the higher brain center of the motor control to maintain the set point, we may expect long-term correlations in the PRP deviations. The DFA coefficients may characterize this phenomenon as being equal to 0.5 for uncorrelated fluctuations (like white noise) and 1.0 for a purely self-similar process like the 1/f noise. Thus, a decrease in the DFA coefficient might reflect a lower motor control of the higher brain center. In our fractal model of motor control, the multifractal analysis allows to separately describe interacting factors with different fractal dimensions; and the multiscale analysis might reveal shifts of the correlated fluctuations among the temporal scales, possibly providing further clues on the motor adaptation strategies during a dual-task performance.

We estimated the multifractal multiscale structure of the PRP series by the MFMS-DFA algorithm described and downloadable in [12]. Briefly, we calculated the cumulative sum, y_i, of each PRP series and split y_i into M maximally overlapped blocks of n samples. Then we detrended each block with a least-square polynomial regression and calculated the variance of the residuals in each k-th block, $\sigma^2{}_n(k)$. The variability function $F_q(n)$ is the q-th moment of $\sigma^2{}_n$ [6]:

$$\begin{cases} F_q(n) = \left(\frac{1}{M} \sum\limits_{k=1}^{M} \left(\sigma_n^2(k)\right)^{q/2} \right)^{1/q} & \text{for } q \neq 0 \\ F_q(n) = e^{\frac{1}{2M} \sum\limits_{k=1}^{M} \ln\left(\sigma_n^2(k)\right)} & \text{for } q = 0 \end{cases} \quad (1)$$

We calculated $F_q(n)$ for $-5 \leq q \leq 5$ and $6 \leq n \leq L/4$, with L the length in samples of the PRP series. We mapped the scale units from number of pedal revolutions, n, to time τ, in seconds, with the transformation:

$$\tau = n \times \mu_{PRP} \quad (2)$$

where μ_{PRP} is the mean PRP, in seconds. We spline-interpolated the log τ axis evenly [21] between 8 and 128 s to consider the same scales in all the participants. Scales $\tau < 8$ s were excluded to avoid the large estimation bias for negative q at shorter scales [12]. We calculated $F_q(\tau)$ twice, for detrending polynomials of first and second order (Figure 2) because the second-order polynomial removes the longer trends more efficiently but it over-fits block sizes shorter than 12 beats [12,22,23].

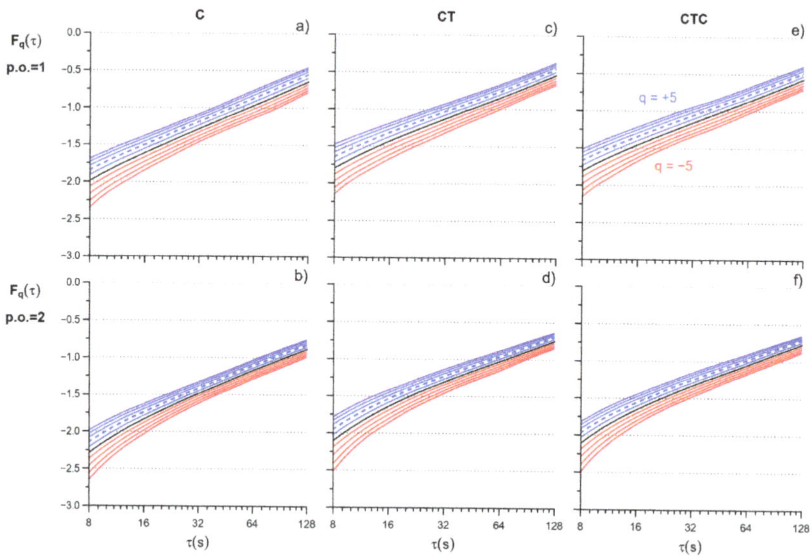

Figure 2. MF-DFA variability functions. $F_q(\tau)$ after polynomial detrending of order 1 (p.o. = 1) in C (**a**), CT (**b**), and CTC (**c**) and order 2 (p.o. = 2) in C (**d**), CT (**e**), and CTC (**f**): average of 36 participants. $F_q(\tau)$ in red for $q < 0$, blue for $q > 0$, and black for $q = 0$; the dashed line is $q = 2$ (moment order of the monofractal DFA). Note the larger effects of overfitting at the shorter scales for the second-order polynomial, which, however, is expected to better remove trends at the larger scales.

We calculated the multifractal coefficients as the derivative of log $F_q(\tau)$ vs. log τ separately for the two detrending orders. The final estimate, $\alpha(q,\tau)$, combined the coefficients of the two detrending orders with a weighted average that progressively weights order 1 more at the shorter than longer scales, as described in [12]. To check the influence of possibly present long-term drifts in the recorded series, we recalculated $\alpha(q,\tau)$ twice, after removing a linear or a quadratic drift from the original PRP series. We did not find substantial effects of drift removal when the scale coefficients were estimated by combining the two detrending orders (see Appendix A); thus, the analysis was performed without drift removal so as not to introduce an additional pre-elaboration stage.

2.2.2. Statistical Comparison with the Reference Condition

We compared the $\alpha(q,\tau)$ coefficients in "CT" and "CTC" vs. the reference "C" by the Wilcoxon signed-rank test for each q and τ. In this way, we obtained the Wilcoxon V signed-rank statistics as a function of both the moment order and scale: $V(q,\tau)$. The significance of observing a given $V(q,\tau)$ value was represented by a color map. The map allows visualizing the regions of the q-τ space with the more significant differences with the reference condition.

To stratify the results by skill levels, the group was subdivided into tertiles according to the individual Tetris score, and the comparisons were repeated in each tertile separately.

2.3. *Degree of Multifractality Scale by Scale*

2.3.1. Cumulative Multifractality Function

We quantified the degree of multifractality at each scale by the cumulative function of the squared increments of scale exponents, $\alpha_{CF}(\tau)$:

$$\alpha_{CF}(\tau) = \sum_{q=-4}^{5} [\alpha(q,\tau) - \alpha(q-1,\tau)]^2 \tag{3}$$

For empirical physiological time series, this index of multifractality is more stable than the width of the singularity spectrum based on the Legendre transform [13].

2.3.2. Surrogate Data Analysis

We employed two generators of surrogate data [24]: (1) the "random phase" (RP) algorithm that preserves the second-order statistics (e.g., the power spectrum) but not the amplitude distribution by shuffling the phases of the Fourier spectrum; and (2) the "iterative amplitude adjusted Fourier transform" (IAAFT) algorithm, which tries to preserve both the power spectrum and the amplitude distribution through an iterative procedure of gaussianization, phase randomization, and de-gaussianization. We created 100 surrogate series for each PRP recording, both for the RP- and the IAAFT generator.

2.3.3. Statistical Comparison with the Surrogate Data

We tested the presence of multifractality at each τ by comparison with the surrogate data. For each participant j, with $1 \leq j \leq 36$, we first calculated the cumulative function of the original series "O", $\alpha_{CF}^{O,j}(\tau)$. Then we calculated the cumulative function of each of 100 RP-surrogate series i, $\alpha_{CF}^{i,j}(\tau)$, with $1 \leq i \leq 100$, and their median values, $\alpha_{CF}^{M,j}(\tau)$. Finally, we tested the presence of multifractality evaluating whether the cumulative function of the original series, $\alpha_{CF}^{O,j}(\tau)$, was statistically greater than the median cumulative function of the surrogate series, $\alpha_{CF}^{M,j}(\tau)$, at each scale τ, over the group of N = 36 participants, by the Wilcoxon one-tailed paired test. The test was repeated for the IAAFT-surrogate series.

All the statistics were conducted using R (R Core Team, 2023) and RStudio (Posit Team, 2023).

3. Results

3.1. MFMS-DFA and Cognitive Tasks

Figure 3 compares the $\alpha(q,\tau)$ coefficients in the reference cycling condition C with cycling playing Tetris alone, CT, or collaboratively, CTC. Color maps show the statistical significance at $p < 1\%$ in bright yellow and point out marked alterations induced by the cognitive tasks. Playing Tetris alone produces changes at two scales, around τ = 16 s and τ = 64 s, mainly for positive moment orders (the differences quickly lose significance for $q < -1$). The upper left panel shows that for $q > 0$, $\alpha(q,\tau)$ decreases from C to CT. In particular, for $q = 2$ (moment order of the monofractal DFA, dashed lines in Figure 3), the scale coefficient at τ = 16 s decreases from the value of a fractional Brownian motion process ($\alpha > 1$) to the value of pink noise ($\alpha = 1$) and at τ = 64 s decreases from pink noise to the value of a fractional Gaussian noise ($\alpha < 0.8$).

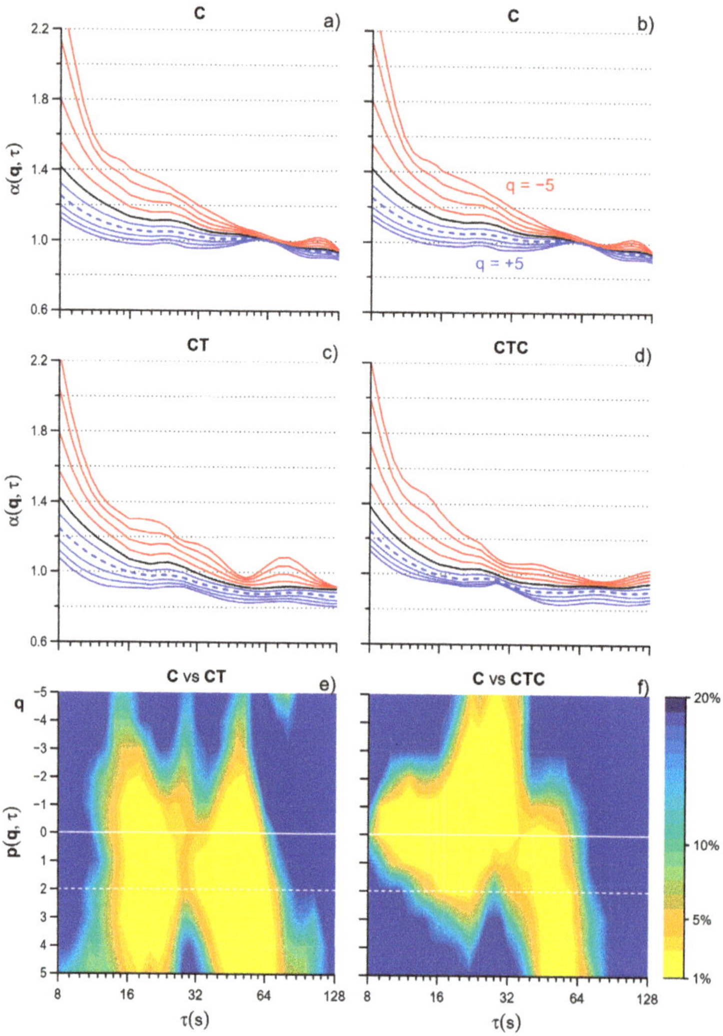

Figure 3. MFMS-DFA coefficients in C (**a**,**b**), CT (**c**), and CTC (**d**) and color maps of the statistical significance of the Wilcoxon p comparing C vs. CT (**e**) and CTC (**f**); $\alpha(q,\tau)$ in red for $q < 0$, blue for $q > 0$, black for $q = 0$; dashed lines indicate $q = 2$ (moment order of the monofractal DFA).

Playing Tetris collaboratively (CTC) produces marked changes too. While at $\tau = 64$ s the color map of C vs. CTC highlights the same α decrease we found in the C vs. CT comparison, the decrease is less pronounced at $\tau = 16$ s. Furthermore, playing Tetris collaboratively slightly but significantly decreases α around $\tau = 32$ s for $q < 0$.

3.2. Multifractality

Figure 4 compares the $\alpha(q,\tau)$ functions for the original and surrogate series. Visually, a lower dispersion of the multifractal coefficients is apparent at all the scales τ.

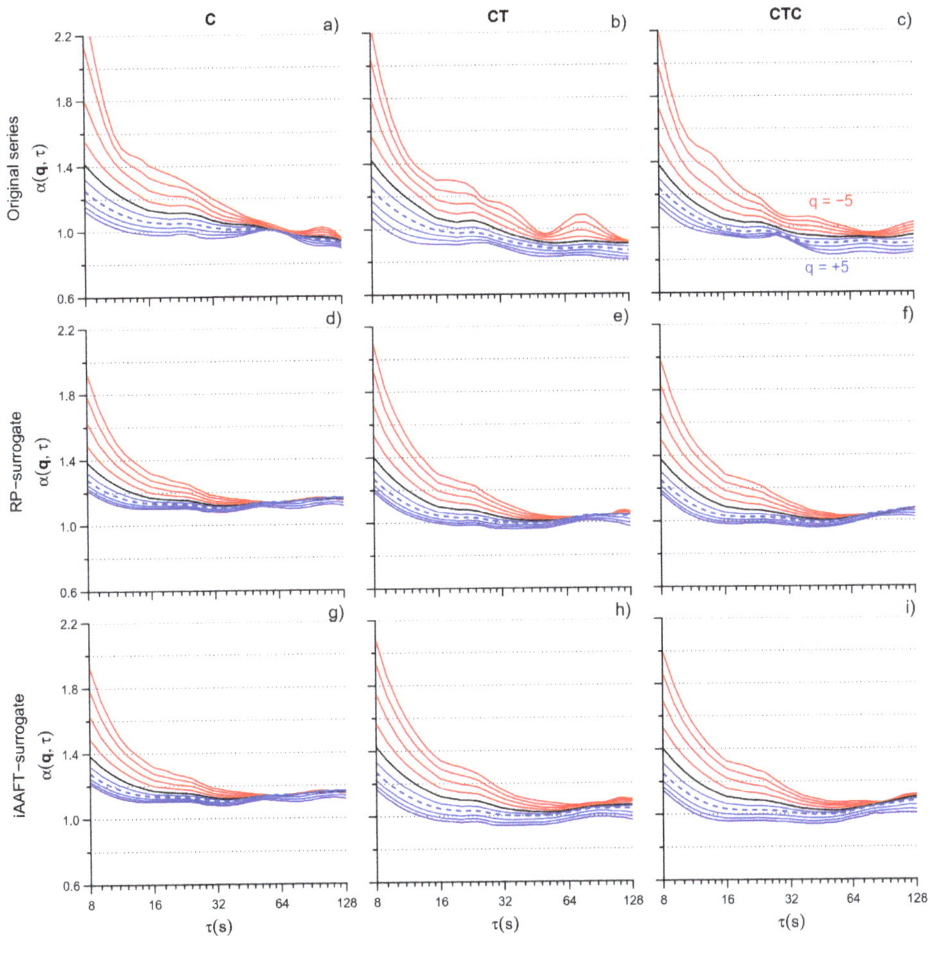

Figure 4. MFMS-DFA coefficients for original and surrogate series. $\alpha(q,\tau)$ of the original series in C (**a**), CT (**b**), and CTC (**c**); RP-surrogate series in C (**d**), CT (**e**), and CTC (**f**); and IAAFT-surrogate series in C (**g**), CT (**h**), and CTC (**i**). For the original series, the figure shows the average over N = 36 participants; for the surrogate series, the median over 100 surrogates was calculated for each of the N = 36 participants and the figure shows the average over N = 36 medians.

The visual trend is confirmed by Figure 5 which shows the degree of multifractality of the original and surrogate series and their statistical comparison. The cumulative function $\alpha_{CF}(\tau)$ is significantly lower for the surrogate data over most of the scales, indicating an almost ubiquitous presence of multifractality at all τ's. It is worth noting a greater significance of multifractality in C than in CT or CTC at $\tau < 32$ s for both the RP and IAAFT surrogates and a systematically larger p (i.e., lower significance) when the comparison is performed against the IAAFT than the RP surrogate.

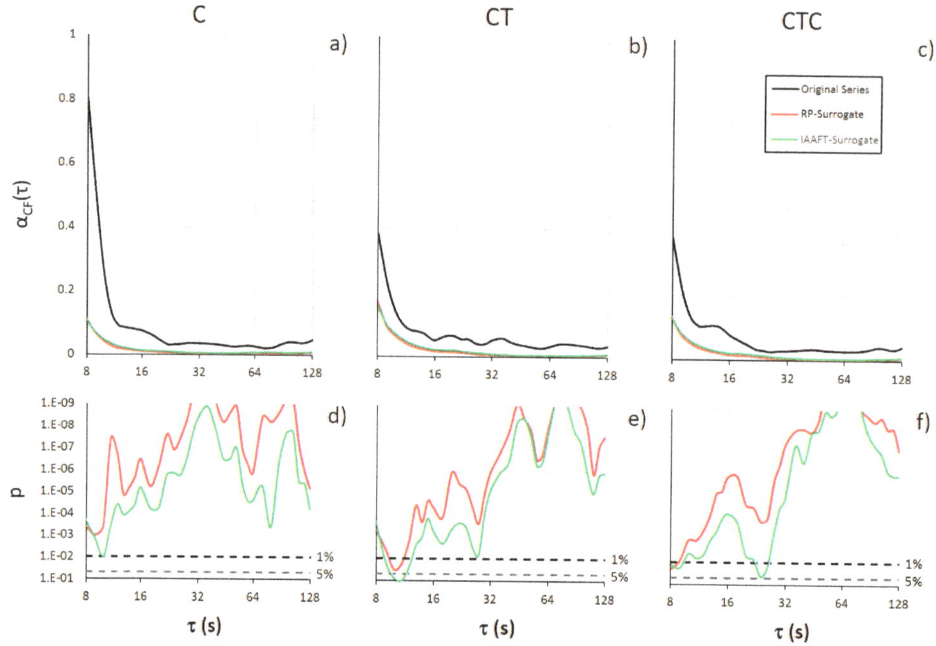

Figure 5. Multifractality cumulative function $\alpha_{CF}(\tau)$. Averages over 36 participants for the original series and its RP- and IAAFT surrogates in C (**a**), CT (**b**), and CTC (**c**). Statistical significance p of the difference between original and surrogate series in C (**d**), CT (**e**), and CTC (**f**): When p is above the dashed lines at 5% and 1% levels, α_{CF} is significantly lower for the surrogate than the original series at the 0.05 and 0.01 statistical significance.

3.3. Stratification by Skill Level

Participants were subdivided into tertiles according to their Tetris best score. The I tertile included 12 participants with a best score between 1400 and 7900; the II tertile 12 participants with a best score greater than 7900 up to 10,967; and the III tertile the remaining 12 participants, with a best score from 10,967 to 32,200. Table 1 reports the general characteristics of participants in each tertile separately. The three subgroups had similar age and sex composition. They also reported similar scores for the perceived workload of the three tasks, with greater scores for the dual tasks.

Table 1. Participants' characteristics by tertiles of Tetris score.

	I Tertile	II Tertile	III Tertile
Females/Males	4/8	5/7	3/9
Age (yoa)	34 (15)	32 (13)	28 (9)
NASA-Task Load Index of Total Workload			
C	46.3 (27.8)	43.3 (21.2)	47.7 (11.3)
CT	64.7 (23.8)	60.3 (12.3)	66.7 (18.7)
CTC	65.3 (20.8)	67.0 (18.7)	63.3 (10.2)

Values as mean (standard deviation) for age, as median (interquartile range) for NASA-TLX questionnaire scores.

The effects of playing Tetris, alone or collaboratively, are evident in the I tertile (participants with the lower skill): By contrast, almost no effects of playing Tetris collaboratively appear in the tertile with the highest skill level (Figure 6). The same high-skill participants, however, when they played Tetris alone, that is, in the experimental condition where they reached their best scores, showed a clear alteration in the $\alpha(q,\tau)$ coefficients (CT vs. C) at the scales around τ = 64 s and $q > 0$.

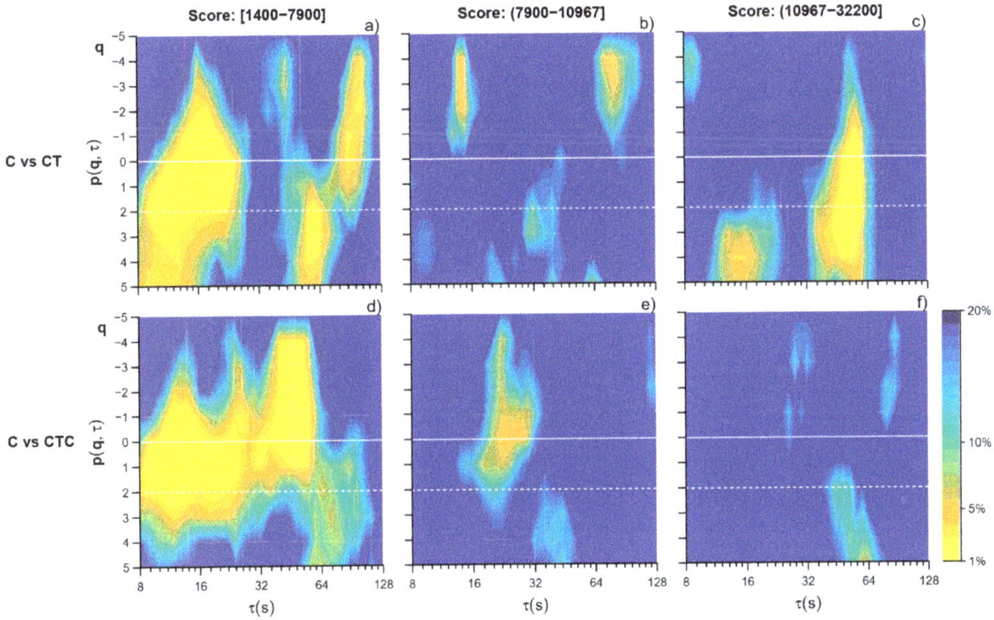

Figure 6. Wilcoxon p significance of the comparison of the MFMS-DFA coefficients between the reference C vs. CT and CTC by tertile of Tetris score. Color codes as in Figure 3 compare C vs. CT in the I (**a**), II (**b**), and III (**c**) tertile and C vs. CTC in the I (**d**), II (**e**), and III (**f**) tertile.

4. Discussion and Conclusions

This work applied the recently proposed MFMS-DFA to quantify for the first time the scale-by-scale profile of the multifractal exponents of motor time series. Multiscale DFA is common in studies on cardiovascular dynamics, and the MFMS-DFA has been already used on heart rate and blood pressure beat-to-beat recordings. By contrast, multifractal analyses of motor series never exploited the temporal structure of scale exponents, and our work provides evidence supporting the feasibility of the MFMS-DFA also in these studies. In particular, we considered an experimental design where concurrent cognitive tasks during cycling limited the duration of the recordings to no more than 10 min. Such short recordings make particularly important the statistical consistency of the $\alpha(q,\tau)$ estimates that characterize the MFMS-DFA method, thanks to its strategy of maximizing the overlapping between consecutive blocks. There are two other innovative solutions for the assessment of fractal analysis that can be applied using the MFMS-DFA and that the present work employs for the first time in a study on motor control. The first solution is the possibility of progressively shifting the order of the detrending polynomials fitting each block of data according to the block size n: The progressive increase from the first- to the second order by mixing the scale coefficients with a weighted average allowed us to efficiently remove second-order trends from the longer data blocks without introducing important polynomial overfitting in the shorter blocks (see also the Appendix A). The second solution is the use of the multifractal cumulative function for quantifying the degree of multifractality scale by scale: This index of multifractality is more stable compared to the traditional width of the Legendre spectrum, which often exhibits "zigzag shapes rather than the expected parabolic shape" on empirical series, as reported by other authors [9,25].

Our results highlighted the usefulness of a multiscale approach to multifractality also for motor time series because the MFMS-DFA identified alterations in the PRP fractal structure induced by the cognitive task, i.e., playing Tetris, separately at short ($\tau \approx 16$ s) and long ($\tau \approx 64$ s) scales (Figure 3e). The alterations occurring at short and long scales may have a different nature because the short-term alterations prevail in the less skilled participants compared to the other volunteers (Figure 6a–c), while long-term alterations appear with high statistical significance in the subgroup with the best Tetris scores in which, by contrast, short-term alterations are almost absent. A multifractal structure presenting different α coefficients at positive and negative moment orders q suggests that fractal processes with different amplitudes are acting simultaneously, being the influence of the fractal processes with lower amplitude amplified by $q < 0$ and decreased by $q > 0$ while the opposite is true for the fractal processes with higher amplitude. Thus, our results suggest that playing Tetris influences significantly the fractal processes with larger amplitude.

When Tetris was played in collaboration with another participant, the MFMS-DFA suggested a reorganization of the interactions among neural networks that may alter the fractal components with a lower amplitude (Figure 3d). The stratification by skill level provides further insights. While the multifractal structure of the PRP time series was again importantly altered in the less skilled subgroup (Figure 6d), we did not find significant alterations induced by the collaborative task in the subgroup with the best Tetris score, over the whole range of scales and for all the moment orders (Figure 6f). The present work, aimed at evaluating the applicability of the MFMS-DFA for assessing the fractal dynamics of human movements, is not designed to identify the mechanisms underlying the adaptations of brain motor centers during dual-task performance. However, our results appear consistent with the hypothesized phenomenon of dual-task interference, meaning that the performance of a specific task is impaired when the task is executed simultaneously with another task [15]. An explanation of dual-task interference is a limitation in the cognitive capacity of processing the flow of information. The latter has been experimentally associated with the limited recruitment of the neurons in the lateral prefrontal cortex: These neurons may be overloaded by the task-relevant information that increases proportionally with the demands of the concurrent tasks [26]. In our participants, we found significant changes in the MFMS-DFA of the PRP series at different scales and in all moment orders when cycling was performed while playing Tetris (Figure 3). These changes support the hypothesis that playing Tetris interferes with cycling. Actually, the dual-task performance is perceived as more demanding than the single task, according to the scores of the NASA-TLX questionnaire (Table 1). Interestingly, the NASA-TLX scores appear similarly high in CT and CTC compared to the single task in all the tertiles of Tetris best score. This indicates that the dual task was perceived as equally challenging by both more skilled and less skilled players. However, CTC is expected to be less cognitively demanding than CT, since the player can focus on one of the two actions of the Tetris game only: block rotation or block shift. Thus, we may hypothesize that in the less skilled players, the higher brain centers involved in the dual tasks were overloaded by the cognitive task of playing Tetris even when Tetris was played together with another participant. We may also hypothesize that in contrast the task of playing Tetris collaboratively was much less demanding for more skilled players. Thus, in the higher tertile of participants, the lower flow of information to be processed to play Tetris collaboratively allowed the higher brain centers to better elaborate the information required for the motor task of cycling. This would explain why in terms of the multifractal multiscale dynamics, the PRP series of the more skilled players did not differ substantially when cycling was executed as the single task or while playing Tetris collaboratively.

A final comment regards the assessment of multifractality by the MFMS-DFA approach. The multifractal cumulative function, α_{CF}, was able to significantly identify the multifractality, which is expected in cyclic motor time series, over a wide range of scales. By describing the degree of multifractality as a function of τ, α_{CF} showed a stronger statistical significance at the larger scales ($\tau > 32$ s, Figure 5), suggesting the superposition of fractal processes with long dynamics. In the comparison with the surrogate data, α_{CF} was also sufficiently sensitive to quantify more significant differences with the RP than the IAAFT surrogates, indicating that nonlinear dynamical components affect not only the Fourier phase but also the amplitude distribution of the PRP values [24].

In conclusion, the recently proposed MFMS-DFA provides statistically consistent estimates of the multifractal dynamics when applied on the relatively short time series typical of the studies on the interactions among the neural networks involved in motor control and cognitive processes. Its ability to quantify the multifractal structure separately over different ranges of scales may help to better identify the underlying mechanisms, with promising perspectives in the diagnosis of motor disorders and on the course monitoring of diseases affecting the neural networks involved in motor control.

Limitations

Since the aim of the study was to evaluate the feasibility of the MFMS-DFA approach, the results should be considered carefully before proposing physiological interpretation or if they should be the reference in healthy controls for future comparison with diseased conditions. First, we may expect gender differences that require a larger population than the number of participants, 2/3 males and 1/3 females, enrolled in the present work. Second, our participants are healthy, young and middle-aged adults, and for future evaluations of the effects of specific diseases, likely to affect older ages, healthy volunteers properly matched by age with the patients are required.

Author Contributions: Conceptualization and methodology, P.C., L.M.A., V.D.-A. and A.F.; software, A.F.; validation and formal analysis, A.F. and P.C.; investigation, P.C., L.M.A., V.D.-A. and A.F.; resources, L.M.A. and V.D.-A.; data curation, L.M.A. and V.D.-A.; writing—original draft preparation, P.C. and A.F.; writing—review and editing, P.C., L.M.A., V.D.-A. and A.F.; visualization, A.F.; supervision, P.C.; project administration and funding acquisition, P.C., L.M.A., V.D.-A. and A.F. All authors have read and agreed to the published version of the manuscript.

Funding: A.F. was funded by the Italian Ministry of Health (Ricerca Corrente); Fondazione Don Gnocchi was supported by #NEXTGENERATIONEU (NGEU) and funded by the Ministry of University and Research (MUR), National Recovery and Resilience Plan (NRRP), project MNESYS (PE0000006)—a multiscale integrated approach to the study of the nervous system in health and disease (DN. 1553 11.10.2022).

Institutional Review Board Statement: The study was conducted in accordance with the Declaration of Helsinki and approved by the Institutional Review Board "Faculte des STAPS".

Data Availability Statement: The data supporting the main findings of this study are available on reasonable request with access granted on justified request to researchers meeting the criteria for access to confidential data.

Conflicts of Interest: The authors declare no conflict of interest. The funders had no role in the design of the study; in the collection, analyses, or interpretation of data; in the writing of the manuscript; or in the decision to publish the results.

Appendix A

Possibly present long-term drifts in the recordings may influence the DFA coefficients, and trend removal is generally recommended [27]. Drifts can be removed by subtracting linear or quadratic polynomials least-square fitting the series. However, to calculate $F_q(n)$ in Equation (1), the DFA already detrends each block of size of n samples by a polynomial fitting. In our approach, we used both first- and second-order detrending polynomials, which may have mitigated differently the effects of long-term drifts. Thus, to evaluate how importantly long-term drifts may have influenced our results, we calculated the MFMS DFA coefficients, $\alpha(q,\tau)$, without drift removal and removing linear or quadratic drifts. This was carried out using both detrending polynomials of first and second order (Figure A1). The results show that when the second-order polynomial detrending is used within each block of size n, the $\alpha(q,\tau)$ coefficients without or with drift removals are virtually identical. This means that long-term linear or quadratic drifts are automatically removed at all the scales by the DFA method itself which employs the second-order detrending.

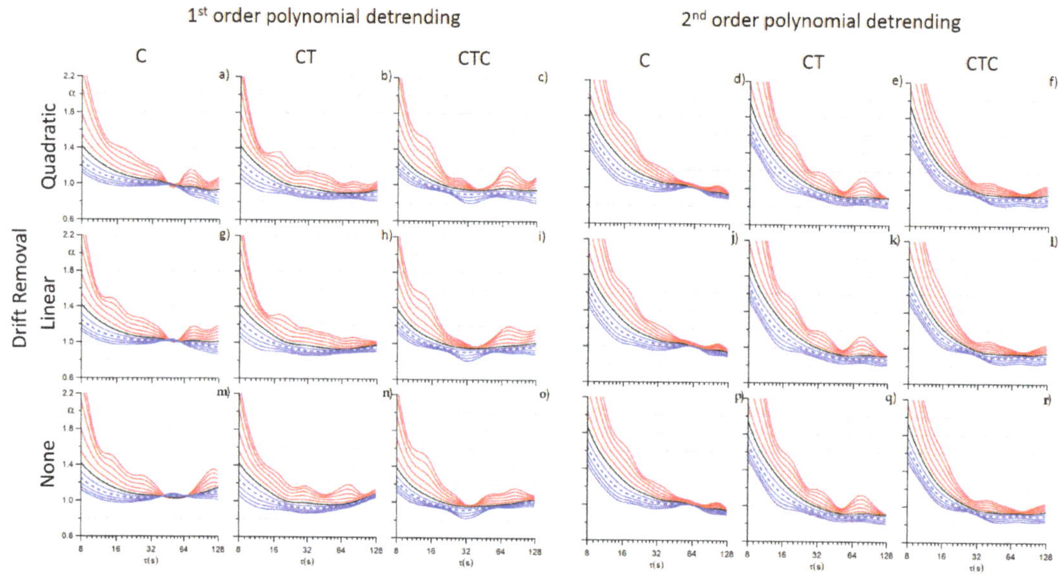

Figure A1. Scale coefficients after quadratic (**a–f**) or linear (**g–l**) drift removal or without drift removal (**m–r**) calculated with first- or second-order detrending in each of the three experimental conditions. Colors as in Figure 2.

We also calculated $\alpha(q,\tau)$ coefficients using the first-order detrending to avoid overfitting effects. Figure A1 shows that in this case only the coefficients estimated at the larger scales ($\tau > 30$ s) are influenced by the drift removal.

Our final estimate mixes the two detrending orders with a weighted average of the $\alpha(q,\tau)$ coefficients after first- and second-order detrending. In the weighted average, estimates with first-order detrending prevail at the shorter scales (to remove the overfitting effect) but have no weight at the larger scales to fully detrend the larger blocks with a parabolic fitting. Thus, we do not expect any substantial influence of drift removal on our estimates. This is confirmed by Figure A2: The mixed coefficients, calculated either without or with drift removal, are virtually identical.

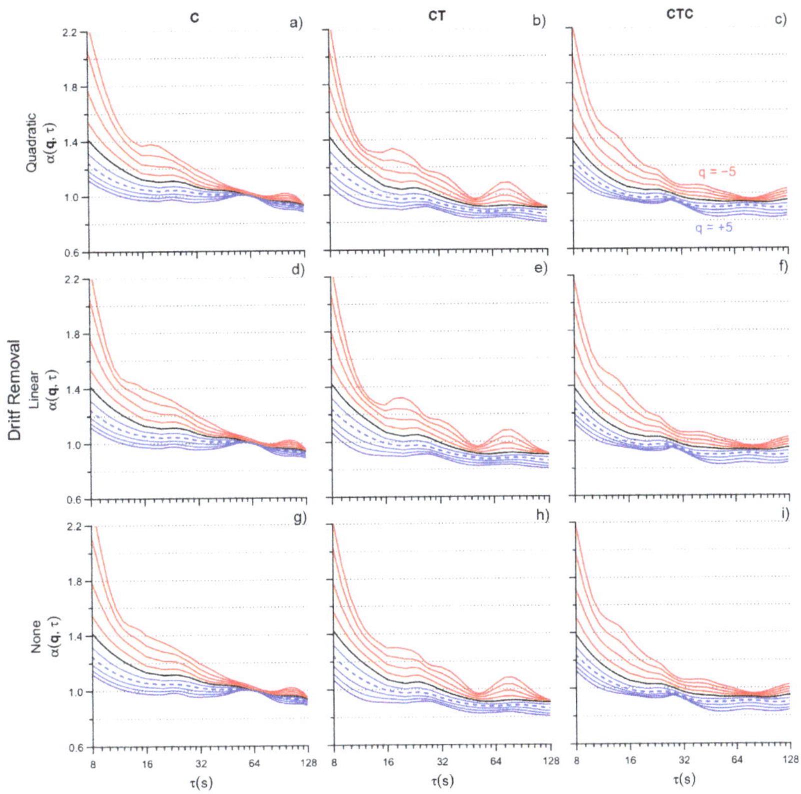

Figure A2. MFMS DFA coefficients estimated with a weighted average of the scale coefficients obtained by first- and second-order detrending: values calculated after quadratic (**a–c**) and linear (**d–f**) drift removal and without drift removal (**g–i**). Colors as in Figure 2.

References

1. Ihlen, E.A.F.; Vereijken, B. Multifractal Formalisms of Human Behavior. *Hum. Mov. Sci.* **2013**, *32*, 633–651. [CrossRef] [PubMed]
2. Dixon, J.A.; Holden, J.G.; Mirman, D.; Stephen, D.G. Multifractal Dynamics in the Emergence of Cognitive Structure. *Top. Cogn. Sci.* **2012**, *4*, 51–62. [CrossRef] [PubMed]
3. Shimizu, Y.; Thurner, S.; Ehrenberger, K. Multifractal spectra as a measure of complexity in human posture. *Fractals* **2002**, *10*, 103–116. [CrossRef]
4. Wilson, T.J.; Mangalam, M.; Stergiou, N.; Likens, A.D. Multifractality in Stride-to-Stride Variations Reveals That Walking Involves More Movement Tuning and Adjusting than Running. *Front. Netw. Physiol.* **2023**, *3*, 1294545. [CrossRef] [PubMed]
5. Peng, C.K.; Havlin, S.; Hausdorff, J.M.; Mietus, J.E.; Stanley, H.E.; Goldberger, A.L. Fractal Mechanisms and Heart Rate Dynamics. Long-Range Correlations and Their Breakdown with Disease. *J. Electrocardiol.* **1995**, *28*, 59–65. [CrossRef] [PubMed]
6. Kantelhardt, J.W.; Zschiegner, S.A.; Koscielny-Bunde, E.; Havlin, S.; Bunde, A.; Stanley, H.E. Multifractal Detrended Fluctuation Analysis of Nonstationary Time Series. *Phys. A Stat. Mech. Appl.* **2002**, *316*, 87–114. [CrossRef]
7. Mukli, P.; Nagy, Z.; Eke, A. Multifractal Formalism by Enforcing the Universal Behavior of Scaling Functions. *Phys. A Stat. Mech. Appl.* **2015**, *417*, 150–167. [CrossRef]
8. Pratviel, Y.; Deschodt-Arsac, V.; Larrue, F.; Arsac, L.M. Fast Hand Movements Unveil Multifractal Roots of Adaptation in the Visuomotor Cognitive System. *Front. Physiol.* **2021**, *12*, 713076. [CrossRef]
9. Delignières, D.; Almurad, Z.M.H.; Roume, C.; Marmelat, V. Multifractal Signatures of Complexity Matching. *Exp. Brain Res.* **2016**, *234*, 2773–2785. [CrossRef]
10. Gilfriche, P.; Deschodt-Arsac, V.; Blons, E.; Arsac, L.M. Frequency-Specific Fractal Analysis of Postural Control Accounts for Control Strategies. *Front. Physiol.* **2018**, *9*, 293. [CrossRef]
11. Delignières, D.; Marmelat, V. Strong Anticipation and Long-Range Cross-Correlation: Application of Detrended Cross-Correlation Analysis to Human Behavioral Data. *Phys. A Stat. Mech. Appl.* **2014**, *394*, 47–60. [CrossRef]

12. Castiglioni, P.; Faini, A. A Fast DFA Algorithm for Multifractal Multiscale Analysis of Physiological Time Series. *Front. Physiol.* **2019**, *10*, 115. [CrossRef] [PubMed]
13. Faini, A.; Parati, G.; Castiglioni, P. Multiscale Assessment of the Degree of Multifractality for Physiological Time Series. *Phil. Trans. R. Soc. A.* **2021**, *379*, 20200254. [CrossRef] [PubMed]
14. Castiglioni, P.; Omboni, S.; Parati, G.; Faini, A. Day and Night Changes of Cardiovascular Complexity: A Multi-Fractal Multi-Scale Analysis. *Entropy* **2020**, *22*, 462. [CrossRef] [PubMed]
15. Pashler, H. Dual-Task Interference in Simple Tasks: Data and Theory. *Psychol. Bull.* **1994**, *116*, 220–244. [CrossRef] [PubMed]
16. Strobach, T. Cognitive Control and Meta-Control in Dual-Task Coordination. *Psychon. Bull. Rev.* **2023**. [CrossRef] [PubMed]
17. Wing, A.M.; Kristofferson, A.B. The Timing of Interresponse Intervals. *Percept. Psychophys.* **1973**, *13*, 455–460. [CrossRef]
18. Arsac, L.M. Multifractal Dynamics in Executive Control When Adapting to Concurrent Motor Tasks. *Front. Physiol.* **2021**, *12*, 662076. [CrossRef]
19. Breukelaar, R.; Demaine, E.D.; Hohenberger, S.; Hoogeboom, H.J.; Kosters, W.A.; Liben-Nowell, D. Tetris is hard, even to approximate. *Int. J. Comput. Geom. Appl.* **2004**, *14*, 41–68. [CrossRef]
20. Hart, S.G.; Staveland, L.E. Development of NASA-TLX (Task Load Index): Results of Empirical and Theoretical Research. In *Advances in Psychology*; Elsevier: Amsterdam, The Netherlands, 1988; Volume 52, pp. 139–183. ISBN 978-0-444-70388-0.
21. Almurad, Z.M.H.; Delignières, D. Evenly Spacing in Detrended Fluctuation Analysis. *Phys. A Stat. Mech. Appl.* **2016**, *451*, 63–69. [CrossRef]
22. Kantelhardt, J.W.; Koscielny-Bunde, E.; Rego, H.H.A.; Havlin, S.; Bunde, A. Detecting Long-Range Correlations with Detrended Fluctuation Analysis. *Phys. A Stat. Mech. Appl.* **2001**, *295*, 441–454. [CrossRef]
23. Bunde, A.; Havlin, S.; Kantelhardt, J.W.; Penzel, T.; Peter, J.-H.; Voigt, K. Correlated and Uncorrelated Regions in Heart-Rate Fluctuations during Sleep. *Phys. Rev. Lett.* **2000**, *85*, 3736–3739. [CrossRef] [PubMed]
24. Lancaster, G.; Iatsenko, D.; Pidde, A.; Ticcinelli, V.; Stefanovska, A. Surrogate Data for Hypothesis Testing of Physical Systems. *Phys. Rep.* **2018**, *748*, 1–60. [CrossRef]
25. Makowiec, D.; Rynkiewicz, A.; Wdowczyk-Szulc, J.; Zarczynska-Buchowiecka, M.; Galaska, R.; Kryszewski, S. Aging in Autonomic Control by Multifractal Studies of Cardiac Interbeat Intervals in the VLF Band. *Physiol. Meas.* **2011**, *32*, 1681–1699. [CrossRef]
26. Watanabe, K.; Funahashi, S. Neural Mechanisms of Dual-Task Interference and Cognitive Capacity Limitation in the Prefrontal Cortex. *Nat. Neurosci.* **2014**, *17*, 601–611. [CrossRef]
27. Hu, K.; Ivanov, P.C.; Chen, Z.; Carpena, P.; Stanley, H.E. Effect of Trends on Detrended Fluctuation Analysis. *Phys. Rev. E Stat. Nonlin. Soft. Matter. Phys.* **2001**, *64*, 011114. [CrossRef]

entropy

MDPI

Article

A Spike Train Production Mechanism Based on Intermittency Dynamics

Stelios M. Potirakis [1,2,*], Fotios K. Diakonos [3] and Yiannis F. Contoyiannis [1]

1 Department of Electrical and Electronics Engineering, University of West Attica, Ancient Olive Grove Campus, 12241 Egaleo, Greece; yiaconto@uniwa.gr
2 Institute for Astronomy, Astrophysics, Space Applications and Remote Sensing, National Observatory of Athens, Metaxa and Vasileos Pavlou, Penteli, 15236 Athens, Greece
3 Department of Physics, University of Athens, 15874 Athens, Greece; fdiakono@phys.uoa.gr
* Correspondence: spoti@uniwa.gr

Abstract: Spike structures appear in several phenomena, whereas spike trains (STs) are of particular importance, since they can carry temporal encoding of information. Regarding the STs of the biological neuron type, several models have already been proposed. While existing models effectively simulate spike generation, they fail to capture the dynamics of high-frequency spontaneous membrane potential fluctuations observed during relaxation intervals between consecutive spikes, dismissing them as random noise. This is eventually an important drawback because it has been shown that, in real data, these spontaneous fluctuations are not random noise. In this work, we suggest an ST production mechanism based on the appropriate coupling of two specific intermittent maps, which are nonlinear first-order difference equations. One of these maps presents small variation in low amplitude values and, at some point, bursts to high values, whereas the other presents the inverse behavior, i.e., from small variation in high values, bursts to low values. The suggested mechanism proves to be able to generate the above-mentioned spontaneous membrane fluctuations possessing the associated dynamical properties observed in real data. Moreover, it is shown to produce spikes that present spike threshold, sharp peak and the hyperpolarization phenomenon, which are key morphological characteristics of biological spikes. Furthermore, the inter-spike interval distribution is shown to be a power law, in agreement with published results for ST data produced by real biological neurons. The use of the suggested mechanism for the production of other types of STs, as well as possible applications, are discussed.

Keywords: spike train; artificial neural networks; biological neurons; intermittency; criticality; tricriticality; phase transitions

1. Introduction

Spike structures appear in the natural world in several phenomena, for example in astrophysical phenomena (diffraction spikes) [1–3], in laser devices [4–6], in biological nervous systems [7–10], and in biological structures, such as spike proteins and the coronavirus (e.g., COVID-19) [11–15]. Especially in the class of the dynamical-in-time spikes (spike train), spikes play a very important role because these structures carry information in time. The most typical example of a spike train (ST) is the action potential of biological neurons that carries the stimulus information in the biological neural network. The membrane potential of a biological neuron consists of two parts: (a) the action potential (voltage spikes), and (b) the relaxation time intervals (characterized by high-frequency voltage fluctuations). Figure 1a depicts a typical segment of a membrane potential time series, where the ST and the relaxation time intervals (high-frequency fluctuations) between the spikes are shown. Figure 1b presents a zoom-in to one of the spikes of Figure 1a, along with the pre- and after-spike high-frequency fluctuations. The experiment through which the specific membrane potential time series was acquired is described in detail in [16], whereas the time series was first presented and analyzed in [17]. Figure 1b is the same as Figure 3a of [17].

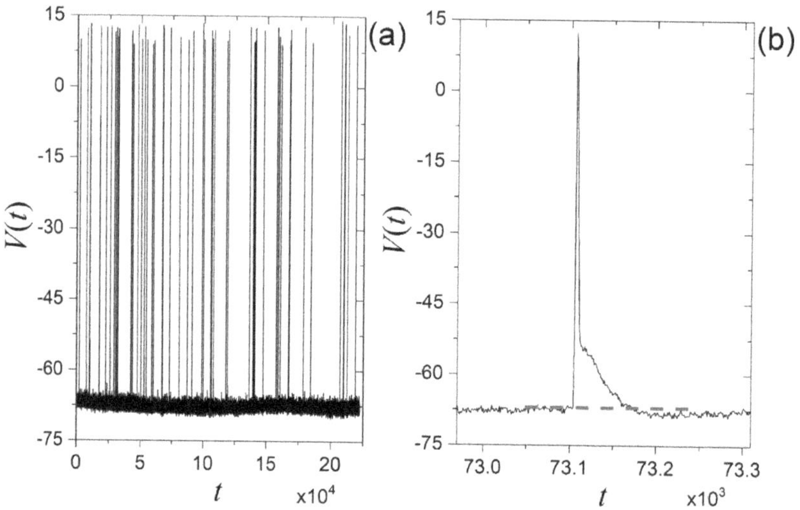

Figure 1. (**a**) Biological membrane potential $V(t)$ of neuron 089s1c1 from in vitro intracellular recordings of CA1 pyramidal neurons of Wistar male rats (adopted from [17]). (**b**) Zoom-in to a spike from Figure 1a, along with pre- and after-spike high-frequency fluctuations. The horizontal dashed red line, denoting the firing threshold, highlights the fact that the after-spike mean level is lower than the pre-spike one (hyperpolarization phenomenon).

When the membrane potential reaches a threshold (denoted by the red line in Figure 1b), the biological neuron fires and generates a signal (spike). This is a temporal process. The individual spikes of Figure 1 present specific morphological characteristics. They present a sharp peak, a firing threshold and the biological phenomenon of the hyperpolarization; that is, the fall part of spike goes lower than the pre-spike relaxation values as in Figure 1b. Moreover, as shown in [16,17], the zone of high-frequency fluctuations between the spikes obeys dynamics that are met in the description of magnetization fluctuations near the critical point [16,18,19] (see also Section 2).

In the present work, we focus on STs of the biological neuron type, as in Figure 1. The main scope of this work is to investigate what mechanism could create such an ST. Several models have already been proposed that capture aspects of the biological neuron functioning or suggest simplified, computationally efficient alternatives, e.g., refs. [7,20–23]. However, these cannot reproduce the above-mentioned dynamics of the high-frequency fluctuations found in the relaxation intervals. This is important because it means that these cannot reproduce the spontaneous emergence of membrane fluctuations [16]. If there is really an ST production mechanism that could fill this gap, understanding it could help us design an artificial neural network (ANN) closer to the functioning of a biological neural network.

Although ANNs were inspired by the functioning of the nervous system (and brain) of living organisms, they have so far embodied a subset of biological neural systems' features.

Thus, despite their extended and particularly successful use in pattern recognition and machine learning in general (including the relatively recent example of deep learning), they need to be further developed to fully emulate biological neural systems functioning.

In this direction, in recent years, significant efforts have been made to enrich the ANNs with additional features of the biological neural systems, such as with the creation of STs. It is reminded that when the membrane potential reaches a threshold in biological neurons, the biological neuron fires and generates a (time-evolving) ST signal. Thus, the effort to introduce time into ANNs through the threshold mechanism has recently begun with the aim of creating spikes. An ANN which does not transmit information at each propagation circle but transmits information only when a certain threshold is crossed is called a spiking neural network (SNN) [20,24–29]. Until now, the most prominent SNN seems to be the leaky integrate-and-fire model [30]. After firing, the state variable is reset to a lower value (as in the biological neurons hyperpolarization phenomenon). Spike-based activation of SNNs is not differentiable, thus, making it difficult to develop gradient-descent-based training methods to perform error backpropagation [31,32]. One way to overcome this problem is to use the sigmoid function, which is differentiable, as an approximation of the Heaviside step function [17,33]. Various techniques have been developed in order to equip SNNs with improved learning capabilities, elevating computational efficiency [33–37]. However, learning mechanisms for SNNs, including SNNs that present inhibition, cannot yet be considered a solved issue [38].

Our approach to the topic of ANNs and the challenge of simulations as close as possible to biological reality is quite different from the predominant research directions in this field. First, in [16], we studied the biological neuron and showed that the high-frequency fluctuations which appear in the relaxation intervals follow the critical dynamics as expressed through the critical intermittency type I map (see Section 2). Then, we introduced the so-called Hybrid Model (HM) [17]. This model is based on key concepts of the stochastic Hopfield ANNs with an important difference: instead of using the classical Boltzmann probability to calculate the statistical weights that spread the information in the ANN from layer to layer, HM uses Fermi Statistics because this interprets the $\{+1, -1\}$ duality as fermions. The term "hybrid" highlights the mixing of ANN formalism and basic principles of Physics. Although HM succeeded in producing the fractal structure of inter-spike intervals and the dynamics of the high-frequency fluctuations found in the relaxation intervals, the morphology of individual spikes was not always close enough to that of real biological ones. HM spikes presented high fluctuations in spike heights and stepped morphology at the repolarization part (fall part), while some individual HM spikes presented flat tops.

Continuing in the above-presented direction, in the present work, we suggest a mechanism for the production of STs in the form of a rigorous formalism based on the phenomenon of intermittency. Thus, the aim of the work is to propose a mechanism that is capable of producing STs that resemble the biological ST shown in Figure 1, in terms of the key morphological characteristics of individual spikes, the fractal structure of inter-spike intervals, and the dynamics of the high-frequency fluctuations observed in the relaxation intervals.

The rest of the paper is organized as follows: Section 2 presents key information about intermittency, a critical intermittency type I map and a tricritical intermittency map, focusing on the fact that they behave as "complementary" repellors. Section 3 presents the proposed ST production mechanism based on the appropriate coupling of critical and tricritical intermittency. Its generalized form and the details on its operation using the abovementioned maps are presented in Section 3.1. Section 3.2 demonstrates its application through a numerical experiment that produces an ST whose individual spikes share key morphological characteristics with the real biological spikes of Figure 1. The dynamical and fractal characteristics of the produced ST are analyzed in Section 4 and proven similar to those of the real biological spikes of Figure 1. Section 4.1 focuses on the dynamics of the relaxation intervals' high-frequency fluctuations, whereas Section 4.2 focuses on the fractal structure of inter-spike intervals. A discussion on the future further investigation of the proposed mechanism, its connection to biological neuron functions/processes and its possible applications is provided in Section 5. Finally, Section 6 summarizes the conclusions.

2. Intermittent Dynamics

Based on previous works [16,17], which demonstrate that the high-frequency amplitude fluctuations of the time intervals between spikes obey critical intermittency dynamics, we examine the idea that a mechanism for ST production can be proposed based on the dynamics of intermittency. The phenomenon of intermittency consists in the temporal alternation of regions of low amplitude fluctuations, called "laminar regions", with regions, which demonstrate high amplitude fluctuations (bursts).

It has been demonstrated that the fluctuations of the order parameter in simulated thermal systems close to the critical point, which signals the transition from paramagnetic to ferromagnetic behavior in magnetic systems, are mathematically described by the critical intermittency type I map [18,19]:

$$\phi_{n+1} = \phi_n + u_1 \phi_n^{z_1} + \varepsilon_n, \tag{1}$$

where $u_1 > 0$, $z_1 > 0$ and ε_n is a uniform "noise", $\varepsilon_n \in [-\varepsilon_1, +\varepsilon_1]$. Note that in case that the exponent is not an integer, the nonlinear term should be used as $+u_1|\phi_n|^{z_1}$. The map of Equation (1) will hereafter be referred to as "$map1$".

As shown in [18,19], if ϕ_n denotes the fluctuations of the order parameter, then the exponent z_1 is connected to the isothermal critical exponent δ, which characterizes the above-mentioned transition, with the relation:

$$z_1 = \delta + 1. \tag{2}$$

Figure 2 depicts a time series constructed using Equation (1) that presents the above-mentioned alternation between laminar regions and bursts. In Figure 2, the start of the laminar region is $\phi_0 = 0$, whereas the end of the laminar region has been considered to be $\phi_L = 0.2$. This means that the time series remains within the laminar region as long as $\phi_n \in [\phi_0, \phi_L]$; in this example it is $\phi_n \in [0, 0.2]$. The way to construct a numerical simulation that produces an intermittent time series, as in Figure 2, is the following. One starts from a random value inside the laminar region. Each next value is calculated from the previous value using an appropriate recurrence relation, e.g., the intermittent map of Equation (1), until a burst ends. Then the trajectory randomly returns to the laminar region, and so on. A burst ends when the next value calculated by the recurrence relation exceeds a predefined high threshold, e.g., the value 0.65 for the time series of Figure 2. At this point, instead of using the calculated as the next value, a new value inside the laminar region, e.g., inside the interval $[0, 0.2]$ for the time series of Figure 2, is used in its place. In this return mechanism, if the burst ends are considered as the input values, then the laminar region values are the output values. In this sense, the return from the burst region back to the laminar region is a kind of "feedback" mechanism.

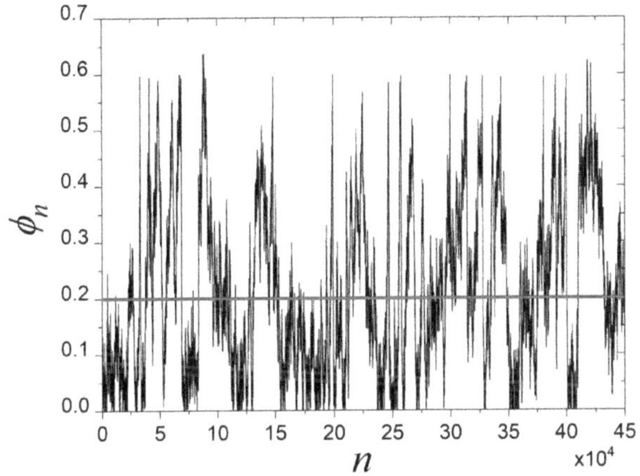

Figure 2. An intermittent time series in which it has been considered that the region from 0 up to 0.2 (bounded upwards by the red horizontal line) is the laminar region and all values above this zone correspond to bursts.

In nature, many phenomena present intermittent dynamics [39], producing intermittent time series. In other words, intermittency presents universality. Could the universality of intermittency lead to a universal ST production mechanism? The waveform of a spike comprises a rise part and a fall part (see Figure 1b). The map of Equation (1) can produce the rise part, but what about the fall part?

It is known from the theory of critical phenomena [40] that there is a region where a second-order phase transition, characterized by critical behavior, meets a first-order one, characterized by an abrupt change. This is accomplished around the so-called Griffiths tricritical point. As has been shown in [41], the thermal fluctuations of the order parameter near this point are described by another intermittent map of the form:

$$\phi_{n+1} = \phi_n - u_2 \phi_n^{-z_2} + \varepsilon_n, \tag{3}$$

where $u_2 > 0$, $z_2 > 0$ and ε_n is a uniform "noise", $\varepsilon_n \in [-\varepsilon_2, +\varepsilon_2]$. Note that, as in Equation (1), in case the exponent is not an integer, the nonlinear term should be used as $-u_2|\phi_n|^{-z_2}$. The map of Equation (3) will hereafter be referred to as "$map2$". The negative sign of the nonlinear term and the negative exponent ensure that the values fall. The reader should keep in mind that, according to the previous discussion, $map1$ can be called the critical intermittency type I map and $map2$ the tricritical intermittency map.

The two maps of Equations (1) and (3), i.e., $map1$ and $map2$, are repellors with respect to their fixed points. The fixed point of $map1$ is at zero (marginally unstable) and its nonlinear term leads the trajectory to higher values (rise), departing from its fixed point and allowing the trajectory to approach the fixed point of $map2$, which lies at a high value (theoretically infinity). Similarly, the fixed point of $map2$ is also marginally unstable and its nonlinear term leads the trajectory to lower values (fall), departing from its fixed point and allowing the trajectory to approach the fixed point of $map1$. Figure 3 presents an example of return plots of these two maps: on the left is the critical intermittency type I ($map1$) and on the right is the tricritical intermittency ($map2$). As can be seen from Figure 3, the nonlinear term $u_1\phi_n^{z_1}$ of $map1$ drives the trajectory away from the laminar region, marked as "$laminar1$" ($\phi_n \in [(\phi_0)_1, (\phi_L)_1]$), by leading to higher values, moving it away from the bisector. This can be considered an "excitatory" process in biological neuron terms. On the contrary, the nonlinear term $-u_2\phi_n^{-z_2}$ of $map2$ drives the trajectory away from the laminar region, marked as "*laminar*2" ($\phi_n \in [(\phi_L)_2, (\phi_0)_2]$), by leading to lower values, moving it away from the bisector. This can be considered an "inhibitory" process in biological neuron terms. In that sense, these two intermittent maps, *map*1 and *map*2, can be seen as "complementary" repellors.

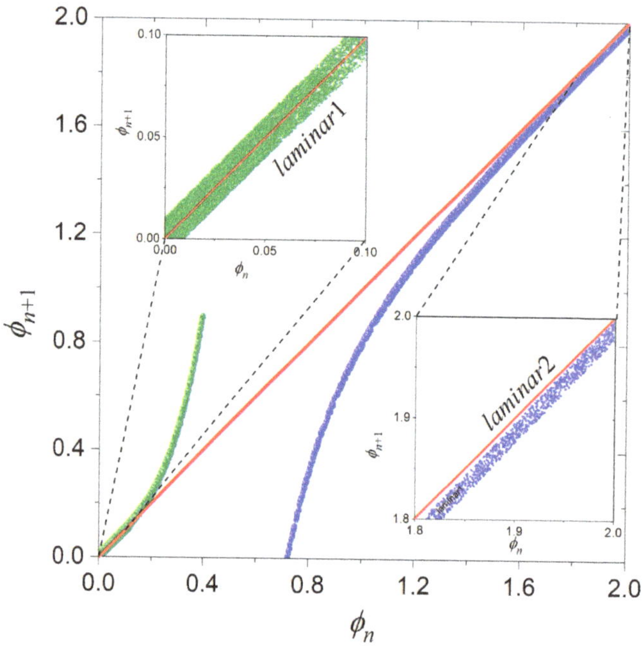

Figure 3. Typical examples of the return plots of the *map*1, given by Equation (1) (green), and the *map*2, given by Equation (3) (blue). The laminar region in both maps is the part of the trajectory that closely follows (is almost parallel to) the bisector (red line). A zoom-in to each laminar region, *laminar*1 and *laminar*2, is presented in the corresponding insets. As soon as the trajectory begins to move away from the bisector, the map has entered the bursts region. Such plots can be constructed for various combinations of the parameters' values in Equations (1) and (3), moving the two laminar regions closer or further away from each other.

3. Spike Train Production Mechanism Based on Intermittency Dynamics

3.1. Suggested Spike Train Production Mechanism

Based on the complementary repellor behavior of the critical and tricritical intermittent maps presented in Section 2, it is expected that an appropriate coupling of these two maps, which would allow their temporal alternation, "*map*1 ⇄ *map*2", each time one of the maps crosses a certain threshold value—ϕ_{Th1} (for *map*1) and ϕ_{Th2} (for *map*2)—at the end or outside the corresponding laminar region, could indeed lead to ST production. Generalizing this suggestion, an appropriate coupling of any critical intermittency "model" with any tricritical intermittency "model", which would allow a corresponding "*critical intermittency* ⇄ *tricritical intermittency*" temporal alternation with the switching criterion being the crossing of the matching thresholds, could also lead to ST production (Figure 4). By the terms critical intermittency "model" and tricritical intermittency "model," we mean any methods of generating time series that exhibit the corresponding dynamics, such as mathematical expressions, ANNs, or even appropriately using parts of real recorded time series.

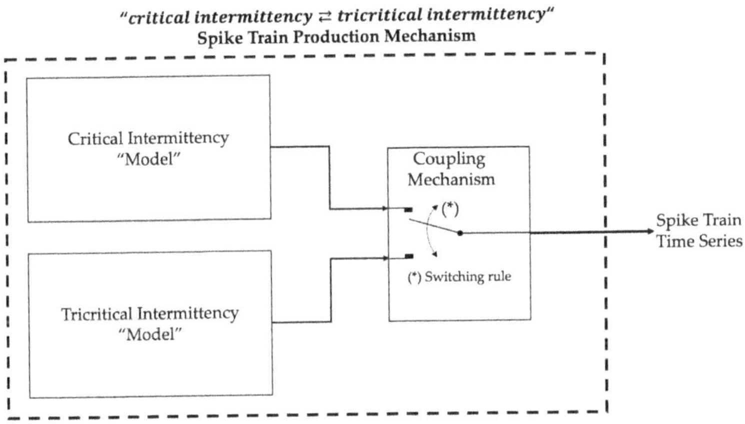

Figure 4. Suggested spike train production mechanism based on intermittency dynamics.

Let us see how the suggested spike train production mechanism can be defined in detail, using *map*1 and *map*2 as critical intermittency and tricritical intermittency "models", respectively.

As already mentioned, the "*critical intermittency* ⇄ *tricritical intermittency*" temporal alternation is performed with a switching criterion: the crossing of a threshold value at the end or outside the corresponding laminar region. For the "*map*1 ⇄ *map*2" case, let us suppose that the procedure starts with *map*1. For *map*1, the initial conditions are a low value $\phi_{n=0} \in laminar1$ and $\varepsilon_{n=0} \in [-\varepsilon_1, +\varepsilon_1]$. As long as *map*1 produces values $\phi_{n+1} \in laminar1$, the switch in Figure 4 remains at the upper position that connects *map*1 output to the system's output. As soon as the trajectory crosses upwards the threshold ϕ_{Th1}, i.e., as soon as a next value $\phi_{n+1=k} > \phi_{Th1}$ is calculated (and before this reaches the system's output), this triggers the switch in Figure 4 to change to the lower position. At this position, the switch connects the system's output to the *map*2 output, which starts at a high value $\phi_k \in laminar2$. As long as the next values produced by *map*2 remain within *laminar*2, i.e., as long as $\phi_{n+1 \geq k} \in laminar2$ holds, the switch in Figure 4 remains at the lower position that connects the *map*2 output to the system's output. As soon as the trajectory crosses downwards the threshold ϕ_{Th2}, i.e., as soon as a next value $\phi_{n+1=m>k} < \phi_{Th2}$ is calculated (and before this reaches the system's output), this triggers the switch in Figure 4 to change back to the upper position. At this position, the switch connects the *map*1 output to the system's output, starting at a low value $\phi_m \in laminar1$ and so on.

The two thresholds, ϕ_{Th1} and ϕ_{Th2}, controlling the switch operation and, consequently, the coupling mechanism, depend on the considered critical intermittency and tricritical intermittency "model" parameters as well as on the targeted ST. In the case of STs of the biological neuron type, $\phi_{Th1} = (\phi_L)_1$, which is the end of the laminar region of *map*1, while $\phi_{Th2} = (\phi_0)_1$, which is the fixed point of *map*1. It is noted that ϕ_{Th1} signifies the firing threshold and, therefore, the start of each spike. Setting $\phi_{Th2} = (\phi_0)_1$ ensures that after each spike the trajectory returns to a value lower than the firing threshold, thus, reproducing the hyperpolarization phenomenon.

In the following, we will examine whether the suggested mechanism, using *map*1 and *map*2, can indeed produce an ST resembling a biological neuron ST, such as the one shown in Figure 1.

3.2. Spike Train Production Example

By setting specific values for all the parameters of *map*1 (Equation (1)) and *map*2 (Equation (3)), as well as for ϕ_{Th1} and ϕ_{Th2} (see Section 3.1), a numerical experiment can be performed that demonstrates the effectiveness of the coupling mechanism suggested in Section 3.1. A set of such values is (a) $map1$ $\{z_1 = 4, u_1 = 0.011, \varepsilon_1 = 0.0175\}$, (b) $map2$ $\{z_2 = 5, u_2 = 17, \varepsilon_2 = 0.07\}$, and (c) switching thresholds $\{\phi_{Th1} = 0.31, \phi_{Th2} = 0\}$. The specific ϕ_{Th1} and ϕ_{Th2} values are, respectively, the end of the laminar region and the fixed point of $map1$ for the above-mentioned $map1$ parameters, following the rule mentioned in Section 3.1 for the production of STs of the biological neuron type. The return plots of each map for the above-mentioned parameters, without any coupling between them, are shown in Figure 5.

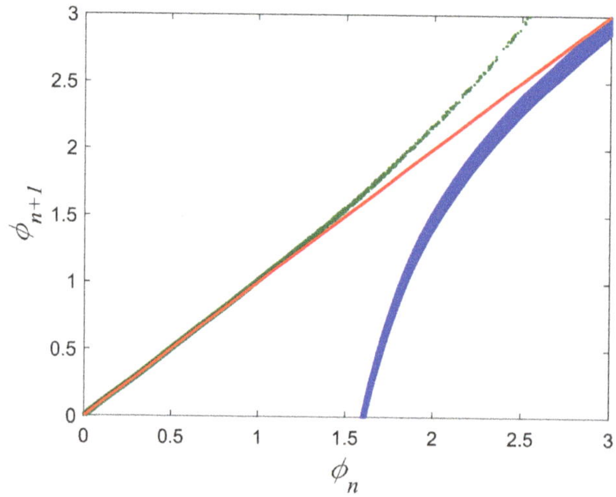

Figure 5. Return plots of $map1$ (green) and $map2$ (blue), uncoupled, for the maps parameters values used in our numerical experiment (see text). Red line denotes the bisector.

In our example, the high value of the laminar region of $map2$ at which the trajectory starts upon switching to $map2$ has been set to be $\phi_k = 3 \in laminar2$. Thus, the switch from $map1$ to $map2$ means a jump to the value 3. Correspondingly, both the initial value and the low value upon switching to $map1$ have been set to be $\phi_{n=1} = \phi_m = 0 \in laminar1$, i.e., the fixed point of $map1$. Therefore, the switch from $map2$ to $map1$ means a jump to the value 0. Although in our experiment the values at which each map starts upon switching were fixed, the randomness in the coupling between the two maps is ensured by the random number generators, which are used to generate the "noise" terms ε_n of $map1$ and $map2$ (see Equations (1) and (3)). Thus, each time the trajectory enters the laminar region of each map, a different path is followed due to the term ε_n included in both Equations (1) and (3), and, consequently, different laminar lengths L, i.e., waiting times within the laminar region (see also Section 4.1), are produced. This randomness is enough and there is no need to introduce additional randomness at the entry points for each laminar trajectory. However, this certainly does not rule out sending the trajectory at a random point of the corresponding laminar region upon switching, which is not affecting the results.

The experiment was run until the algorithmic time $n = N = 3,000,000$, i.e., until the length of the time series reaches a target value of $N = 3,000,000$ points. The pseudocode that produces the ST time series is shown in Figure 6, whereas the Matlab v. 2024a source code for the numerical experiment is provided in the online available Supplementary Material of this article. The produced ST time series is shown in Figure 7a. It is noted that such a long length for the time series was selected for two reasons. Primarily, to achieve a population of spikes sufficient for inter-spike intervals statistical analysis (see Section 4.2).

But also, to tighten the 95% confidence intervals (95%CI) of the estimated values during the analysis presented in Sections 4.1 and 4.2.

```
% Definition of parameters
    map1 parameters: z_1 = 4, u_1 = 0.011, ε_1 = 0.0175
    map2 parameters: z_2 = 5, u_2 = 17, ε_2 = 0.07
    switching thresholds: ϕ_Th1 = 0.31, ϕ_Th2 = 0
    trajectory start values upon switching: (map1)_start = 0, (map2)_start = 3
    length of the ST time series: N = 3,000,000

% Uniform noise production
    (ε_{n=1:N})_1 = 2 · ε_1 · (Uniform Random Noise in the interval [0,1])_{n=1:N} − ε_1
    (ε_{n=1:N})_2 = 2 · ε_2 · (Uniform Random Noise in the interval [0,1])_{n=1:N} − ε_2

% Production of the ST time series {ϕ_n}, n = 1,2,...,N according to Section 3.1

% Set the first value of the ST time series and the first value of the current switch
% status flag (map1)_check [possible values: 1→ map1, 0→ map2]
    ϕ_{n=1} = (map1)_start
    (map1)_check = 1

% Production of the rest of the ST time series values
    FOR n = 2:N
            IF {(ϕ_{n−1} < ϕ_Th1) AND ((map1)_check = 1)}
                    ϕ_n = ϕ_{n−1} + u_1 ϕ_{n−1}^{z_1} + (ε_{n−1})_1
                    (map1)_check = 1
                    IF ϕ_n ≥ ϕ_Th1
                            ϕ_n = (map2)_start
                            (map1)_check = 0
                    END IF
            ELSEIF {(ϕ_{n−1} > ϕ_Th2) AND ((map1)_check = 0)}
                    ϕ_n = ϕ_{n−1} − u_2 ϕ_{n−1}^{−z_2} + (ε_{n−1})_2
                    (map1)_check = 0
                    IF ϕ_n ≤ ϕ_Th2
                            ϕ_n = (map1)_start
                            (map1)_check = 1
                    END IF
            END IF/ELSEIF
    END FOR
```

Figure 6. Pseudocode for the production of the ST time series presented in Figure 7.

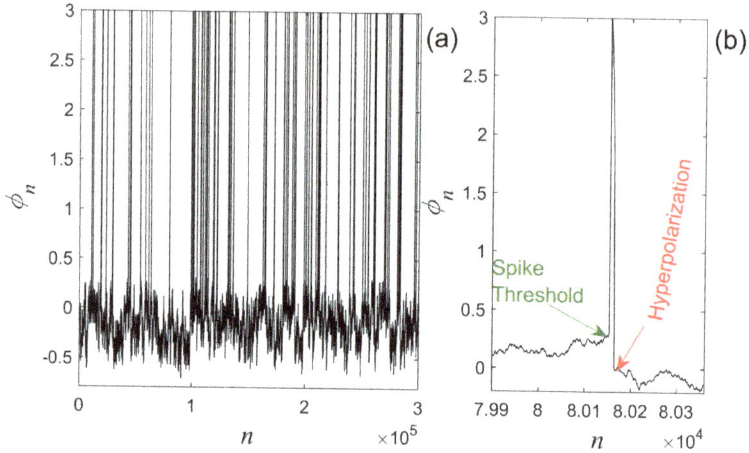

Figure 7. (**a**) The time series produced by the ST production mechanism suggested in Section 3.1, using the parameters: $map1$ $\{z_1 = 4,\ u_1 = 0.011,\ \varepsilon_1 = 0.0175\}$, $map2$ $\{z_2 = 5,\ u_2 = 17,\ \varepsilon_2 = 0.07\}$, switching thresholds $\{\phi_{Th1} = 0.31,\ \phi_{Th2} = 0\}$. Only the first 300,000 points of the produced 3,000,000-points-long time series are shown, so that the spike pattern is clearly visible. (**b**) A spike from the time series of Figure 7a where the spike threshold and the hyperpolarization phenomenon are marked.

As can be seen from Figure 7a, the time series produced by the mechanism suggested in Section 3.1 is an ST. From Figure 7b, it is clear that the generated spikes present sharp peaks, the spike threshold, and the phenomenon of hyperpolarization, which are signatures of biological neuron-type spikes (see Figure 1b). In our suggested ST production mechanism, the external excitation parameter (external field) is absent. The intermittency is an "internal" dynamic with feedback mechanisms between the region of smooth fluctuations (laminar region) and the region of strong fluctuations (bursts). Therefore, the imposition of external fields would destroy this internal self-organization.

The coupling of the two intermittent maps successfully generates STs of the biological neuron type, as they exhibit key characteristics of STs produced by biological neurons. Beyond the morphological resemblance of the individual spikes—sharp peaks, spike threshold, and hyperpolarization—, both the dynamics of the relaxation intervals fluctuations and the fractal nature of biological neuron STs are also reproduced by the suggested spike train production mechanism, as shown in Section 4.

4. Dynamical and Fractal Characteristics of the Produced Spike Train

4.1. On the Dynamics of the Relaxation Intervals' High-Frequency Fluctuations

In this section, we study the dynamics of the relaxation intervals' high-frequency fluctuations for the ST presented in Section 3.2, produced by the mechanism suggested in Section 3.1. In [16], it was found that the high-frequency fluctuations of Figure 1a for the biological neuron obey the critical dynamics as expressed through the intermittency map type I ($map1$, Equation (1)). The use of $map1$ in the suggested ST production mechanism serves the reproduction of this dynamics. This dynamics is determined from the distribution of the laminar lengths L (see below) of the intermittency. According to [39], for the $map1$, this distribution is a power law of the form:

$$P(L) \sim L^{-p}, \tag{4}$$

where $p = {}^{z_1}/_{(z_1-1)}$. Due to Equation (2) we have that:

$$p = 1 + (1/\delta). \quad (5)$$

Given the fact that $1 < \delta < \infty$, from Equation (5), it results that the critical intermittency exists for exponent values $p \in [1,2)$. According to [41], the $map2$ laminar lengths' distribution is also given by the Equation (4), where it is now:

$$p = \frac{z_2}{z_2 + 1} = \frac{\delta + 1}{\delta + 2}. \quad (6)$$

Given the fact that $1 < \delta < \infty$, from Equation (6), it results that the tricritical intermittency exists for exponent values $p \in [0.66, 1)$.

The method of critical fluctuations (MCF) [18,19,42] is used for the estimation of the exponent p in a time series. According to MCF, it has been found [16] that the relaxation intervals' high-frequency fluctuations of the biological neuron of Figure 1a obey critical intermittency with exponent $p = 1.37$. This exponent is very close to the value 1.33, which results from Equation (5) for $\delta = 3$ and indicates the mean field theory (MFT) universality class of critical phenomena [40].

In the following, we apply the MCF to the relaxation intervals' high-frequency fluctuations of the 3,000,000-points-long ST time series produced in the numerical experiment presented in Section 3.2. The application of MCF to an ST is demonstrated with the help of Figure 8. The red-colored level marked in Figure 8a denotes the lower value ϕ_{red} of the ST, whereas the blue-colored level denotes a higher value $\phi_{blue} : \phi_{red} < \phi_{blue} < \phi_{Th1}$ and is treated as a free parameter. The levels ϕ_{red} and ϕ_{blue} delimit a "laminar region" for the ST time series, which is actually a subset of $laminar1$ of $map1$. The laminar lengths L are the waiting times of the ST time series inside this zone. Thus, the laminar lengths are the number of successive ϕ-values obeying the condition $\phi_{red} < \phi < \phi_{blue}$. As already mentioned, the ϕ_{blue} value is a free parameter. This means that all values $\phi_{red} < \phi < \phi_{Th1}$ are examined as possible ϕ_{blue} values, while the examination is performed exhaustively by progressively increasing the number of equally spaced values covering the whole amplitude range until the best power-law distribution of laminar lengths is found. As soon as the laminar lengths distribution is calculated, the following fitting function is used in order to estimate the power-law exponent:

$$g(L) \sim L^{-p_2} e^{-p_3 L}. \quad (7)$$

The specific fitting function is a truncated power-law function, where p_2 is the exponent of the power-law factor, and p_3 is the exponent of the exponential corrective term. If the estimated p_3 exponent is close to zero, the distribution tends to the scaling form of Equation (4); thus, p_2 is an estimate of the exponent p. As already mentioned, if $p \in [1,2)$, then the time series has been produced by a system that is in its critical state. Therefore, the exponent p_3 is a measure of how close the system is to criticality if simultaneously $p_2 \in [1,2)$.

We applied the fitting function of Equation (7) to the distribution of laminar lengths of Figure 8b, calculated for $\phi_{red} = -0.64$ and $\phi_{blue} = 0.07$, and estimated the values of the exponents p_2 and p_3 by nonlinear least squares optimization using the trust-region-reflective algorithm [43]. We found $p_3 = 0.001$ with a 95% $CI : (-0.001,\ 0.002)$ and $p_2 = 1.372$ with a 95% $CI : (1.364,\ 1.381)$. Note that in this experiment, $\phi_{blue} = 0.07$ is not the only ϕ_{blue} value leading to $p_3 \approx 0$ (here considered as any $p_3 \in [0, 001]$). However, using any of these ϕ_{blue} values lead essentially to the same results. The obtained p_2 and p_3 exponents indicate the critical dynamics of the relaxation intervals' high-frequency fluctuations of the ST produced by the mechanism suggested in Section 3.1. Moreover, the value of this exponent p_2, on the one hand, is very close to the theoretical one for the MFT ($\delta = 3$), which is $p = 1.33$, and on the other hand, it is practically identical to the value of the exponent $p = 1.37$ that has been determined in [17] for the membrane potential relaxation intervals' high-frequency fluctuations of the biological neuron ST of Figure 1. Note that running the same numerical experiment again with different seeds for the pseudorandom number generators may lead to different ϕ_{blue} values that lead to $p_3 \approx 0$. However, as soon as the ϕ_{blue} value leading to $p_3 \approx 0$ is determined, the p_2 value, resulting from the application of the fitting function of Equation (7) to the distribution of laminar lengths, remains essentially the same.

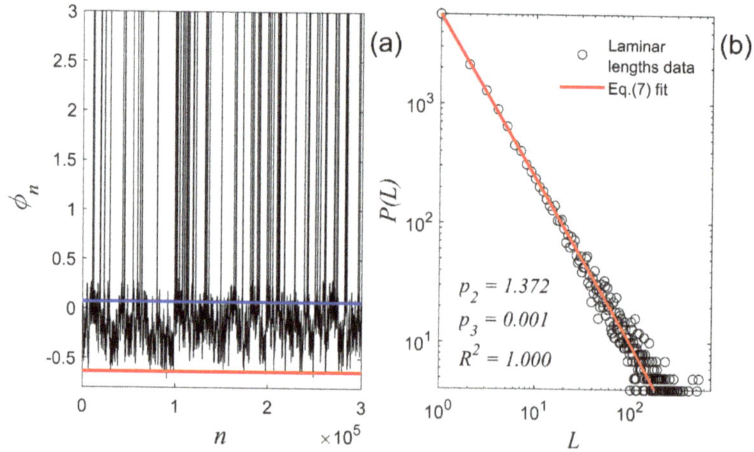

Figure 8. (**a**) The ST time series produced by the numerical experiment of Section 3.2 (also depicted in Figure 7a). The laminar region of the high-frequency fluctuations of the relaxation intervals was found to be bound between $\phi_{red} = -0.64$ (red horizontal line) and $\phi_{blue} = 0.07$ (blue horizontal line). (**b**) The distribution of laminar lengths resulting from the laminar region marked in Figure 8a. The estimated exponent values by fitting Equation (7) are: $p_2 = 1.372$, $p_3 = 0.001$, with goodness of fit $R^2 = 1.000$. Although only the first 300,000 points of the analyzed time series are shown in Figure 8a, the distribution was calculated using the total length of the time series (3,000,000 points).

4.2. On the Distribution of Inter-Spike Intervals

In this section, we calculated the distribution of the inter-spike intervals (distances between successive spikes) for the time series produced by the numerical experiment of Section 3.2. The distribution of inter-spike intervals for the 3,000,000-points-long ST time series is a power law of the form $P(s) \sim s^{-1.372}$ (Figure 9), proving the fractal structure of the portrait of the ST produced by the mechanism suggested in Section 3.1. This is yet another similarity to the STs produced by biological neurons, such as the one presented in Figure 1, which also exhibit a similar fractal structure [17]. It should be noted that the exponent of the inter-spike intervals power law is dependent on the selected parameters to conduct the numerical experiment. It is noted that the distribution was also fit to the truncated power-law function of Equation (7), also by nonlinear least squares optimization using the trust-region-reflective algorithm [43], and the result was identical to the fitting of the laminar lengths' distribution of Figure 8b, i.e., $p_3 = 0.001$ with a 95% $CI : (-0.001,\ 0.002)$ and $p_2 = 1.372$ with a 95% $CI : (1.364,\ 1.381)$ with goodness of fit $R^2 = 1.000$.

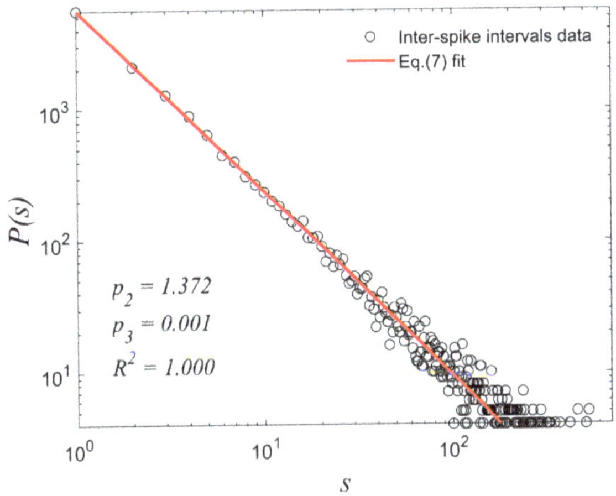

Figure 9. The distribution of the inter-spike intervals of the 3,000,000-points-long ST time series produced in the numerical experiment of Section 3.2 is a power law of the form $P(s) \sim s^{-1.372}$.

5. Discussion

Criticality in nature is a universal phenomenon. Since critical dynamics are a key constituent element of the ST production mechanism suggested in Section 3.1, which is clearly reflected on the produced ST time series, it is expected that the suggested mechanism is universal. So, in future work, we intend to study additional critical and tricritical intermittency "models" (other than *map*1 and *map*2) within the framework of the ST production mechanism suggested in Section 3.1, both as alternatives to the production of STs of the biological neuron type and to investigate the production of other ST types found in different phenomena.

It has been shown in the pioneering work of Hodgkin et al. [7], that an ST, presenting the morphological characteristics of an ST produced by a biological neuron, can successfully be produced by the solution of the differential equations of an electrical circuit, known as the Hodgkin–Huxley model. As already mentioned in the Introduction Section, several other models have been proposed since its introduction, trying either to capture aspects of the biological neuron functioning or to suggest simplified, computationally efficient alternatives [20–23]. However, these cannot reproduce the dynamics of the high-frequency fluctuations found in the relaxation intervals [16,17]. But why does the coupling of the two types of dynamic intermittency succeed in producing STs that exhibit similarity to biological neurons in terms of the morphology of individual spikes, the power-law distribution of the inter-spike intervals and the dynamics of the relaxation intervals' high-frequency fluctuations? Can this coupling of critical and tricritical intermittency be attributed to biological neuron functions/processes and under which conditions? The answer to these questions is more of a neuroscience problem and, thus, beyond the scope of the present work. However, we would like to propose, as a working hypothesis, that the two types of dynamic intermittency may be "hidden" in the sodium–potassium pump mechanism and possible feedback mechanisms between ionic transportation through the membrane and the membrane potential under specific conditions [17].

It has been suggested that biological neuron ST patterns serve as a kind of time encoding to pass messages across the nervous system and brain [44], while several efforts have been made to mimic this encoding with artificial systems, e.g., Refs. [44–46], as well as to achieve brain–machine interface [47] and to exploit it for cryptography applications, e.g., Refs. [48–50]. The STs produced by the mechanism proposed in Section 3.1 can produce

STs with a variety of ST patterns, depending on the intermittency "models" parameters (e.g., $map1$ and $map2$ parameters), switching rules and pseudorandom number generator parameters. Thus, the proposed mechanism may be useful as an alternative for encryption and cryptography applications.

6. Conclusions

The findings of the present work prove that a coupling between the critical intermittency and the tricritical intermittency can lead to the production of STs. The individual spikes of the produced STs present the key morphological characteristics of biological spikes (sharp peak, spike threshold and hyperpolarization phenomenon). Moreover, the inter-spike interval distribution is a power law, whereas the high-frequency fluctuations found in the relaxation intervals present critical dynamics that belong to the Mean Field Theory universality class. The universal character of the phenomenon of intermittency implies that the suggested ST production mechanism may be able to produce STs that appear in different phenomena, as well as that different critical and tricritical intermittency "models" (other than $map1$ and $map2$) may be able to produce similar STs within the framework of the suggested mechanism. In future works, we intend to study various critical and tricritical time series generators, or real-world time series, in place of the critical and tricritical "model" blocks of the suggested ST production mechanism, with the aim of producing both the biological type and other types of STs to test mechanism's universality.

Supplementary Materials: The following supporting information can be downloaded at: https://www.mdpi.com/article/10.3390/e27030267/s1, Matlab source code for the production of the ST time series.

Author Contributions: Conceptualization, S.M.P., F.K.D. and Y.F.C.; methodology, S.M.P., F.K.D. and Y.F.C.; software, S.M.P., F.K.D. and Y.F.C.; validation, S.M.P., F.K.D. and Y.F.C.; formal analysis, S.M.P., F.K.D. and Y.F.C.; investigation, S.M.P., F.K.D. and Y.F.C.; writing—original draft preparation, S.M.P. and Y.F.C.; writing—review and editing, F.K.D. and Y.F.C.; All authors have read and agreed to the published version of the manuscript.

Funding: This research received no external funding.

Data Availability Statement: The original contributions presented in this study are included in the article. Further inquiries can be directed to the authors.

Conflicts of Interest: The authors declare no conflicts of interest.

References

1. Lendermann, M.; Tan, J.S.Q.; Koh, J.M.; Cheong, K.H. Computational imaging prediction of starburst-effect diffraction spikes. *Sci. Rep.* **2018**, *8*, 16919. [CrossRef]
2. Armatas, S.; Bouratzis, C.; Hillaris, A.; Alissandrakis, C.E.; Preka-Papadema, P.; Moussas, X.; Mitsakou, E.; Tsitsipis, P.; Kontogeorgos, A. Detection of spike-like structures near the front of type-II bursts. *Astron. Astroph.* **2019**, *624*, A76. [CrossRef]
3. Nemiroff, R.; Bonnell, J. (Eds.) (15 April 2001). "Diffraction Spikes Explained". Astronomy Picture of the Day. NASA. Available online: https://apod.nasa.gov/apod/ap010415.html (accessed on 8 December 2024).
4. Mekemeza-Ona, K.; Routier, B.; Charbonnier, B. All optical Q-switched laser based spiking neuron. *Front. Phys.* **2022**, *10*, 1017714. [CrossRef]
5. Prucnal, P.R.; Shastri, B.J.; Ferreira de Lima, T.; Nahmias, M.A.; Tait, A.N. Recent progress in semiconductor excitable lasers for photonic spike processing. *Adv. Opt. Photon.* **2016**, *8*, 228–299. [CrossRef]
6. Mu, P.; Wang, K.; Liu, G.; Wang, Y.; Liu, X.; Guo, G.; Hu, G. Spike Dynamics Analysis in Semiconductor Ring Laser. *Electronics* **2024**, *13*, 260. [CrossRef]
7. Hodgkin, A.L.; Huxley, A.F.; Katz, B. Measurement of current-voltage relations in the membrane of the giant axon of Loligo. *J. Physiol.* **1952**, *116*, 424–448. [CrossRef] [PubMed]
8. Brunel, N. Dynamics of sparsely connected networks of excitatory and inhibitory spiking neurons. *J. Comput. Neurosci.* **2000**, *8*, 183–208. [CrossRef]

9. Nagel, K.; Wilson, R. Biophysical mechanisms underlying olfactory receptor neuron dynamics. *Nat. Neurosci.* **2011**, *14*, 208–216. [CrossRef]
10. Jolivet, R.; Rauch, A.; Lüscher, H.R.; Gerstner, W. Predicting spike timing of neocortical pyramidal neurons by simple threshold models. *J. Comput. Neurosci.* **2006**, *21*, 35–49. [CrossRef]
11. Deng, X.; Baker, S.C. Coronaviruses: Molecular biology (Coronaviridae). *Encycl. Virol.* **2021**, *2*, 198–207. [CrossRef]
12. Zhu, C.; He, G.; Yin, Q.; Zeng, L.; Ye, X.; Shi, Y.; Xu, W. Molecular biology of the SARs-CoV-2 spike protein: A review of current knowledge. *J. Medic. Virol.* **2021**, *93*, 5729–5741. [CrossRef] [PubMed]
13. Li, F. Structure, function, and evolution of Coronavirus spike proteins. *Ann. Rev. Virol.* **2016**, *3*, 237–261. [CrossRef]
14. Walls, A.C.; Park, Y.-J.; Tortorici, M.A.; Wall, A.; McGuire, A.T.; Veesler, D. Structure, function, and antigenicity of the SARS-CoV-2 spike glycoprotein. *Cell* **2020**, *181*, 281–292.e6. [CrossRef] [PubMed]
15. Ke, Z.; Oton, J.; Qu, K.; Cortese, M.; Zila, V.; McKeane, L.; Nakane, T.; Zivanov, J.; Neufeldt, C.J.; Cerikan, B.; et al. Structures and distributions of SARS-CoV-2 spike proteins on intact virions. *Nature* **2020**, *588*, 498–502. [CrossRef] [PubMed]
16. Kosmidis, E.K.; Contoyiannis, Y.F.; Papatheodoropoulos, C.; Diakonos, F.K. Traits of criticality in membrane potential fluctuations of pyramidal neurons in the CA1 region of rat hippocampus. *Eur. J. Neurosc.* **2018**, *48*, 2343–2353. [CrossRef]
17. Contoyiannis, Y.F.; Kosmidis, E.K.; Diakonos, F.K.; Kampitakis, M.; Potirakis, S.M. A hybrid artificial neural network for the generation of critical fluctuations and inter-spike intervals. *Chaos Solitons Fractals* **2022**, *159*, 112115. [CrossRef]
18. Contoyiannis, Y.F.; Diakonos, F.K.; Malakis, A. Intermittent dynamics of critical fluctuations. *Phys. Rev. Lett.* **2002**, *89*, 035701. [CrossRef]
19. Contoyiannis, Y.; Diakonos, F. Criticality and intermittency in the order parameter space. *Phys. Lett. A* **2000**, *268*, 286–292. [CrossRef]
20. Gerstner, W.; Kistler, W.M. *Spiking Neuron Models: Single Neurons, Populations, Plasticity*; Cambridge University Press: Cambridge, UK, 2002.
21. Torres, J.J.; Varona, P. Modeling Biological Neural Networks. In *Handbook of Natural Computing*; Rozenberg, G., Bäck, T., Kok, J.N., Eds.; Springer: Berlin/Heidelberg, Germany, 2012. [CrossRef]
22. Almog, M.; Korngreen, A. Is realistic neuronal modeling realistic? *J. Neurophys.* **2016**, *116*, 2180–2209. [CrossRef]
23. Ma, J.; Tang, J. A review for dynamics in neuron and neuronal network. *Nonlin. Dyn.* **2017**, *89*, 1569–1578. [CrossRef]
24. Maass, W. Networks of spiking neurons: The third generation of neural network models. *Neur. Net.* **1997**, *10*, 1659–1671. [CrossRef]
25. Stanojevic, A.; Woźniak, S.; Bellec, G.; Cherubini, G.; Pantazi, A.; Gerstner, W. High-performance deep spiking neural networks with 0.3 spikes per neuron. *Nat. Commun.* **2024**, *15*, 6793. [CrossRef] [PubMed]
26. Lee, D.; Lee, G.; Kwon, D.; Lee, S.; Kim, Y.; Kim, J. Flexon: A flexible digital neuron for efficient spiking neural network simulations. In Proceedings of the 2018 ACM/IEEE 45th Annual International Symposium on Computer Architecture (ISCA), Los Angeles, CA, USA, 1–6 June 2018; pp. 275–288. [CrossRef]
27. Gerstner, W. Spiking neurons. In *Pulsed Neural Networks*; Maass, W., Bishop, C.M., Eds.; MIT Press: Cambridge, MA, USA, 2001; pp. 3–53. [CrossRef]
28. Yamazaki, K.; Vo-Ho, V.-K.; Bulsara, D.; Le, N. Spiking neural networks and their applications: A Review. *Brain Sci.* **2022**, *12*, 863. [CrossRef] [PubMed]
29. Nunes, J.D.; Carvalho, M.; Carneiro, D.; Cardoso, J.S. Spiking neural networks: A survey. *IEEE Access* **2022**, *10*, 60738–60764. [CrossRef]
30. Ganguly, C.; Bezugam, S.S.; Abs, E.; Payvand, M.; Dey, S.; Suri, M. Spike frequency adaptation: Bridging neural models and neuromorphic applications. *Commun. Eng.* **2024**, *3*, 22. [CrossRef]
31. De Wilde, P. *Neural Network Models: Theory and Projects*, 2nd ed.; Springer: London, UK, 1997.
32. Bohte, S.M.; Kok, J.N.; La Poutré, J.A. SpikeProp: Backpropagation for networks of spiking neurons. In Proceedings of the ESANN 2000, 8th European Symposium on Artificial Neural Networks, Bruges, Belgium, 26–28 April 2000.
33. Ribeiro, B.; Antunes, F.; Perdigão, D.; Silva, C. Convolutional spiking neural networks targeting learning and inference in highly imbalanced datasets. *Patt. Recogn. Lett.* **2024**, *in press*. [CrossRef]
34. Furber, S. Large-scale neuromorphic computing systems. *J. Neural Eng.* **2016**, *13*, 051001. [CrossRef]
35. Ballard, D.H. Modular learning in neural networks. In Proceedings of the Sixth National Conference on Artificial Intelligence, Seattle, WA, USA, 13–17 July 1987; Volume 1, pp. 279–284.
36. Pfeiffer, M.; Pfeil, T. Deep learning with spiking neurons: Opportunities and challenges. *Front. Neurosci.* **2018**, *12*, 774. [CrossRef]
37. Tavanaei, A.; Ghodrati, M.; Kheradpisheh, S.R.; Masquelier, T.; Maida, A. Deep learning in spiking neural networks. *Neur. Net.* **2019**, *111*, 47–63. [CrossRef]
38. Kiselev, M.; Ivanitsky, A.; Larionov, D. A purely spiking approach to reinforcement learning. *Cogn. Sys. Res.* **2005**, *89*, 101317. [CrossRef]
39. Schuster, H.G. *Deterministic Chaos: An Introduction*; VCH: Weinheim, Germany, 1995.

40. Huang, K. *Statistical Mechanics*, 2nd ed.; Wiley: New York, NY, USA, 1987.
41. Contoyiannis, Y.; Potirakis, S.M.; Eftaxias, K.; Contoyianni, L. Tricritical crossover in earthquake preparation by analyzing preseismic electromagnetic emissions. *J. Geodyn.* **2015**, *84*, 40–54. [CrossRef]
42. Potirakis, S.M.; Contoyiannis, Y.; Schekotov, A.; Eftaxias, K.; Hayakawa, M. Evidence of critical dynamics in various electromagnetic precursors. *Eur. Phys. J. Spec. Top.* **2021**, *230*, 151–177. [CrossRef]
43. Rojas-Dueñas, G.; Riba, J.-R.; Moreno-Eguilaz, M. Nonlinear least squares optimization for parametric identification of DC–DC converters. *IEEE Trans. Power Electron.* **2021**, *36*, 654–661. [CrossRef]
44. Auge, D.; Hille, J.; Mueller, E.; Knoll, A. A Survey of encoding techniques for signal processing in spiking neural networks. *Neur. Process Lett.* **2021**, *53*, 4693–4710. [CrossRef]
45. Subbulakshmi Radhakrishnan, S.; Sebastian, A.; Oberoi, A.; Das, S.; Das, S. A biomimetic neural encoder for spiking neural network. *Nat. Commun.* **2021**, *12*, 2143. [CrossRef] [PubMed]
46. Yu, Q.; Tang, H.; Tan, K.C.; Yu, H. A brain-inspired spiking neural network model with temporal encoding and learning. *Neurocomputing* **2014**, *138*, 3–13. [CrossRef]
47. Wen, S.; Yin, A.; Tseng, P.H.; Itti, L.; Lebedev, M.A.; Nicolelis, M. Capturing spike train temporal pattern with wavelet average coefficient for brain machine interface. *Sci. Rep.* **2021**, *11*, 19020. [CrossRef]
48. Dong, P.; Dang, K.N.; Nguyen, D.; Tran, X. A light-weight neuromorphic controlling clock gating based multi-core cryptography platform. *Microproc. Microsyst.* **2024**, *106*, 105040. [CrossRef]
49. Balamurugan, V.; Karthikeyan, R.; Sundaravadivazhagan, B.; Cyriac, R. Enhanced Elman spike neural network based fractional order discrete Tchebyshev encryption fostered big data analytical method for enhancing cloud data security. *Wireless Net.* **2023**, *29*, 523–537. [CrossRef]
50. Almomani, A.; Alauthman, M.; Alweshah, M.; Dorgham, O.; Albalas, F. A comparative study on spiking neural network encoding schema: Implemented with cloud computing. *Clust. Comput.* **2019**, *22*, 419–433. [CrossRef]

MDPI AG
Grosspeteranlage 5
4052 Basel
Switzerland
Tel.: +41 61 683 77 34

Entropy Editorial Office
E-mail: entropy@mdpi.com
www.mdpi.com/journal/entropy

www.ingramcontent.com/pod-product-compliance
Lightning Source LLC
Chambersburg PA
CBHW061443210926
51850CB00008B/1602
* 9 7 8 3 7 2 5 8 6 3 6 6 2 *